PLANT TISSUE CULTURE

Academic Press Rapid Manuscript Reproduction

PROCEEDINGS OF A SYMPOSIUM BASED ON THE UNESCO TRAINING COURSE ON PLANT TISSUE CULTURE: METHODS AND APPLICATIONS IN AGRICULTURE, SPONSORED BY UNESCO AND HELD IN CAMPINAS, SAO PAULO, BRAZIL, ON NOVEMBER 8–22, 1978

PLANT TISSUE CULTURE

Methods and Applications in Agriculture

Edited by

TREVOR A. THORPE

Department of Biology
University of Calgary
Calgary, Alberta, Canada

ACADEMIC PRESS 1981

A Subsidiary of Harcourt Brace Jovanovich, Publishers

New York London Toronto Sydney San Francisco

ACADEMIC PRESS, INC.
111 Fifth Avenue, New York, New York 10003

United Kingdom Edition published by
ACADEMIC PRESS, INC. (LONDON) LTD.
24/28 Oval Road, London NW1 7DX

Library of Congress Cataloging in Publication Data
Main entry under title:

Plant tissue culture.

Includes bibliographies and index.
1. Plant tissue culture--Congresses. 2. Plant cell culture--Congresses. 3. Plant-breeding--Congresses.
I. Thorpe, Trevor A.
SB123.6.P53 631.5'3 81-1850
ISBN 0-12-690680-7

PRINTED IN THE UNITED STATES OF AMERICA

82 83 84 9 8 7 6 5 4 3 2

CONTENTS

CONTRIBUTORS

Numbers in parentheses indicate the pages on which the authors' contributions begin.

Eugênio Aquarone (359), *Faculdade de Ciências Farmacêuticas, Universidade de São Paulo, 05508 São Paulo, SP, Brazil*

Stefania Biondi* (1), *Department of Biology, University of Calgary, Calgary, Alberta T2N 1N4, Canada*

Linda S. Caldas (349), *Departmento de Botanica, Universidade de Brasilia, 70.000, Brasilia, DF, Brazil*

Otto J. Crocomo (359), *Escola Superior de Agricultura "Luíz de Queiroz," Universidade de São Paulo, 13400 Piracicaba, SP, Brazil*

David A. Evans (45, 213), *Pioneer Research, Campbell Institute for Agricultural Research, 2611 Branch Pike, Cinnaminson, New Jersey 08077*

Christopher E. Flick† (45), *Center for Somatic Cell Genetics and Biochemistry Research, State University of New York, Binghamton, New York 13901*

Oluf L. Gamborg (21, 115), *International Plant Research Institute, 887 Industrial Road, San Carlos, California 94070*

Otto R. Gottlieb (359), *Instituto de Química, Universidade de São Paulo, 05508 São Paulo, SP, Brazil*

Walter Handro (155), *Departmento de Botanica, Caxia Postal 11,230, Universidade de São Paulo, 01.000, São Paulo, SP, Brazil*

C. John Jensen (253), *Department of Agricultural Research, Research Establishment, Risø, DK-400 Roskilde, Denmark*

K. K. Kartha (181), *Prairie Regional Laboratory, National Research Council of Canada, Saskatoon, Saskatchewan, S7N OW9, Canada*

Ming-Chin Liu (299), *Taiwan Sugar Research Institute, 154 Shen Chan Road, Tainan 700, Taiwan, China*

L. C. Monaco (325), *Institute Agronomic, 28, 13.100, Campinas, SP, Brazil*

Colette Nitsch (241), *C.N.R.S., Genetique et Physiologie du Developpement des Plantes, 91190 Gif-sur-Yvette, France*

Kiyoharu Oono (273), *Division of Genetics, National Institute of Agricultural Sciences, 3-1-1 Kannondai, Yatabe, Tsukuba, Ibaraki-ken 305, Japan*

*Present address: Istituto Botanico, Università di Bologna, Via Irnerio 42, 40126 Bologna, Italy

†Present address: Pioneer Research, Campbell Institute for Agricultural Research, 2611 Branch Pike, Cinnaminson, New Jersey 08077

Sandra M. Reed (213), *Pioneer Research, Campbell Institute for Agricultural Research, 2611 Branch Pike, Cinnaminson, New Jersey 08077*

Elias A. Shahin (115), *International Plant Research Institute, 887 88 Industrial Road, San Carlos, California 94070*

William R. Sharp (45, 325), *Pioneer Research, Campbell Institute for Agricultural Research, 2611 Branch Pike, Cinnaminson, New Jersey 08077*

Jerry P. Shyluk (21, 115), *Prairie Regional Laboratory, National Council of Research, Saskatoon, Saskatchewan, S7N OW9, Canada*

Harry E. Sommer (349), *School of Forest Resources, University of Georgia, Athens, Georgia 90601*

M. R. Sondahl (325), *Departmento de Genetica, Instituto Agronomico, Caxia Postal 28, 13.100, Campinas, SP, Brazil*

Trevor A. Thorpe (1, 253), *Department of Biology, University of Calgary, Calgary, Alberta T2N 1N4, Canada*

Edward C. Yeung (253), *Department of Biology, University of Calgary, Calgary, Alberta T2N 1N4, Canada*

PREFACE

Plant tissue culture techniques have become a powerful tool for studying basic and applied problems in plant biology. Furthermore, in the last ten years these techniques have found wide commercial application in the propagation of plants, mainly horticultural species, and in the elimination of pathogens. The interest in plant tissue culture and its potential for plant improvement is worldwide, as the increase in members and participating countries in the International Association for Plant Tissue Culture would attest. In addition, the Plant Biology Panel of the International Cell Research Organization (ICRO) of UNESCO has designated the teaching of plant tissue culture techniques as one of its high priorities.

The Plant Biology Panel of ICRO organized its first International Training Course entitled Plant Tissue Culture Methods and Applications in Agriculture at the Agronomic Institute in Campinas, Brazil, in 1978. The course was cosponsored by UNESCO and research organizations in Brazil. It was during this course that the idea for a book arose. As originally planned, it was to have been edited by Drs. Oluf L. Gamborg, Maro R. Sondahl and William R. Sharp, who were principals in the course and who selected the authors and topics. I became editor when the project was underway and when nearly half of the chapters had been submitted.

Our common goal was to produce a laboratory manual of plant tissue culture technology. Protocols are presented in detail, and the rational behind each method is implicitly or explicitly stated. There are two major sections: Part A, which emphasizes methodology, and Part B, which emphasizes the applications. We recognize that a certain degree of overlap exists, and trust this will be beneficial, because it reinforces and unifies underlying concepts.

The first chapters present the requirements for a tissue culture facility and then discuss nutrition, media, and characteristics of cultured plant cells and their growth and behavior *in vitro,* particularly with reference to embryogenesis and organogenesis. Protoplasts, mutagenesis and *in vitro* selection, meristem culture, freeze preservation, and cytogenetic techniques complete Part A. In Part B, androgenesis, *in vitro* fertilization, and embryo culture are discussed. These are followed by chapters outlining the application of *in vitro* methodology to selected crops such as rice, sugar cane, coffee, and forest trees. The final chapter deals with the potential of tissue culture in the biosynthesis of secondary products. It is clear that tissue culture methods are playing a significant role in plant improvement, either directly or as an adjunct to more traditional methods. Furthermore, there is the implicit conviction that tissue culture methods have an even greater role to play in the future.

This volume is designed for scientists who wish to apply *in vitro* methods to the management and breeding of plants of economic importance, i.e., those involved in modification and improvement of plants, for the production of food, fiber, or fuel. Melchers[1] has pointed out that although basic research cannot be constrained by "planning" in the true sense of the word, to be efficient, applied research must be thoroughly planned. Part of this planning must include a very exact understanding of the principles of the methodology to be used and its limitations, as well as the proper approach to the problem. It is my hope that this volume will prove useful to those who must thoroughly plan their research in tackling problems in agriculture that are amenable to the tissue culture approach. With the increase in world population, the continued loss of prime agricultural land to housing and industry, the use of more marginal land for agriculture, and the increase in problems such as salinity, new approaches are needed. Tissue culture offers one such possibility.

It gives me great pleasure to acknowledge with gratitude those who assisted in the editing and preparation of this volume: first, Laura Bentley, who served as my assistant, and competently organized the chapter sections and the references, proofread the typed articles, and assisted with the indexing; second, Erin Smith who did most of the typing and skillfully made all the required changes; and third, Win Packer and Susan Tait, who typed some of the chapters. The proficiency and dedication of these assistants made my task easier.

Preparing this volume has been a valuable learning experience; I hope that those of you who use it will likewise benefit.

[1]Melchers, G. (1980). In "Perspectives in Plant Cell and Tissue Culture" (I. K. Vasil, ed.). Int. Rev. Cytol. Suppl. 11B, p. 24. Academic Press, New York.

REQUIREMENTS FOR A TISSUE CULTURE FACILITY

Stefania Biondi
Trevor A. Thorpe

Department of Biology
University of Calgary
Calgary, Alberta, Canada

I. INTRODUCTION

Today, it is generally accepted that the term "plant tissue culture" broadly refers to the cultivation *in vitro* of all plant parts, whether a single cell, a tissue or an organ, under aseptic conditions; although Street has recommended a more restricted use of the term (20). Plant tissue culture is a technique which has great potential as a means of vegetatively propagating economically important species; a potential which is being realized commercially at present. However, a tissue culture system is also very often a "model" system which allows one to investigate physiological, biochemical, genetic and structural problems related to plants and the technique is being used also as an adjunct to more traditional means in plant modification. Many of these approaches being used at present are described in succeeding chapters in this book. It is mainly in view of using tissue culture as a tool in basic and applied research that the requirements of a plant tissue culture facility will be examined in this chapter.

The underlying principles involved in plant tissue culture are very simple. Firstly, it is necessary to isolate a plant part from the intact plant and its inter-organ, inter-tissue and inter-cellular relationships. Secondly, it is necessary to provide the plant part with an appropriate environment in which it can express its intrinsic or induced potential. This means that a suitable culture medium and proper culture conditions must be provided. Finally, the above must be carried out aseptically. In practice, this means that the

ISBN 0-12-690680-7

culture must be free of bacterial, algal, fungal and other contaminants. Contamination by such microorganisms is a very real problem in tissue culture and one which demands a great deal of skill, care and organization because the media used to support higher plant cell growth also supports the growth of these microorganisms. If their growth is not prevented, they may overgrow the plant cells, inhibit their development and interfere with the physiology and biochemistry of the system by the release of metabolic products. Secondly, we will see that much of the equipment used in a tissue culture laboratory is aimed at careful control of all the components pertaining to the physical (and to some extent as a consequence, to the physiological) environment of the system (e.g., media components, gaseous atmosphere, types of vessels used, light and temperature regimes, etc.). All this is aimed at ensuring that the system is as defined as possible. Nonetheless, it is important to realize that as de Fossard has pointed out, it is very rare for any of us to attain optimal, fully defined, fully reproducible culture conditions (4). No doubt a rational yet imaginative use of laboratory equipment coupled with a knowledge of which factors may or may not influence the system and how, plays an essential role toward achieving this somewhat elusive goal.

The following topics are covered in this chapter: basic organization and facilities, glassware, instruments and miscellaneous equipment, water, controlled environments and liquid cultures.

II. BASIC ORGANIZATION AND FACILITIES

The cultivation of a plant tissue *in vitro* does not *per se* require complex or expensive equipment. It has been said that all one really requires is a pressure cooker and a few jam jars! The extent to which more sophisticated apparatuses are necessary depends on the nature of the research undertaken. For example, one may wish to investigate the ultrastructural changes occurring in the course of the growth and differentiation of a particular system. In that case an electron microscope and darkroom facilities will be required. Biochemical studies carried out on a tissue culture model system may necessitate the acquisition of a high-speed centrifuge, a spectrophotometer, a freeze-dryer, etc. Such requirements will not be reviewed here.

The basic organization and facilities of most tissue culture laboratories today can be summarized as follows:

(1) A general laboratory area with provisions for either independent or common working spaces or both. Some equipment and materials will necessarily be communal and should be easily accessible to all workers.

(2) Large sinks (some lead-lined to resist acids and alkalis) and draining areas. Washing machines to wash glassware in bulk and hot-air cabinets or ovens for drying washed glassware are useful in most cases.

(3) Cabinet and shelf space for safe storage of chemicals and dust-free storage space for clean glassware.

(4) Transfer areas for aseptic manipulations. Such a facility can be provided in several ways and will be reviewed later.

(5) An autoclave and/or oven for sterilizing media, solutions, water, culture vessels and instruments.

(6) Culture rooms or incubators where cultures can be incubated under controlled light, temperature, and, if possible, humidity regimes.

(7) Essential services such as electricity, water, gas, and, if possible, compressed air and vacuum in the working areas.

(8) A supply of distilled, double-distilled and deionized (demineralized) water; and

(9) Various instruments and appliances.

A. *Sterilization of Equipment and Plant Material*

As mentioned earlier, the rigorous exclusion of contaminating microorganisms is an essential feature of tissue culture. All nutrient media, culture vessels, and instruments used in handling the tissue as well as the plant material itself must be sterile. An obvious precaution is not to share areas where tissue culture work is being carried out with microbiologists and pathologists. Secondly, cleanliness and a certain amount of efficient organization will contribute toward reducing the risk of widespread as well as occasional contamination.

The sterilization of glassware and metallic instruments can be carried out in dry heat, i.e., an oven. The unusual resistance of bacteria, particularly spores, to dry heat has long been recognized. However, a systematic study on the temperatures and exposure times which are lethal to microorganisms is lacking and consequently so is a general consensus as to the temperature and exposure time required for efficient dry heat sterilization. It has been shown that even within one bacterial genus (*Bacillus*), these requirements varied considerably from species to species (e.g., 120°C for

40 mins to 140°C for 180 mins) (Sterilizers Operating and Instruction Manual, Despatch Oven Company, Minneapolis, Minn., ME-143, 11/5/70). They have suggested that 60-120 min exposure at 160°C should be sufficient. However, it is important to realize that these conditions are dependent on the material which is being sterilized, particularly on the length of time it takes to reach the set sterilization temperature. Furthermore, microorganisms surrounded by organic matter, such as grease, are greatly protected against the action of dry heat. Thus, materials should be thoroughly cleaned beforehand. Finally, as an indication of the lack of consensus on this matter, Sarles and co-workers (quoted in 4) suggest a minimum exposure of 3 hr at 160-180°C.

An exposure of 160°C dry heat for 1-2 hr is regarded as being approximately equivalent to moist heat (steam) sterilization at 121°C for 10-15 mins (Despatch Oven Co.). Indeed, these are the conditions encountered during steam sterilization in an autoclave. Modern autoclaves, such as the AMSCO[1] "General Purpose" and "Vacamatic A", operate at a temperature of 121° or 132°C (the former is by far the most common) with a pressure of 15 p.s.i. (103.4×10^3 Pa). Liquid sterilization (media, water) is followed by a slow exhaust. Exposure time is dependent on the volume of liquid to be sterilized (Table I).

Domestic pressure cookers can be used for small volumes of liquid. Dry goods (instruments, empty glassware) can either be "wrapped" (in aluminum foil, brown paper or sealed metal containers), or "unwrapped". Clearly glassware must be stop-

TABLE I. Suggested Minimum Sterilization Times for Media

Volume per container (ml)	*Minimum sterilization time at 121°C*
20-50	*15*
75	*20*
250-500	*25*
1000	*30*
1500	*35*
2000	*40*

[1]*Mention of a specific make, brand or trade name in this book does not imply that only those indicated are suitable, or that other comparable substitutes may not be available.*

pered and instruments wrapped if sterility is to be maintained up to and beyond cooling. The "dry goods" program in the autoclave is followed by a rapid exhaust; unwrapped goods can generally be sterilized for 15 mins and wrapped goods for 20 mins. In addition, unwrapped goods are subjected to a longer drying period. In locations like Calgary where low ambient humidity reduces the risk of contamination, instruments used in handling sterile tissue need not be autoclaved. Instead, soaking in 70-80% (v/v) ethanol followed by flaming is the routine procedure and is carried out repeatedly while aseptic manipulations are in progress. This procedure is essential even if the instruments have been subjected to prior dry or steam sterilization.

It is important to realize that not all plastic labware, which is more and more frequently replacing glassware, is resistant to heat sterilization. Only polypropylene, polymethylpentene, polyallomer, Tefzel^R ETFE and Teflon^R FEP may be repeatedly autoclaved at 121°C (Nalgene Labware and Thermolyne Apparatus, 1979 Catalog). Polycarbonate shows some loss of mechanical strength with repeated autoclaving and cycles should be limited to 20 min. Linear polyethylene is not usually considered autoclavable, but if the container is small and the temperature does not at any time exceed 120°C, no major problems arise. Polystyrene, polyvinyl chloride, styrene-acrylonitrile, acrylic and conventional polyethylene are not autoclavable under any conditions. Only Teflon^R FEP may be dry-heat sterilized.

To avoid sticking of the plastic sealing rings in flask or test-tube closures, these should be placed with threads totally disengaged during autoclaving. Furthermore, it should be noted that plastic containers may take longer to reach the sterilization temperature than glass and therefore longer cycles may be necessary.

Some growth factors such as zeatin, abscisic acid, urea and certain vitamins are heat-labile. They cannot then be autoclaved with the rest of the nutrient medium. Filter-sterilization is the technique which is applied instead. For small quantities, bacteria-proof membrane filters (Millipore^R by the Millipore Corp., Bedford, Massachusetts; Gelman Triacetate Metrical^R by the Gelman Instrument Co., Ann Arbor, Michigan; Sartorius Regular^R via V.A. Howe and Co., London, U.K., Canlab, Canada, etc.) can be fitted in 1-10 ml graduated syringes (B-D^R, Cornwall^R and Luer-Lok^R by Becton, Dickinson and Co., Rutherford, New Jersey), with a Swinny^R filter adapter and a needle or cannula. Only the adapter and the needle need to be sterilized. The filter-sterilized solution is added to the rest of the medium, which has been previously autoclaved, just before it sets at about 40°C (if agar is present).

A less common method reported by Street consists of treating the dry thermolabile compound with high grade (Analar) ethyl ether, removing the ether at a temperature below 30°C, dissolving the solid in sterile distilled water and adding the solution aseptically to the sterile medium (20).

For large volumes, filter-sterilization can be carried out using a Millipore® vacuum filtering set-up. This consists of a pyrex funnel and base with a coarse frit-glass support for the membrane filter. These are held together by a spring clamp. When used with side-arm flasks, a vacuum can be applied and also automatic transfer to a burette can be done. The whole apparatus can be autoclaved before use. For more information about the effects of autoclaving on culture media, refer to Bragt *et al.* (3).

Although one should avoid touching the sterile part of instruments and sterile plant material with one's hands, it is essential that the worker's hands be relatively aseptic during manipulations. A wash with an anti-bacterial detergent followed by spraying with 40-70% ethanol is quite effective. In fact, by repeatedly dipping fingers in ethanol, we have found that the seed coat of sterile *Pinus radiata* seeds can be removed manually without any ensuing contamination.

Plant tissues do not generally harbor microorganisms internally. If they do, an antibiotic can be added to the medium. However, this is not usually a reliable way of ensuring asepsis.

Plant material can be surface-sterilized by a variety of sterilants. The type and concentration of sterilant to be used and exposure time must be decided upon empirically. Street offers some guidelines on this question (20). Some of the more commonly utilized surface-sterilants are:

(1) A solution of calcium or sodium hypochlorite (which releases chlorine as the active sterilant),
(2) Hydrogen peroxide,
(3) Bromine water,
(4) Silver nitrate,
(5) Commercial bleach containing hypochlorite: chlorinated lime ("bleaching powder") or aqueous solutions (Clorox®, Javex®, etc.).

It is important that the plant surface be properly wetted by the sterilizing solution. A 30 sec immersion in 70% (v/v) ethanol or the addition of a few drops of a liquid detergent ("wetting agent") such as Teepol^R (British Drug Houses, Ltd.) or Tween-80^R (Fisher Scientific Co. Ltd.) to the sterilizing solution will ensure that this happens. Sterilants must be removed by thoroughly rinsing with several changes of sterile distilled water.

B. *Transfer Areas for Aseptic Manipulation*

It is generally accepted that even in a very clean laboratory with a dry atmosphere, some protection against air-borne microorganisms must be sought during the manipulation of *in vitro* systems.

When a large number of cultures must be routinely manipulated or when large culture equipment is handled, walk-in transfer rooms can be useful. Such rooms are fitted with a unit consisting of a fan and bacteria-proof filters which forces in sterile air. Overhead ultra-violet lighting, although considered to be dangerous to both humans and plants, is a standard feature. The walls and floor should be smooth so that dirt does not accumulate and so they can be thoroughly cleaned and disinfected. Sliding doors, airtight windows and the absence of any radiators ensure that there are no drafts or convection currents to blow dust around. Ventilation must be provided by an air-conditioning unit.

The simplest sterile transfer cabinet is a bench-top plastic shield which can either be enclosed on all sides except the front (with a sloping face to afford some protection from the worker's breath) or entirely enclosed with 2 rubber-circled apertures in the front through which the worker's arms go. Such "glove boxes" (Germfree Laboratories, Inc., Miami, Florida; Scientific Products, Evanston, Ill.) are generally made of fiberglass or Plexiglass and can be simple or have several accessories built into them (an air-filtering system, a lighting system, etc.). Glove boxes can be used within a transfer room but this is not generally necessary.

The laminar air-flow cabinet (Envirco^R by ViroMart Ltd., Burlington, Ontario; Scientific Products, Evanston, Ill.; John Bass Ltd., Crawley, Sussex, U.K.; Microflow Ltd., Fleet, Hants., U.K.) is probably the most common accessory used today for aseptic manipulations. It is cheaper and easier to install than a transfer room and allows a better view, more space and more comfort than a glove box. Here, too, air is forced into the cabinet through a bacterial filter; it flows outward (forward) over the working bench at a uniform rate.

The constant air-flow prevents particles from settling on the working bench. These cabinets are available in different sizes commercially. They can also be constructed by using Hepa^R filters and appropriate motors.

Aseptic glass bells (Bellco Glass Inc.) are useful when a liquid has to be transferred to a sterile container in the absence of a transfer room or cabinet or a glove box. Aseptic bells are designed either to be attached to a reservoir with rubber tubing, or with a rubber stopper. Alternatively, micro amounts of solution can be dispensed by pushing a syringe needle through a special (Luer^R tip) adaptor on the bell.

III. GLASSWARE

Tissue culture work does not normally require more glassware or instruments than is found in a well-stocked laboratory. The requirements are flexible and almost anything can be adapted for use. Culture vessels should be preferably made of Pyrex or boro-silicate glass. The latter material is becoming expensive and soda glass can be as much as 7-8 times cheaper. However, de Fossard states that soda glass containers must be discarded after 12 months of use, or be treated with Repelcote^R (Hopkins and Williams Ltd.), a dimethyldichloro-silane (4). Disposable and often pre-sterilized plastic petri dishes (usually 100 x 15 mm) and jars are, however, becoming more widespread.

Wide-neck Erlenmeyer flasks (50, 100, 125, 250 ml capacity) are routinely used as culture vessels. Larger flasks are also used for medium preparation. Test tubes (often 25 x 150 mm culture tubes), petri dishes, Universal bottles or jars and a vast assortment of other glass containers (see 23) can be adapted to fit the needs of a particular tissue culture system. In essence, any container that can be autoclaved (or is pre-sterilized) and cleaned thoroughly, that does not release any toxic substances and is clear (if the culture requires light) is suitable. In fact, all new culture vessels may release substances which affect the composition of the medium or are toxic to the tissue. Hence, they must be thoroughly washed before use. Street suggests that new vessels should be filled with double-distilled water and autoclaved at least twice (30 min each time) with a detergent wash in between (20). Other glassware items routinely required in tissue culture laboratories include measuring cylinders, pipettes, beakers and volumetric flasks, used essentially for media preparation, and burettes to dispense media. Flasks with break-resistant collars, heavy wall construction and side-arms can be used for filtration (with a Buchner funnel) or for media dispensing (Bellco Glass Inc.). In fact, graduated spring-loaded syringes with valves for uptake and output of a fixed volume of liquid are often used for media dispensing (B-D^R, Cornwall^R by Becton, Dickson and Co.). Plastic containers are particularly useful for freeze-storage of pre-prepared solutions such as inorganic salts, vitamins, growth factors, etc. Such solutions, as well as dry media components (sucrose, agar), can be stored in pre-determined volumes or weights in a series of small containers. Other types of glassware which are used for liquid suspension cultures will be described in a later section.

In laboratories where a considerable number of large-scale experiments are set up, the problem of washing glassware can assume significant proportions. It is advisable to have washing machines and/or sufficient space for storage and washing of dirty glassware as well as draining/drying areas for washed glassware. Automatic washing machines generally ensure that thorough and repeated rinsing takes place after the detergent wash. In most cases, commercially available detergents (e.g., 7-X^R by Linbro Chemicals Co. Inc., New Haven, Connecticut; Jet-Clean^R, FL-70^R and Sparkleen^R by Fisher Scientific Co., Ltd.) can meet the needs of most glassware washing operations. Rinsing is important because detergents are often highly toxic to plant cells. Procedures for cleaning highly contaminated glassware include the following (23):

(1) Boiling the glassware in white-soap solution, rinsing it thoroughly in water and then in 95% ethanol, drying it and finally wrapping and dry sterilizing it.

(2) Cleaning by ultrasonics.

(3) Washing in pyrophosphate solution.

(4) Boiling in metaphosphate (Alconox^R), rinsing, boiling in dilute hydrochloric acid, rinsing in water, then in 95% ethanol.

(5) Acid bath which, despite the inherent danger, some glassware, particularly if it has been in contact with toxic chemicals, may have to be washed in a mixture of potassium dichromate and concentrated sulphuric acid. The glassware can be immersed in such an acid bath, which itself can be contained in polythene or PVC containers, for several hours. The operator must wear a lab coat and acid-resistant gloves at the very least. Acid-resistant plastic clothing and goggles may be advisable. Repeated rinsing in tap, de-ionized,

then distilled water is essential. For glassware coming into direct contact with living tissue, 30 min in the autoclave, immersed in double-distilled water has been suggested.

IV. INSTRUMENTS AND MISCELLANEOUS EQUIPMENT

The instruments most commonly required to handle tissue in preparation for and during culture are scalpels and forceps. The shape and size of scalpel blades can be suited to the material being handled; long forceps (about 15 cm) and fine-point forceps may be useful. In addition, the need for scissors, razor blades, spatulas and mounted needles may arise. Cork borers are useful to obtain a standardized explant size. This can be an important prerequisite for quantitative studies on tissue development. Yeoman and Macleod cut sterile tubers of carrot, turnip, parsnip and Jerusalem artichoke into slices and then removed cylinders of tissue by using a stainless steel cannula (cork borer) (27). The tissue cylinders were then lined up parallel to one another on a special cutter made of a fixed row of parallel razor blades.

Hypodermic syringes, as previously mentioned, are used for filter-sterilization. They can also be used to dispense nutrients in minute quantities to hanging-droplet cultures (see Section VII *C*).

Media dispensing equipment, particularly automatic petri dish filling machines (Petrimat^R, Bellco Glass Inc.) which can pour up to 600 dishes per hour, bubble-free and aseptically by using a dish-handling unit, UV light and a pump unit are available, but this is generally superfluous for standard research. Other equipment that can be regarded as basic to a tissue culture facility are:

(1) A pH meter.

(2) A balance, e.g., (i) electronic analytical to 5 decimal places, weighs up to 30 or 166 gm ± 0.1 mg: Sartorius^R (models 1201 and 2003 MPI/BCD),

(ii) up to 19 gm ± 0.01 mg: Micro-gram-atic^R by R. Mettler, Zurich,

(iii) less accurate balances (to 2 decimal places) are usually sufficient, certainly for weighing out sucrose, agar, etc.: (e.g., Mettler P1200 model - up to 1200 gm ± 10 mg).

(3) Bunsen burners, including a Touch-o-Matic^R (Bellco).

(4) Spirit lamps (for flaming instruments).

(5) Hot plates with magnetic stirrers for melting media.

(6) Microwave oven for rapid melting of large volumes of agar medium.

(7) A microscope and haemocytometer for cell counting.

(8) A low-speed centrifuge to determine packed cell volume.

(9) A dissecting (low-power) microscope for overall observation of tissue development such as counting adventitious organs; stereozoom models are recommended (Kyowa, Olympus, Nikon, Wild Heerbrugg, etc.).

(10) Large (e.g., 10 or 25-liter) plastic carboys to store high quality water.

(11) A fume-hood.

(12) Metal or wooden racks to support test and culture tubes.

(13) Metal trays and carts for transport of culture flasks, racks of tubes, etc.

Stoppers and various closures have to be considered. Again, a certain flexibility exists so that the choice of the closure method can be made individually, based on what is available and what the cost is. Most commonly, flasks (especially if they have a collar) are stoppered with non-absorbent cotton wool (sometimes wrapped in cheese cloth or muslin), autoclavable foam plugs (Identi-plugs^R, Fisher Scientific Co.) or aluminum foil. The latter can also be used to cover cotton wool and foam plugs. Screw-cap Erlenmeyer flasks (Bellco Glass Inc.) and flasks with a Delong^R type neck which can be stoppered with metal (aluminum or stainless steel, Bellco Glass Inc.) or plastic caps are also available. Plastic (polypropylene) caps such as KimKaps^R (Kimble, Division of Owens, Ill.), Kaputs^R (Bellco Glass Inc.) and Bacti-Capalls^R (Fisher Scientific Co.) are autoclavable. Test tubes are generally stoppered with such caps, too. Recently Kimble P.M. (permeable membrane) Cap^R closures were introduced on the market. They are transparent, autoclavable polypropylene caps with a membrane built into the top which prevents contamination yet allows gas to permeate and acts as an effective moisture barrier (water retention is 3 times better than with KimKaps). Culture vessels which are pre-sterilized or have already been autoclaved can be sealed with Parafilm^R (American Can Co., Greenwich, CT.), which is a waxy, non-autoclavable adherent and stretchable sheet, or thin plastic sheeting such as the household kind (e.g., Saran Wrap^R). To seal the edge around petri dishes, these sheets (generally cut into strips) need not be sterilized. In other cases, 70% ethanol can be used (but not on Parafilm^R). Serum caps are required when experiments on the gaseous atmosphere inside the culture vessel are carried out.

V. WATER

Even treated tap water usually contains the following contaminants, often in significant amounts:

(1) Dissolved inorganic material (from tanks, pipes, etc.).
(2) Dissolved organic material (tannins, industrial effluents, etc.).
(3) Micro-organisms (a severe infection can affect ion-exchange resins and allow dissolved inorganics through).
(4) Particulate matter (silt, dust, etc.).

In order to obtain high purity "laboratory grade" water, glass distillation is usually carried out. In distillation, water is boiled and the purified vapor resulting from this is condensed. However, this procedure is relatively expensive in terms of electricity costs and can produce sub-standard water quality if the still is not meticulously maintained. A new technique was introduced a few years ago known as the Milli-RO^R Reverse Osmosis system. Here, water is separated from dissolved solids by applying a pressure differential across a water-permeable membrane. The effluent is pure laboratory grade water which can be delivered at a rate of 4, 15, or 40 liters per hour depending on the model. The level of contaminants remains too high for "reagent grade" water but laboratory grade water of consistent quality is produced. This can be used in association with the Milli-Q^R system for reagent grade quality water (see later).

De-ionized water, which is essentially suitable only for rinsing labware, is produced by running water through two ion-exchange resin columns (one anodic, one cathodic). Water treated in this way has an extremely low specific conductivity (i.e., low content of inorganic salts) but still contains un-ionized volatile materials such as oils, organic gasses, and large molecules. Furthermore, the de-ionizing bed can serve as a substrate for bacterial growth which may release toxic products, the so-called "pyrogens" (23).

It should be noted that many distillation stills have tin-lined copper heat exchangers. Copper can be toxic to living cells and water from such a still should be redistilled in Pyrex^R, Vycor^R, quartz or some non-toxic metal like nickel. The Bellco Glass Co. manufactures a Pyrex still but for some nutrient studies involving boron, Pyrex, which is a borosilicate, cannot be used. Contact with rubber can invalidate experiments on sulphur or zinc requirements.

For ultra-pure water the following ion exchange system is available from Chromatographic Specialties Ltd., Brockville, Ontario: tap water passes through an Organosorb^R Cartridge which removes most of the organics, free chlorine and chloramines, phosphate complexes, and turbidity. The water then passes through a high capacity cartridge which removes all ionized constituents except free CO_2 and silica. Passage through two additional cartridges produces water equivalent to triple distilled in which all ionized minerals including silica and free CO_2 are eliminated down to a level of 4 ppb. The Milli-RO^R/Milli-Q^R systems are similar.

VI. CONTROLLED ENVIRONMENTS

It is generally considered good practice to incubate cultures under controlled temperature, light and humidity regimes, even though systematic studies on the physical/environmental requirements of plant tissue cultures are conspicuously rare. Incubation (culture) rooms and commercially available incubation cabinets (incubators) satisfy this requirement. An incubator room should be air-conditioned because good air circulation is important. Perforated shelves to support culture vessels and a low-speed, large-blade fan are well suited for this purpose.

Fluorescent tubes and a timing device to set light-dark regimes (photoperiods) are standard accessories. In addition a dark area, simply closed off from the rest of the room with thick, black curtains, is necessary if some cultures require continuous darkness.

Incubators ensure that the set environmental conditions (particularly temperature) remain much more constant. Controlled Environments (Pembina, N. Dakota) produces a wide variety of incubators with the following characteristics:

(1) Temperature $\pm$ 0.3^{o}C at one point at any time.
(2) Temperature range 4°-45°C, 2°-40°C, etc.
(3) Uniform air-flow.
(4) Variable air speed.
(5) Variable light intensity.
(6) Adjustable lamp canopy.
(7) 24-hr timer programs for photoperiods.
(8) Humidifier/dehumidifier.

Floor models (92 x 76 x 198 cm; 183 x 75 x 198 cm), bench models (137 x 69 x 132 cm) or walk-in (343 x 234 x 254 cm) models are available. Canlab (Canada) distributes two particular models (Thelco[R] 6M and Precision[R] 815) with temperature ranges of 5° to 70°C and -10° to 50°C respectively. In fact, refrigeration or a cold water circulating coil are necessary for effective temperature control during hot weather. In incubation rooms, slight refrigeration or thermostatically controlled intermittent heat is necessary if the ambient temperature is to be maintained at 25°C or less in hot weather or to counterbalance the heat emanating from lamps and running electric motors. An externally located temperature recorder allows one to monitor temperature fluctuations which, if they are frequent and excessive, usually announce impending thermostat failure. There should also be a safety cut-out device and warning lights in case the temperature controls fail. A stand-by generator or emergency power supply is essential wherever cultures are being incubated and/or shaken.

Freezers and refrigerators are essential for storage of solutions, extracts, chemicals and plant material. Refrigerators, and walk-in refrigerators (cold rooms) for experiments to be carried out below room temperature, usually have a temperature range close to -10° to 10°C. Freezers, depending on the model (e.g., Canlab), can have temperature ranges of -85° or -75° to 0°, or -10° to -1°C. Freeze preservation of plant material can be ensured, for short periods, at -20°C in ordinary mechanical freezers, or at -80°C in two-stage mechanical freezers. Long-term storage, particularly for such purposes as the establishment of gene banks, the preservation of rare genomes, etc., must be carried out in suitable cryostats and freezing units (e.g., LR-33 Biological Freezer-6[R] by Union Carbide Corp.) with controlled rates of cooling. Alternatively, tissue can be frozen at -140°C in the vapor of liquid N_2 or -196°C by immersion in liquid N_2 and then stored in a freezer. This is acceptable if the tissue is being temporarily stored before being analysed (e.g., metabolite extractions, etc.).

For more information on the technology of long-term freeze preservation, the following can be consulted: Bajaj (1); Bajaj and Reinert (2); Withers (25); Withers and Street (26); and Chapter 6 in this book.

VII. LIQUID CULTURES

When plant cells are cultured on a liquid medium as opposed to an agar-solidified one, such cultures are known as liquid cultures. Liquid cultures can either be *stationary* or *agitated* (shaken).

Stationary cultures usually involve the use of filter paper supports and methods have been described for the rapid preparation of such supports (8). Recently Horsch *et al.* (9) have described a "filter paper growth assay" (FPGA). Here, a thin layer of cells is separated from agar-solidified medium by a disc of filter paper. The disc and adhering cells can be removed for weighing or transfer. The authors state that this technique allows continual monitoring of cell growth, less labor is required for inoculation, harvest and clean-up, fewer cells are needed per inoculum, and growth can be assessed at any time without sacrificing or disturbing the cells. Thus, the FPGA appears to combine the benefits of liquid (suspension) culture with the lower cost and simplicity of callus culture. Generally, however, liquid cultures are designed to avoid the substances often present in gelling agents such as agar. This can be critical when nutritional work is undertaken.

Although in some cases, filter paper supports offer an ideal situation, e.g., aseptic seed germination, they do not provide a radically altered environment as compared to semi-solid cultures. In particular, nutrients are still absorbed through the base of the explant, so that gradients can build up. However, total immersion of the tissue into the liquid medium raises the problem of aeration. Hence, agitation is required. In agitated liquid cultures gaseous exchange is facilitated, polarization of the tissue in relation to gravity is eliminated, and nutrient and growth regulator gradients are removed.

A. *Periodic Immersion Cultures*

In such systems, the culture vessel is agitated in such a way that the cells are alternately submerged and exposed to the air inside the vessel. This ensures good contact with the medium yet adequate gaseous exchange. The simplest and oldest way of achieving this is with the roller-tube method first made practical by Gey (7). Culture tubes (16 x 150 mm) filled with 1-2 ml of medium and well stoppered are placed in a perforated drum which is sloped so as to allow the liquid to cover the lower side of the tube to about half its length. The drum is slowly rotated so that the cultures are immersed for about 1 min out of every 10 (23). Although most commonly used in animal cell cultures, this method has been applied to studies on root growth. Bellco Glass Inc. manufactures roller drums with a speed range of 0.1-144 rpm and which can hold 164 or 351 tubes of 16 mm diameter or 156 tubes of 25 mm diameter. The drum can be tilted from 0-7°. The New Brunswick Scientific Co. Inc., Edison, N.J. also manufactures such drums.

A more sophisticated design is that of the Steward apparatus or auxophyton (18, 19). This consists of one to several circular platforms mounted on a shaft inclined at 10°-12° to the horizontal. The platforms rotate (1-2 rpm) and accommodate two types of culture vessels:

(1) Tumble tubes: these are closed at both ends with a side-arm at mid-point for inoculation and are essentially like the tubes described in the preceding paragraph. They are usually 12.5 cm long, 3.5 cm in diameter and carry 10 ml of medium.

(2) Nipple flasks: these are flat-bottomed round flasks with several side projections ("nipples"). They allow for a greater culture volume to be accommodated (1000 ml with 10 nipples, 250 ml with 8 nipples).

In these set-ups, agitation and aeration is provided by the medium flowing from one end of the tube to the other, or in and out of the nipples. Large-scale spinning cultures have been described by Lamport (13) and Short *et al.* (17) in which two 10-liter Pyrex bottles each with 4.5 l of culture are inclined at a 45° angle and rotated at 80-120 rpm.

B. *Continuous Immersion Cultures*

In order to maintain the tissue or cells in constant contact with the medium and yet ensure aeration, these cultures are constantly shaken or stirred. Shaking can be done in reciprocating or rotary (Gyrotory®) platform shakers; forced air input and/or magnetic stirrers ensure that the culture is continuously stirred.

Platform shakers (New Brunswick Scientific Co., New Brunswick, N.J.; L.H. Engineering Co. Ltd., Stoke Pages, Bucks., U.K.; Adolf Kuhner, A.G., Basel, Switzerland) are fitted with clips of various sizes to hold flasks of 25-1000 ml capacity, or with a studded rubber base (for low speeds only). They generally have a speed range of 30-400 rpm but Street (20) has stated that speeds above 150 rpm are unsuitable. Orbital motion will generally have a stroke of 2-4 cm. A 250 ml flask containing 50-70 ml of culture can successfully be shaken with a 4 cm stroke at 100-200 rpm (21). If available, shakers should be located in incubation rooms. If not, shakers with an enclosing incubator can be acquired (Controlled Environments; Incubator Shakers by New Brunswick Scientific Co.; Clim-o-shake® by Adolf Kuhner, A.G., Basel,

Switzerland) in which light intensity, light quality, photoperiod, temperature and humidity can be regulated. In addition, reciprocating and Gyrotory[R] water-bath shakers are available from New Brunswick Scientific Co. A variant on flask clamps has been introduced known as the ErlAngle Flask Clamp (New Brunswick Scientific Co.) which holds the base of the Erlenmeyer flask at a 15° angle from the platform. When shaken, the flask contents rise from the bottom and are vigorously washed around the walls with the consequent exposure of more liquid surface area to the atmosphere.

Agitated cultures are prone to high water loss and must therefore be well sealed. The use of aluminum foil is generally recommended.

To transfer aliquots of the culture to fresh medium, for example at 3-week intervals, a spring-loaded pipetting unit, a hypodermic syringe fitted with a cannula whose bore (up to approximately 2 mm diameter) allows passage of single cells or small cell clumps, or a graduated pipette (with the tip removed if necessary) can be used. Another piece of equipment adapted to cell suspension cultures is a nylon, Teflon[R] or stainless steel screen or sieve. Komamine and co-workers (6, 10, 11) have shown that by sieving the cell population of carrot suspension cultures through nylon screens with 81-, 47- and 31-μm pores, they could collect cell clusters of a particular size and significantly increase synchronized somatic embryogenesis.

Modifications to standard Erlenmeyer flasks for the purposes of suspension cultures have been described (21). For example, the addition of a center well (with a gas absorbent) or of side-arms allows the gas phase in the closed system to be changed; two-tier vessels, such as a 100 ml flask projecting through the bottom of a 250 ml flask, can be used to grow two cultures in the same gaseous atmosphere. Large fermenter-type vessels based on those used for growing large quantities of microorganisms, have entered the field of plant cell culture too. They generally consist of cylindrical, conical or circular vessels with inlet and outlet ports for aeration, transfer and inoculation. They may be connected to electronic regulating and sampling devices; constant agitation is maintained by magnetic stirrers and/or forced aeration.

Three basic systems are in use today: batch propagation, semi-continuous culture and continuous culture (see also Chapter 14, this volume). With *batch propagation*, there is a finite volume of medium in which cell growth ceases when essential nutrients are exhausted. In *semi-continuous cultures*, the medium is periodically drained and replaced with fresh medium. *Continuous cultures* can be either of the *open* or *closed* type. In the latter, the inflow of fresh medium is balanced by an outflow of the same volume of culture (i.e., spent medium + cells). There are two ways of monitoring this process:

(1) In a *chemostat* the growth rate of the cells is regulated by a controlled inflow of growth-limiting nutrient(s). The inflow rate depends on the doubling time of the cells. Generally it is between 25-35 hr (12).

(2) In a *turbidostat*, inflow of fresh medium occurs when the increased turbidity of the culture caused by the increase in cell number reaches a set threshold.

These large batch cultures (1.5 - 10.0 l) are fixed and can therefore be connected to reservoirs, gas supplies, thermostats and other electronic devices. A variety of set-ups have been described by Street (20). However, Kurz has pointed out that most plant cells form aggregates in suspension cultures which are agitated by conventional methods (12). While this may be acceptable in some cases (e.g., large-scale production of secondary metabolites), for some physiological or metabolic investigations, single cells are required. Fermenters which work on the air-lift principle have proven best for plant cells (22), with the result that scaling-up of culture vessel size is taking place. (For more details see refs. 12, 14, 20-22, 24 and Chapter 14, this volume).

C. Small Scale Suspension Cultures

Watch glasses have long been used as small scale culture vessels. For example, a chemical watch glass 5 cm in diameter set in a closed petri dish and surrounded by a ring of moistened cotton to support it and maintain a humid environment was used in 1929 by Fell and Robinson (5). White listed a number of other watch glass designs which could be used such as small Syracuse[R] watch glasses (A.H. Thomas) each covered with a 25 mm round cover glass (23). The main objection to watch glass cultures is the difficulty in sealing. A more modern approach is to suspend the cells in a small drop of medium (approximately 10^{-8} m^3 volume) in a microchamber formed by a central cavity or depression on a glass slide. A cover-slip sealed with mineral oil offers effective protection against evaporation. This method is useful for *in situ* staining and/or microscopic observation of individual cells.

The "hanging droplet" or "microdroplet array" (MDA) technique is one that offers the added advantage of large-scale screening of cultures rapidly and with little labor (15, 16). The method consists of dropping 40 μl droplets of liquid culture out of a 1 ml pipette onto the lid of a petri dish in a

7 x 7 array. A solution with osmotic pressure of approximately 60% that of the culture medium is added to the bottom half of the petri dish. The lid is inverted onto the dish so that "hanging droplets" are obtained. The dishes can be sealed with Parafilm^R, stacked and enclosed in plastic containers under high humidity. The hanging droplets contained 400 protoplasts each which formed a monolayer at the meniscus of the droplet. They were able to test 5000 experimental treatments using only 0.25 m^3 of culture room or incubator space and less than 1 g of tissue.

In conclusion, we have outlined the requirements for a tissue culture facility. We have also attempted to give the theoretical and practical basis for our recommendations. The actual requirements will vary with the type of operation to be carried out. The references cited in this chapter should be used for more detailed information. More specific use of the equipment, etc., as well as specialized requirements for various purposes are described in the appropriate sections in this volume. We also draw your attention to the brochure "General Procedures for the Cell Culture Laboratory" which is available from Corning Glass Works, Oneonta, New York.

REFERENCES

1. Bajaj, Y.P.S., *In* "Plant Cell and Tissue Culture. Principles and Applications" (W.R. Sharp, P.O. Larsen, E.F. Paddock, and V. Raghavan, eds.), p. 745. Ohio State Univ. Press, Columbus, (1979).
2. Bajaj, Y.P.S., and Reinert, J., *In* "Plant Cell, Tissue and Organ Culture" (J. Reinert and Y.P.S. Bajaj, eds.), p. 757. Springer-Verlag, Berlin, (1977).
3. Bragt, van J., Mossel, D.A.A., Pierik, R.L.M., and Veldstra, H., (eds.) "Effects of Sterilization on Components in Nutrient Media", H. Veenman and Zonen, N.V., Wageningen, (1971).
4. de Fossard, R.A., "Tissue Culture for Plant Propagators" Univ. of New England Printery, Armidale, N.S.W., (1976).
5. Fell, H.B., and Robinson, R., *Biochem. J. 23*, 767 (1929).
6. Fujimura, T., and Komamine, A., *Plant Sci. Lett. 5*, 359 (1975).
7. Gey, G.O., *Am. J. Cancer 17*, 752 (1933).
8. Heller, R., *In* "Tissue Culture. Methods and Applications". (P.F. Kruse, Jr. and M.K. Patterson, Jr., eds.), p. 387. Academic Press, New York, (1973).
9. Horsch, R.B., King, J., and Jones, G.E., *Can. J. Bot.* (in press) (1980).

10. Komamine, A., *Plant Physiol.* *64*, 162 (1979).
11. Komamine, A., *Z. Pflanzenphysiol.* *95*, 13 (1979).
12. Kurz, W.G.W., *In* "Tissue Culture. Methods and Applications". (P.F. Kruse, Jr. and M.K. Patterson, Jr., eds.), p. 359. Academic Press, New York, (1973).
13. Lamport, D.T.A., *Exp. Cell Res.* *33*, 195 (1964).
14. Miller, R.A., Shyluk, J.P., Gamborg, O.L., and Kirkpatrick, J.W., *Science* *159*, 540 (1968).
15. Potrykus, I., Harms, C.T., and Lörz, H., *In* "Cell Genetics in Higher Plants" (D. Dudits, G.L. Farkas and P. Maliga, eds.), p. 129. Akadémiai Kiado, Budapest, (1976).
16. Potrykus, I., Lörz, H., and Harms, C.T., *In* "Plant Tissue Culture and its Bio-technological Application" (W. Barz, E. Reinhard and M.H. Zenk, eds.), p. 323. Springer-Verlag, Berlin, (1977).
17. Short, K.C., Brown, E.G., and Street, H.E., *J. Exp. Bot.* *20*, 579 (1969).
18. Steward, F.C., Caplin, S.M., and Millar, F.K., *Ann. Bot.* *16*, 58 (1952).
19. Steward, F.C., and Shantz, E.M., *In* "The Chemistry and Mode of Action of Plant Growth Substances" (R.L. Wain and F. Wightman, eds.), p. 165. Butterworths Ltd., London, (1965).
20. Street, H.E. (ed.), "Plant Tissue and Cell Culture", 2nd edition. Blackwell Scientific Publications, Oxford, (1977).
21. Thomas, E., and Davey, M.R., "From Single Cells to Plants". Wykeham Publications (London) Ltd., London, (1975).
22. Wagner, F., and Vogelman, H., *In* "Plant Tissue Culture and its Bio-technological Application" (W. Barz, E. Reinhard and M.H. Zenk, eds.), p. 27. Springer-Verlag, Berlin, (1977).
23. White, P.R., "The Cultivation of Animal and Plant Cells", 2nd edition. The Ronald Press Company, New York, (1963).
24. Wilson, S.B., King, P.J., and Street, H.E., *J. Exp. Bot.* *21*, 177 (1971).
25. Withers, L.A., *In* "Frontiers of Plant Tissue Culture 1978" (T.A. Thorpe, ed.), p. 297. Univ. of Calgary Press, Calgary, (1978).
26. Withers, L.A., and Street, H.E., *In* "Plant Tissue Culture and its Bio-technological Application" (W. Barz, E. Reinhard and M.H. Zenk, eds.), p. 226. Springer-Verlag, Berlin, (1977).
27. Yeoman, M.M., and MacLeod, A.J., *In* "Plant Tissue and Cell Culture", 2nd edition. (H.E. Street, ed.), p. 31. Blackwell Scientific Publications, Oxford, (1977).

NUTRITION, MEDIA AND CHARACTERISTICS OF PLANT CELL AND TISSUE CULTURES

Oluf L. Gamborg

International Plant Research Institute
San Carlos, California

Jerry P. Shyluk

Prairie Regional Laboratory
National Research Council of Canada
Saskatoon, Saskatchewan, Canada

The historical development of plant tissue culture is described in reviews and books on plant tissue culture (63, 65) and will not be dealt with at this time. The present discussions will be concerned with two major aspects:

I. Nutrition and Media
II. Growth Characteristics and Behaviour of Cultured Cells, followed by a section on methodology.

I. NUTRITION AND MEDIA

A. *Introduction*

For convenience the plant tissue culture technology can be divided into five classes. These classes are based primarily on the type of materials used:

(1) Callus Culture. The culture of cell masses on agar media and produced from an explant of a seedling or other plant source.

(2) Cell Culture. The culture of cells in liquid media in vessels which are usually aerated by agitation.

ISBN 0-12-690680-7

(3) Organ Culture. The aseptic culture on nutrient media of embryos, anthers (microspores), ovaries, roots, shoots, or other plant organs.

(4) Meristem Culture and Morphogenesis. The aseptic culture of shoot meristems or other explant tissue on nutrient media for the purpose of growing complete plants.

(5) Protoplast Culture. The aseptic isolation and culture of plant protoplasts from cultured cells or plant tissues.

These five classes encompass most investigations on plant tissue culture.

In a discussion on methods it is important to recognize that the choice of culture procedures and conditions are based primarily on the objective or purpose of the investigation. The subjects of organ culture (anther, embryo), meristem culture and protoplasts will be discussed in detail in separate chapters.

Cells from any plant species can be cultured aseptically on or in a nutrient medium. The cultures are started by planting a sterile tissue section on an agar medium. Within 2-4 weeks, depending upon plant species, a mass of unorganized cells (callus) is produced. Such a callus can be subcultured indefinitely by transferring a small piece to fresh agar medium.

If a liquid suspension culture is desired, a callus mass, preferably friable in texture, is transferred to liquid medium and the vessel incubated on a shaker. Gradually over several weeks, and by subculturing, a liquid suspension culture can be obtained. The time required to obtain a callus tissue and a suspension culture varies greatly and depends primarily on the plant species, the origin of the explant tissue, and the composition of the culture medium.

Juvenile tissues are generally most likely to produce a callus. However, callus cultures have been obtained from seedlings, young shoots or buds, root tips or developing embryos: fruits, floral parts, tubers and bulbs. Any plant tissue with living cells can be tested. Under the influence of growth regulators, which are discussed under nutrition, the cells may be induced to divide. Callus tissues have been obtained from wood phloem, leaf materials, fruit and other tissues in which the cells are highly specialized (differentiated) and not meristematic.

The details of materials, equipment and procedures will be discussed in connection with the laboratory section. This also includes methods of obtaining sterile tissues (also see Chapter 1).

B. *Nutritional Requirements*

Success in the technology and application of *in vitro* methods is due largely to a better understanding of the nutritional requirements of cultured cells and tissues (20, 42, 63). The two factors which most frequently determine the success of cell cultures are explant origin and the general nutritional milieu. The subject of explant origin has been mentioned earlier and it will be discussed later.

The nutritional milieu consists of essential and optional components. The essential nutrients consists of inorganic salts, a carbon and energy source, vitamins and phytohormones (growth regulators) (Table I). Other components, including organic nitrogen compounds, organic acids, and complex substances, can be important but are optional.

1. *Inorganic Salts.* The inorganic nutrients of a plant cell culture are those required by the normal plant. The following are required in millimole quantities: N, K, P, Ca, S and Mg. The optimum concentration of each nutrient for achieving maximum growth rates varies considerably. For most purposes a nutrient medium should contain at least 25 and up to 60 mM inorganic nitrogen. The cells may grow on nitrate alone but often there is a distinct beneficial effect and occasionally a requirement for ammonium or another source of reduced nitrogen. Nitrate is commonly used in the range from 25-40 mM. The amount of ammonium varies between 2 and 20 mM. Possibly an optimum concentration would be 2-8 mM. Amounts in excess of 8 mM can result in reduced growth. Cells can be grown with ammonium as the sole nitrogen source provided citrate, succinate, malate or another TCA cycle acid is present at ca. 10 mM. It has also been possible to grow cells on other nitrogen sources such as urea, glutamine or casein hydrolyzate. Potassium is required and must be supplied at concentrations of 20 mM or higher. Sodium cannot be used as a substitute. Potassium is generally supplied as the nitrate or as the chloride. The optimum concentrations of P, Mg, Ca and S under conditions where other requirements are satisfied for cell growth vary from 1-3 mM.

The essential nutrients required in micromolar concentrations include Fe, Mn, Zn, B, Cu and Mo. Fe and sometimes Zn are supplied as the chelate with versene (ethylenediaminetetraacetic acid). It may also be important to add Co; I is used in several media but can be omitted.

Plant cells tolerate high concentrations of chloride and sodium, and they have no apparent effect on growth rate.

TABLE I. *Composition of Murashige-Skoog and Gamborg's Media for Plant Tissue and Cell Culture*

	MS(43)		B5(17)	
Macronutrients	mg/l	mM	mg/l	mM
NH_4NO_3	1650	20.6	-	-
KNO_3	1900	18.8	2500	25
$CaCl_2 \cdot 2H_2O$	440	3.0	150	1.0
$MgSO_4 \cdot 7H_2O$	370	1.5	250	1.0
KH_2PO_4	170	1.2	-	-
$(NH_4)_2SO_4$	-	-	134	1.0
$NaH_2PO_4 \cdot H_2O$	-	-	150	1.1
Micronutrients	mg/l	μM	mg/l	μM
KI	0.83	5	0.75	4.5
H_3BO_3	6.2	100	3.0	50
$MnSO_4 \cdot 4H_2O$	22.3	100	-	-
$MnSO_4 \cdot H_2O$	-	-	10	60
$ZnSO_4 \cdot 7H_2O$	8.6	30	2.0	7.0
$Na_2MoO_4 \cdot 2H_2O$	0.25	1.0	0.25	1.0
$CuSO_4 \cdot 5H_2O$	0.025	0.1	0.025	0.1
$CoCl_2 \cdot 6H_2O$	0.025	0.1	0.025	0.1
Fe-Versenate (EDTA)	43.0	100	43.0	100
Vitamins & Hormones	mg/l		mg/l	
Inositol	100		100	
Nicotinic acid	0.5		1.0	
Pyridoxine·HCl	0.5		1.0	
Thiamine·HCl	0.1		10.0	
IAA	1-30		-	
Kinetin	0.04-10		0.1	
2,4-D	-		0.1-2.0	
Sucrose	30000		20000	
pH	5.7		5.5	

The chlorides and sodium salts are therefore useful sources of nutrients for nutritional studies, since they in effect permit the addition of a single nutrient.

2. *Carbon and Energy Source.* The standard carbon source is sucrose or glucose. Fructose also can be used but is

generally much less suitable. Other carbohydrates which have been tested include lactose, maltose, galactose and starch, but these compounds are generally much inferior to sucrose or glucose as a carbon source. Sorbitol has been used for cells of apple and other species of the Rosaceae. Sucrose is generally used at a concentration of 2-3%. Most media contain m-inositol. There is no absolute requirement, but the inclusion of ca. 100 mg per liter improves cell growth.

3. *Vitamins.* Normal plants synthesize the vitamins required for growth and development. When cells of higher plants are grown in culture, some vitamins may become limiting. There is an absolute requirement for thiamine (47). The biochemical basis for this requirement has not been determined. Growth is also improved by the addition of nicotinic acid and pyridoxine. Some media also contain panthothenate and biotin, but these and other vitamins are generally not considered growth limiting factors. The only exception could be in cases when it would be desirable to grow cells at very low concentrations. The following may then be included: p-amino-benzoic acid, folate, choline chloride, riboflavin and ascorbic acid.

4. *Phytohormones.* The cytokinins and auxins are the two classes of compounds which are of special importance in plant tissue culture. Some of these are naturally occurring hormones, e.g., indoleacetic acid (IAA) and zeatin; others are synthetic growth regulators such as 2,4-dichlorophenoxyacetic acid (2,4-D) and benzyladenine (BA). An auxin and sometimes a cytokinin are required for inducing cell division and the formation of callus. The compound most frequently used and the most effective is 2,4-D. Other auxins in use include naphthaleneacetic acid (NAA), IAA, indolebutyric acid (IBA) and p-chlorophenoxyacetic acid (pCPA).

The cytokinins are derivatives of adenine. Several cytokinins occur in cells of all organisms, but the hormone activity is detectable only in plants. The compounds which are used most frequently are kinetin, BA, isopentenyl adenosine (IPA) and zeatin. Other types of plant hormones such as gibberellic acid (GA_3) and abscisic acid have never been demonstrated to be essential for growth. GA_3 is essential for meristem culture of some species.

5. *Organic Nitrogen.* Common sources of organic nitrogen in nutrient media include amino acids, glutamine, asparagine and adenine. A source of organic nitrogen should not be necessary, but it is often beneficial to include a protein digest or L-glutamine. A common approach is to add 0.02-0.1%

casein hydrolyzate or vitamin-free casamino acids to the medium. In some cases L-glutamine may replace the protein digest. Plant cells can tolerate up to 8 mM of L-glutamine. The amino acids when added singly should be used with caution, since they can be inhibitory. Examples of amino acids included in media and amount in mg/l include: glycine 2, asparagine 100, L-tyrosine 100 (only in agar media for morphogenesis), L-arginine and cysteine 10. An approach which has been successful is to group the amino acids for separate testing to ascertain their effect on growth. The inclusion of organic nitrogen compounds is generally necessary only when a callus is being initiated, although there may be some benefit in fortifying the medium for established callus and suspension cultures. Nucleotides have not been established as limiting factors. Some benefit has been derived by adding adenine sulfate to agar media used for morphogenesis.

6. Organic Acids. Plant cells are not able to utilize organic acids as sole carbon sources. Relatively little research has been done to establish if organic acids can be growth limiting. Addition of TCA cycle acids such as citrate, malate, succinate or fumarate permits growth of plant cells on ammonium as the sole nitrogen source. The cells tolerate a concentration of up to 10 mM of the acids. Pyruvate also may enhance growth of cells cultured at low density.

7. Complex Substances. A variety of extracts have been tested. These include protein hydrolyzates, yeast extracts, malt extracts and a variety of plant preparations including coconut and corn endosperm, orange juice and tomato juice. With the exception of protein hydrolyzates and coconut liquid endosperm most other substances are used as a last resort and frequently have been found to be replaceable by defined nutrients. The complex substance, if used at higher concentrations, may adversely affect cell growth. It is advisable to make a preliminary test in a range of 0.1 to 1 g per liter to assess effect on growth. Coconut milk is commonly used at 2-15% (v/v).

8. Environmental Factors. Several factors influence the growth of callus and cell suspension cultures. The pH is an important variable in the culture milieu. Plant cells in culture require an acidic pH, and an initial pH of 5.5-5.8 is optimum. The pH changes during the growing cycle of a cell suspension culture. Initially there is a decrease to below pH 5. Subsequently the pH increases and may reach 6 or higher.

The growth rate is closely related to the prevailing temperature. Growth may occur at temperatures below 20°C. A temperature between 26-28°C is considered optimum for achieving maximum growth. Some cell lines can grow at 32-33°C.

Plant cells in culture are aerobic and aeration is an essential requirement for suspension cultures Shakers are operated at rates which vary from 100-150 rpm.

Light is generally not essential for growing of callus or cell cultures. The cultures may grow equally well in the dark and light. However, light may have a profound effect on the metabolism of the cells A "dark" room is frequently an area which is kept dark but lights are turned on for inspection of the cultures at the time of subculturing. In general, light is provided by Gro Lux[R] or similar sources or white fluorescent lamps at intensities of ca. 300-10,000 lux ($\simeq$ 0.2-5 mW cm^{-2}) at the culture level. The light is given continuously or in photoperiods of 8-16 hr.

C. Media Considerations for Different Culture Systems

1. Callus and Cell Suspension Culture. A wide variety of media compositions have been used with success. For most purposes the salt compositions of media such as Murashige-Skoog, or B5 seem to be adequate. Variations and modifications of these media are widely used. To these are added the B-vitamins, inositol and sucrose in a variety of quantities. The hormones added to the medium induce the cells to divide. 2,4-D at about 1-5 μM is perhaps the most efficient. The addition of 1 μM kinetin may be beneficial. It is also advisable to add 0.05% protein hydrolyzate. For particularly difficult tissues it may be beneficial also to add 2% coconut milk and 2-4 mM L-glutamine.

In starting a callus it is advisable to set up a large number of replicates (see Chapter 12 for an example). The growth rate and friability of callus produced can vary widely between explants and even within replicates on the same medium.

The callus can be maintained on agar or a liquid suspension culture can be obtained. For the latter a B5 type salt composition may be more suitable than Murashige-Skoog. The B51 which is a modification of the B5 medium compensates for weaknesses of the B5 (55). The amounts of Ca, P and NH_4 are higher. In the early stages of establishing cell suspension cultures it is important to use a friable callus and maintain a relatively high callus to liquid ratio.

2. *Morphogenesis.* The Murashige-Skoog (MS) salt mixture is the most suitable basic medium for plant regeneration from tissues and callus (42, 43). The basic composition of the Linsmaier-Skoog medium is identical to the MS. The hormones are the important compounds in a plant regenerating medium. The capacity for plant regeneration of tissues vary widely. In some species the process is readily induced while in others morphogenesis fails to occur. The cytokinins induce shoot formation in plant species which have the capacity for organogenesis. The shoots may be formed from stem, leaf, cotyledon or other organs. In some cereal plants the cultured immature embryos form a callus which regenerates shootlets. When the shootlets are transferred to an auxin enriched medium complete plants are obtained (also see Chapter 3).

In addition to cytokinins, the concentration of sucrose as well as the inclusion of adenine sulfate, L-tyrosine and charcoal also can influence the efficiency of plant regeneration. The auxins which are used to promote root formation on the shoots include NAA, IBA and IAA. 2,4-D represses formation of both shoots and roots and should not be included in media for plant regeneration, except in the case of cereal plants.

The cultured cells of some species have the capacity to differentiate and produce embryos which develop into complete plants (e.g., carrot, celery, coffee). Such cultures are initiated and maintained in 2,4-D containing media. When embryo development and plant regeneration is desired the cells are washed and incubated in hormone-free liquid or agar medium (18). The cell density should be kept very low during embryo development. Any growth regulator added to the medium adversely affects embryogenesis except perhaps zeatin, which reportedly may stimulate the process.

3. *Protoplasts.* Protoplasts are basically plant cells without the walls. The nutritional requirements are generally the same as for cultured plant cells. However, the removal of the cell wall makes it necessary to include osmotic stabilizers and additional nutritional ingredients to preserve the protoplasts and ensure their viability. The principal difference between protoplast and cell culture media is the inclusion of osmotic stabilizers such as sorbitol, mannitol or glucose. Glucose rather than sucrose may be the main carbon source, and a pentose is often added. The mineral salt composition is altered to contain a considerably higher concentration of calcium salt.

The success in culturing protoplasts is closely related to the conditions for their isolation as well as to the physiological state of the tissue origin. Protoplasts from cell suspension cultures generally are more likely to regenerate the cell walls and divide than those from plant tissues (see Chapter 4).

D. *Choice of Media*

A wide variety of salt mixtures has been reported (see 50). Several have been tested with callus and cell cultures of a wide variety of plant species. A medium is identified by its mineral salt composition. Vitamins, hormones and other organic supplements vary widely with respect to composition and concentration. The choice depends on the plant species and to a degree upon the intended use of the culture.

Tables I and II list two media. The Murashige-Skoog medium is widely used for general plant tissue cultures but

TABLE II. Concentrations of Inorganic Nutrients in Murashige-Skoog and Gamborg's Media for Plant Tissue Culture

Macronutrients (mM)	*MS (43)*	*B5 (17)*
K	20	25
N(NO_3)	40	25
N(NH_4)	20	2.0
Mg	1.5	1.0
P	1.25	1.1
Ca	3.0	1.0
S	1.5	1.0
Na	-	1.1
Cl	6.0	2.0
Micronutrients (μM)		
I	5.0	4.5
B	100	50
Mn	100	60
Zn	30	7.0
Mo	1.0	1.0
Cu	0.1	0.15
Co	0.1	0.1
Fe	100	100

especially for morphogenesis, meristem culture and plant regeneration (41, 42). The medium is characterized by high concentrations of mineral salts. The B5 medium contains lower amounts of mineral salts, which appears to be preferred by cells of some species (16, 17). A comprehensive list of culture media is reported by Street and Shillito (63a).

II. GROWTH CHARACTERISTICS

Callus grown on solid agar media achieves a relatively slow rate of growth. The new cells are formed on the periphery of the existing callus mass. Consequently the callus will consist of cells which vary considerably in age. Since nutrients gradually are depleted from the agar, a "vertical" nutrient gradient is formed.

Because of the low degree of uniformity, slower growth rate and the development of nutrient gradients, the usefulness of callus as an experimental system is limited. The principal use of the agar callus method is for purposes of maintaining cell lines and for morphogenesis.

Cell suspension cultures when fully established consist of a nearly homogeneous population. The system has the advantage that the nutrients can be continually adjusted and the cells are surrounded by the medium. The suspension cultures consist of mixtures of single cells and cell clusters. The cells in liquid suspension can be grown in petri dishes, in flasks or on a larger scale in multiliter fermenters (see Chapter 1).

A. *Growth Kinetics and Measurements of Suspension Cultures*

The cells in culture exhibit the expected phases of a growth cycle. When cells are subcultured into fresh medium there is a lag phase. Then follows a period of cell division (the exponential phase) and finally the cell population reaches a stationary phase. Prior to the last phase, the cells expand in size. The cell generation time (doubling time) in suspension cultures may be as low as 18 hours but generally varies from 24 to 48 hours in well established cell cultures. Growth is most often assessed by the dry weight of cell mass or cell count measured at intervals over a period of several days. The tendency for cell aggregation makes it difficult to obtain meaningful cell counts. There are reports of using enzymes or chemicals to achieve single cell cultures.

For general purposes the objective with cell suspension cultures is to achieve rapid growth rates, and uniform cells, with all cells being viable. Frequent subculturing in a suitable medium ensures these qualities. Cells should be subcultured at weekly intervals or less if they are to be used for experimental purposes. The size of the inoculum (dilution rate) will determine the lag period. There is generally a minimum inoculum size below which a culture will not recover. An inoculum corresponding to 0.7-0.9 mg d. wt. per ml is usually sufficient. With any culture the safest approach is to use a relatively heavy inoculum and gradually establish an optimum for a given subculture regime.

Attempts have been made to synchronize the cell division cycle. The objective is to achieve a high percentage of cells passing through the growth cycle in synchrony. Some of the better results have been achieved in a fermenter in which the division cycle was regulated by pulses of ethylene and carbon dioxide. The advantage of having a culture system to obtain synchrony is that it would make it possible to study a large population of cells carrying out the same metabolic activity and function.

B. *Genetic Variation and Chromosome Stability*

Many reports have appeared on the variation in ploidy level of cultured cells. Those reports were often based on observations with callus tissue and slower growing cell lines not fully established. There is considerable evidence that suspension cultures which have achieved a rapid growth rate have a stable and relatively uniform ploidy level (11, 60). However, the ploidy level is often different from that of the original tissues (Table III).

There is apparently no regime which will assure that cells in culture retain the ploidy level of the original tissues.

During callus initiation and particularly in the first weeks of culture a selection process is in effect. It is well known that phenotypic variations exist within a population of cells. The cells which have adapted the best to the chemical (nutrition) and physical environment and have the shortest doubling time will predominate and gradually the entire population will consist of these cells.

The evidence for genetic variability is supported by the reports in recent years of the isolation of cell lines resistant to antibiotics, amino acid analogues and ability to utilize maltose or galactose (37, 66).

Cells in a population also differ in their capacity to

TABLE III. Chromosome Numbers of Cell Cultures of Different Plant Species[a]

		Chromosome number:	
Species	*Cell line*	*Plant tissue*	*Cell culture*
Glycine max	SB-64	2n = 48	37
Nicotiana glauca	NG-478	2n = 24	24
Nicotiana tabacum	SU-677	2n = 48	48
Nicotiana tabacum	NT-775	2n = 48	76
Solanum chacoence	SOC-1175	2n = 25	36
Triticum monococcum	WTM-1066	2n = 14	24
Lycopersicon esculentum	TS-675	2n = 24	104
Zea mays	BM-477	2n = 20	44
Sorghum bicolor	GPR-168	2n = 20	66
Brassica napus	RZ-573	2n = 38	40-44
Datura innoxia	DI-76	n = 12 60%	12
		37%	24

[a]Evans, D. and Gamborg, O.L., *Plant Sci. Lett.* (in press).

accumulate compounds such as pigments or other secondary plant products. Other cell lines have shown resistance to high salt concentrations and herbicides.

The biochemical basis for resistance or tolerance to chemicals has in most cases not been established. Resistance to amino acid analogues is sometimes accompanied by a simultaneous accumulation of the respective natural amino acid.

C. Differentiation and Totipotency

Plant cells have the potential to differentiate on division and develop organs and complete plants (42, 51, 58) (Table IV).

The phenomenon of differentiation can occur in explanted tissues from stems, leaves, corms, tubers, embryos and other organs. Some callus formation may be an intermediate step. In some species plant regeneration can occur from callus or cell suspension cultures. This topic is discussed in greater detail in Chapter 3.

The process may be initiated by cytokinins which induce shoot formation. Roots will generally be produced on the shoots in the presence of a rooting hormone (NAA, IBA or IAA), or in the absence of any growth regulator.

A special application of differentiation is the production of haploid plants from microspores. When combined

TABLE IV. *Examples of Crop Plant Research in Which Tissue Culture was Utilized for Plant Regeneration*

Plant Species	Common Name	References
Hordeum vulgare	barley	8, 14
Zea mays	corn	22
Pennisetum spp.	millet	67
Avena sativa	oat	7
Oryza sativa	rice	45, 46
Sorghum bicolor	sorghum	6, 21
Triticum spp.	wheat	23, 59
Saccharum officinarum	sugar cane	24, 38
Beta vulgaris	sugar beet	53
Solanum tuberosum	potato	34, 62
Manihot esculenta	cassava	27, 44
Ipomea batatas	sweet potato	57
Glycine max	soybean	17, 26
Psophocarpus tetragonolobus	winged bean	4, 23a
Pisum sativum	pea	19, 27
Medicago sativa	alfalfa	54
Trifolium pratense	red clover	48
Stylosanthes hamata	tropical legume	56
Trifolium alexandrinum	beerseem clover	40
Vigna sinensis	cow pea	26
Cicer arietinum	chick pea	26
Brassica napus	rapeseed	28, 31, 64
Brassica campestris	rapeseed	32
Ananas comosus	pineapple	70
Phoenix dactylifera	date palm	52
Carica papaya	papaya	72
Citrus spp.	orange	49, see 51
Malus spp.	apple	35
Pyrus communis	pear	36
Coffea arabica	coffee	61, see 51
Prunus amygdalus	almond	39
Vitis vinifera	grape	3, 33
Pinus pinaster	pine	9, see 58
Picea abies	spruce	69
Eucalyptus spp.	eucalyptus	see 58
Linum usitatissimum	flax	15
Cucumis pepo	pumpkin	25
Asparagus officinalis	asparagus	10
Allium cepa	onion	12
Brassica oleracea	cauliflower	71
Daucus carota	carrot	65
Carum carvi	caraway	1

TABLE IV. (continued)

Plant Species	*Common Name*	*References*
Brassica oleracea	*broccoli*	*2*
Lycopersicon esculentum	*tomato*	*27, 73*
Fragaria	*strawberry*	*5, see 51*

with treatments to induce chromosome doubling, the plants will be homozygous and essentially consist of pure lines (see Chapter 8).

Morphogenesis will be discussed in connection with protoplasts, from which it is also possible to regenerate plants.

There is another aspect of cell differentiation which should be mentioned. Although cultures consist almost entirely of rapidly dividing (meristematic) cells, it is not uncommon to obtain callus which consists of a mixture of pigmented and non-pigmented cells.

The cells accumulate a pigment. The capability of cells to accumulate pigment and secondary metabolites such as alkaloids, steroids and terpenes is a differentiation process and the cells have become specialized. This property of plant cells to produce compounds of medicinal value has been explored for several years (65, 66), and is discussed in Chapter 14.

III. ESTABLISHING CALLUS AND CELL CULTURES (16)

A. Equipment and Facilities

(1) Laminar air-flow sterile cabinets.

(2) Autoclave and dry sterilization oven.

(3) Filter sterilization equipment.

(4) Distillation apparatus or demineralizer for high purity water.

(5) Culture room(s) or cabinets with light and temperature control.

(6) Shakers.

(7) Stainless steel dissecting knives, scalpel or Beaver[R] blades mounted on stainless steel handles, forceps, scissors, and spatulas.

(8) Glass jars with metal lids and autoclavable liners.

(9) Flasks, DeLong[R] flasks or equivalent types with

relatively long neck and without collar.

(10) Petri dishes (100 x 15 mm).

(11) Stoppers.

(12) Pipettes. 5, 10, and 20 ml with tips removed.

Other equipment which should be available are refrigerators, freezers, microscopes, balances, and a pH meter (see Chapter 1).

B. *Reagent Materials*

(1) Water of high purity (see Chapter 1).

(2) Mineral salts and organic compounds. The compounds should be of the highest grade. Growth regulators such as naphthaleneacetic acid and 2,4-dichlorophenoxyacetic acid should be purified by charcoal decolorization followed by one or more recrystalizations from a water-ethanol mixture. Cytokinins can be similarly recrystallized from ethanol. Amino acids should consist of the L-isomer.

(3) Protein hydrolyzates. These are available as enzymatics and as acid hydrolyzates. The enzymatic digests are preferable because all the amino acids are preserved.

(4) Coconut milk. Ripe coconuts are generally used. A hole is drilled through one of the germination pores. The liquid is collected from several nuts, heated to 80°C with stirring, filtered and stored frozen.

(5) Agar. "Bacto" from the Difco Company or "Noble" are frequently used.

(6) Ethanol, 95%, 70%.

(7) Hypochlorite. Solutions of 5% sodium hypochlorite or 20% household bleach such as Chlorox®.

(8) Wetting agent (Tween-80®).

C. *Media Preparation*

The chemicals are dissolved in distilled or demineralized water, the stock solutions added and the pH adjusted. The solution is made to volume and then distributed into flasks. The flasks are stoppered and labeled with Time® tape of different colors. Each color signifies a specific medium. The medium is autoclaved at 121°C for 15 min and the flasks removed for cooling as soon as possible. Agar media are autoclaved in quantities of 500 ml and subsequently poured into sterile containers. All media are stored at 10°C.

1. *Stock Solutions for B5 Medium.* (For preparation of

B5 medium see below; for composition see Table I). *Note:* B5 medium refers to the basic medium with no growth hormone or organic supplements. 1-B5 medium refers to the basic medium plus 1 ppm (1 mg/liter) of 2,4-D.

(1) Micronutrients (store in freezer).

	mg/100 ml
$MnSO_4 \cdot H_2O$	1000
H_3BO_3	300
$ZnSO_4 \cdot 7H_2O$	200
$Na_2MoO_4 \cdot 2H_2O$	25
$CuSO_4 \cdot 5H_2O$	2.5

(2) Vitamins (store in freezer).

	mg/100 ml
Nicotinic acid	100
Thiamine·HCl	1000
Pyridoxine·HCl	100
Myo-Inositol	10000

(3) Calcium chloride
$CaCl_2 \cdot 2H_2O$ 15 g/100 ml

(4) Potassium iodide
KI 75 mg/100 ml
Store in amber bottle in refrigerator.

(5) 2,4-Dichlorophenoxyacetic acid (2,4-D) (2.27 mM). Dissolve 50 mg 2,4-D in 2-5 ml ethanol, heat slightly and gradually dilute to 100 ml with water. Store in refrigerator.

(6) Naphthaleneacetic acid (NAA) (2.68 mM). Prepare the same as 2,4-D above.

(7) Kinetin (1.0 mM). Dissolve 21.5 mg of kinetin in a small volume of 0.15N HCl by heating slightly and gradually diluting to 100 ml with distilled water. Store in refrigerator. Similar procedures can be used for benzyladenine and other cytokinins.

2. *Preparation of B5 Medium with 1 mg/l 2,4-D (1-B5).*

Ingredient	Amount/l
$NaH_2PO_4 \cdot H_2O$	150 mg
KNO_3	2500 mg
$(NH_4)_2SO_4$	134 mg
$MgSO_4 \cdot 7H_2O$	250 mg
Ferric EDTA	43 mg
Sucrose	20 g
$CaCl_2 \cdot 2H_2O$, stock solution	1.0 ml
Micronutrients, stock solution	1.0 ml
Potassium iodide, stock solution	1.0 ml
Vitamins, stock solution	1.0 ml
2,4-D, stock solution	2.0 ml

Final pH adjusted to 5.5 with 0.2N KOH or 0.2N HCl.

3. Modifications to B5.

(1) B5C medium. Add 1.0 g/l of N-Z-Amine[R] Type A. A pancreatic hydrolyzate of casein, obtainable from Humko-Sheffield, New York.

(2) Agar media. Add 6 to 8 g of agar per liter of B5. Dissolve in an autoclave, on a hot plate with magenetic stirrer, or in a microwave oven.

(3) Ammonium citrate medium. Omit KNO_3 and $(NH_4)_2SO_4$ and add 20 mM KCl, 10 mM citric acid and 20 mM NH_4OH.

4. Murashige-Skoog Medium (MS). (For composition see Table I).

(1) MS-Micronutrient stock solution (keep frozen).

Ingredient	mg/100 ml
H_3BO_3	620
$MnSO_4 \cdot 4H_2O$	2230
$ZnSO_4 \cdot 7H_2O$	860
$Na_2MoO_4 \cdot 2H_2O$	25
$CuSO_4 \cdot 5H_2O$	2.5
$CoCl_2 \cdot 6H_2O$	2.5

(2) All other stock solutions the same as for B5.

5. Preparation of MS Medium (1-MS).

Ingredient	Amount/1
NH_4NO_3	1650 mg
KNO_3	1900 mg
$MgSO_4 \cdot 7H_2O$	370 mg
KH_2PO_4	170 mg
Ferric EDTA	43 mg
Sucrose	30 g
$CaCl_2 \cdot 2H_2O$ (B5 stock solution)	2.9 ml
MS-micronutrients (stock solution)	1.0 ml
KI (B5 stock solution)	1.0 ml
Vitamins (B5 stock solution)	1.0 ml
2,4-D (B5 stock solution)	2.0 ml

Adjust pH to 5.8 with 0.2N KOH or 0.2N HCl.

6. Preparation of Agar Media.

(1) Add the growth hormone and supplements.

(2) Adjust the pH.

(3) Add 0.6% Difco Agar and heat solution while stirring until agar is dissolved.

(4) Distribute the medium in glass vessels and plug the

vessels.

(5) Autoclave for 20 min at 20 psi.

Note: Using sterile containers, the agar medium is autoclaved, cooled to ca. 50°C and poured into vessels in a laminar air flow cabinet.

(6) After cooling the media are stored preferably at 4-10°C.

D. Media Concentrates

For larger operations media concentrates may be an advantage. The media without hormones or organic supplements are prepared at ten times final desired concentration and stored frozen in plastic bags with a specified volume.

1. 0-B5 Medium, 10x Concentrate.

Ingredients	To Make 4 Liters
$NaH_2PO_4 \cdot H_2O$	6.0 g
KNO_3	100.0 g
$(NH_4)_2SO_4$	5.36 g
$MgSO_4 \cdot 7H_2O$	10.0 g
Iron Versenate (EDTA)	1.72 g
$CaCl_2 \cdot 2H_2O$ Stock (15 g/100 ml)	40 ml
Micronutrients (stock solution)	40 ml
KI Stock (75 mg/100 ml)	40 ml
Vitamins (stock solution)	40 ml
Sucrose	800 g

Note the following:

(1) Dissolve above ingredients in 3 liters of glass-redistilled H_2O and then adjust volume to 4 liters.

(2) Do not adjust pH.

(3) Dispense into Whirl--Pak[R] or equivalent bags (100 ml into 6 oz bag or 400 ml into 18 oz bag).

(4) Store bags in deep-freeze.

(5) This preparation is 10x the standard B5 medium.

(6) Preparation of 1B5 medium (1 liter): dispense 100 ml concentrate in glass distilled water, add 2,4-D from stock solution, adjust pH and make volume.

2. 0-MS Medium, 10x Concentrate.

Ingredients	To Make 4 Liters
NH_4NO_3	66.0 g
KNO_3	76.0 g
$MgSO_4 \cdot 7H_2O$	14.8 g
KH_2PO_4	6.8 g
Ferric Versenate (EDTA)	1.72 g

$CaCl_2 \cdot 2H_2O$ (stock 15 g/100 ml)		116	ml
Micronutrients (stock solution)		40	ml
	mg/100 ml		
H_3BO_3	620		
$MnSO_4 \cdot 4H_2O$	2230		
$ZnSO_4 \cdot 7H_2O$	860		
$Na_2MoO_4 \cdot 2H_2O$	25		
$CuSO_4 \cdot 5H_2O$	2.5		
$CoCl_2 \cdot 6H_2O$	2.5		
KI (stock solution, 75 mg/100 ml)		40	ml
Vitamins (B5 stock solution)		40	ml
Sucrose		1200	g

Proceed as for 0-B5 concentrate.

E. *Explant Materials*

The choice of plant species and tissues is dictated by the research objective and availability. Almost any part of a plant can be induced to produce a callus and a suspension culture. Juvenile tissues are most likely to succeed. Seedlings refer to young plants living on storage nutrients of the seed.

The materials in order of priority are sections of sterile seedlings, swelling buds, stem or storage organs, leaf materials and, for cereals, immature embryos, mesocotyl or basal stem sections of young plants.

1. *Seed Sterilizing and Germination.*

(1) Transfer seeds to 70% ethanol in a flask and put on a shaker for 2 min.

(2) Discard the alcohol and add 20% commercial bleach containing 5% hypochlorite or use 1-2% sodium hypochlorite. Leave the flask on a shaker for 15-20 min. Discard floating seeds. A second treatment with hypochlorite may be necessary if the seeds are heavily contaminated.

(3) Rinse the seeds in sterile, distilled water.

(4) Place the seeds on double layers of pre-sterilized filter paper in petri dishes. Add sterile distilled water and seal with Parafilm[R].

Note: The seeds may also be germinated aseptically on moist cotton in glass jars with caps or sealed with Parafilm[R] or on sterile moist vermiculite in culture tubes. Most seeds germinate well in the dark at 25-28°C. It is advantageous to use relatively few seeds per container, since a single seed may contaminate the others in a particular container.

(5) When the seeds germinate, any part of the seedling is suitable as explant for callus formation.

2. *Bud, Leaf and Stem Sections.* The aerial portion of plants are sterilized by submerging for 1-3 min in 70% ethanol followed by 2-3 rinses in sterile, distilled water.

3. *Tubers, Bulbs and Roots.* The surface of the organs are sterilized in 70% ethanol followed by immersion in 20% commercial bleach (5% sodium hypochlorite) solution.

F. *Starting Callus and Cell Cultures*

(1) Place 3-5 sections of about 0.5 cm each from sterile seedlings into nutrient medium containing 2,4-D and solidified with 0.6-0.8% Difco Bacto agar. Suitable containers are glass jars, flasks, culture tubes (1 section each), or pre-sterilized plastic jars with screw caps.
(2) Incubate the sections in the dark (or low light) at 26-28°C.
(3) After 3-4 weeks the callus should be about 5 times the size of the explant.
(4) Transfer the callus from 2-3 containers into 20 ml of liquid medium in a 125 ml flask.
(5) Incubate the flasks on a gyratory shaker at 150 rpm in continuous light at 26°C.
(6) Subculture. In the early stages it may be necessary to decant a portion of the spent medium at two week intervals by pipetting and replacing the volume with fresh medium. A cell suspension should form within 4-6 weeks. Initially it is advisable to use a low dilution rate of 1:1 to 1:4 of fresh medium.

G. *Plant Regeneration from Embryo-forming Cell Suspension Cultures*

(1) Aseptically collect 5-10 ml of cell suspension and wash 2-4 times in hormone-free medium by filtration or centrifugation. Either:
(2a) Disperse the cells at a very low density in hormone-free liquid medium in petri dishes. Seal the dish with Parafilm^{R}.
(3a) Incubate on rotary shaker at 30-50 rpm at 25-27°C in light. Or alternatively:
(2b) Disperse the cells on hormone-free agar plates (100 x 15 mm or 60 x 15 mm).

(3b) Incubate at 25-27°C in light. In either case:

(4) Transfer plantlets to hormone-free agar medium in glass tubes or jars and incubate at 25°C in light at 16:8 h photoperiod.

(5) Transfer the rooted plants to Jiffy® pots or peat-vermiculite (or soil) and water with a nutrient solution (see 16).

(6) Grow the plants in a greenhouse or growth room.

REFERENCES

1. Ammirato, P.V., *Bot. Gaz. 135,* 328 (1974).
2. Anderson, W.C., and Carstens, J.B., *J. Amer. Soc. Hort. Sci. 102,* 69 (1977).
3. Barlass, M., and Skene, K.G.M., *Vitis 17,* 335 (1978).
4. Bottino, P.J., Maire, C.E., and Goff, L.M., *Can. J. Bot. 57,* 1773 (1979).
5. Boxus, P.H., *J. Hort. Sci. 49,* 209 (1974).
6. Brar, D.S., Rambold, S., Gamborg, O.L., and Constabel, F., *Z. Pflanzenphysiol. 95,* 377 (1979).
7. Cummings, P.D., Green, C.E., and Stuthman, D.D., *Crop Sci. 16,* 465 (1976).
8. Dale, P.J., and Deambrogio, E., *Z. Pflanzenphysiol. 94,* 65 (1979).
9. David, A., and David, H., *Z. Pflanzenphysiol. 94,* 173 (1979).
10. Dore, C., *Acta Horticulturae 78,* 89 (1977).
11. Evans, D.A., and Gamborg, O.L., *Environ. Expt. Bot. 19,* 269 (1979).
12. Fridborg, G., *Physiol. Plant. 25,* 436 (1971).
13. Gamborg, O.L., *Plant Physiol. 45,* 372 (1970).
14. Gamborg, O.L., and Eveleigh, D.E., *Can. J. Biochem. 46,* 417 (1968).
15. Gamborg, O.L., and Shyluk, J.P., *Bot. Gaz. 137,* 301 (1976).
16. Gamborg, O.L., and Wetter, L.R. (eds.), *In* "Plant Tissue Culture Methods", National Research Council of Canada, Saskatoon, Canada (1975).
17. Gamborg, O.L., Miller, R.A., and Ojima, K., *Exp. Cell Res. 50,* 151 (1968).
18. Gamborg, O.L., Constabel, F., and Miller, R.A., *Planta 95,* 355 (1970).
19. Gamborg, O.L., Constabel, F., and Shyluk, J.P., *Physiol. Plant. 30,* 125 (1974).
20. Gamborg, O.L., Murashige, T., Thorpe, T.A., and Vasil, I.K., *In Vitro 12,* 473 (1976).

21. Gamborg, O.L., Shyluk, J.P., Brar, D.S., and Constabel, F., *Plant Sci. Lett. 10*, 67 (1977).
22. Gengenbach, B.G., Green, C.E., and Donovan, C.M., *Proc. Nat. Acad. Sci. (U.S.A.) 74*, 5113 (1977).
23. Gosh, G., Avivi, L., and Galun, E., *Z. Pflanzenphysiol. 91*, 267 (1979).
23a. Gregory, N.M., Haq, N., and Evans, P.K., *Plant Sci. Lett. 18*, 395 (1980).
24. Heinz, D.J., and Mee, G.W.P., *Crop Sci. 9*, 346 (1969).
25. Jelaska, S., *Physiol. Plant. 31*, 257 (1974).
26. Kartha, K.K., and Gamborg, O.L., *In* "Diseases of Tropical Food Crops" (H. Maraite and J.A. Meyer, eds.), p. 267, Proc. Int. Symp. U.C.L. Louvain-la-Neuve, Belgium (1978).
27. Kartha, K.K., and Gamborg, O.L., *In* "Plant Cell and Tissue Culture, Principles and Applications" (W.R. Sharp, P.O. Larsen, E.F. Paddock, and V. Raghavan, eds.), p. 711, Ohio State Univ. Press, Columbus, Ohio (1977).
28. Kartha, K.K., Gamborg, O.L., Shyluk, J.P., and Constabel, F., *Physiol. Plant. 31*, 217 (1974).
29. Kartha, K.K., Gamborg, O.L., Shyluk, J.P., and Constabel, F., *Z. Pflanzenphysiol. 77*, 292 (1976).
30. Kartha, K.K., Leung, N.L., and Gamborg, O.L., *Plant Sci. Lett. 15*, 7 (1979).
31. Keller, W.A., and Armstrong, K.C., *Z. Pflanzensucht 80*, 100 (1978).
32. Keller, W.A., Rajhathy, T., and Lacapra, J., *Can. J. Genet. Cytol. 17*, 655 (1975).
33. Krul, W.R., and Worley, J.F., *J. Amer. Soc. Hort. Sci. 102*, 360 (1977).
34. Lam, S.L., *Amer. Pot. 54*, 575 (1977).
35. Lane, W.D., *Plant Sci. Lett. 13*, 281 (1978).
36. Lane, W.D., *Plant Sci. Lett. 16*, 337 (1979).
37. Limberg, M., Cress, D., and Lark, K.G., *Plant Physiol. 63*, 718 (1979).
38. Liu, M.C., Shang, K.C., Chen, W.H., and Shih, S.C., *Plant Breeding 29* (1979).
39. Mehra, A., and Mehra, P.N., *Bot. Gaz. 135*, 61 (1974).
40. Mokhtarzadeh, A., and Constantin, M.J., *Crop Sci. 18*, 567 (1978).
41. Murashige, T., *In Vitro 9*, 81 (1973).
42. Murashige, T., *Annu. Rev. Plant Physiol. 25*, 135 (1974).
43. Murashige, T., and Skoog, F., *Physiol. Plant. 15*, 431 (1962).
44. Nair, N.G., Kartha, K.K., and Gamborg, O.L., *Z. Pflanzenphysiol. 95*, 51 (1979).

45. Nakano, H., Tashiro, T., and Maeda, E., *Z. Pflanzenphysiol. 76*, 444 (1975).
46. Niizeki, H., *Japan. Agric. Res. Q. 3*, 41 (1968).
47. Ohira, K., Ikeda, M., and Ojima, K., *Plant Cell Physiol. 17*, 583 (1976).
48. Phillips, G.C., and Collins, G.B., *Crop Sci. 19*, 213 (1979).
49. Rangan, T.S., Murashige, T., and Bitters, W.P., *In* "Proceedings First International Citrus Symposium", Vol. 1, (H.D. Chapman, ed.), p. 225, Univ. Calif. Press (1969).
50. Rechcigl, M., *In* "CRC Handbook Series in Nutrition and Food", Sect. G, Vol. IV, CRC Press Inc., Boca Raton, Florida (1977).
51. Reinert, J., and Bajaj, Y.P.S. (eds.), *In* "Plant Cell, Tissue, and Organ Culture", Springer-Verlag, New York (1977).
52. Reuveni, O., *Hortscience 14*, 457 (1979).
53. Rogozinska, J., Goska, M., and Kuzdowicz, A., *Acta Soc. Bot. Pol. 46*, 471 (1977).
54. Saunders, J.W., and Bingham, E.T., *Crop Sci. 12*, 804 (1972).
55. Savage, A.D., and Gamborg, O.L., *Plant Sci. Lett. 16*, 367 (1979).
56. Scowcroft, W.F., and Adamson, J.A., *Plant Sci. Lett. 7*, 39 (1976).
57. Sehgal, C.B., *Z. Pflanzenphysiol. 88*, 349 (1978).
58. Sharp, W.R., Larsen, P.O., Paddock, E.F., and Raghaven, V. (eds.), *In* "Plant Cell and Tissue Culture", Ohio State Univ. Press, Columbus, Ohio (1979).
59. Shimada, T., Sasakuma, T., and Tsunewaki, K., *Can. J. Gen. Cytol. 11*, 294 (1969).
60. Singh, B.D., Harvey, B.L., Kao, K.N., and Miller, R.A., *Can. J. Genet. Cytol. 14*, 65 (1972).
61. Sondahl, M.R., and Sharp, W.R., *Z. Pflanzenphysiol. 81*, 395 (1977).
62. Sopory, S.K., Jacobsen, E., and Wenzel, G., *Plant Sci. Lett. 12*, 47 (1978).
63. Street, H.E. (ed.), *In* "Plant Tissue and Cell Culture", Botanical Monographs, Vol. II, 2nd Edition, Blackwell, Oxford (1977).
63a Street, H.E., and Shillito, R.D., *In* "CRC Handbook Series, Nutrition and Food, Section G, Vol. IV (M. Rechcigl, Jr., ed.), p. 305, CRC Press, Boco Raton, Florida (1977).
64. Stringam, G.R., *Z. Pflanzenphysiol. 92*, 459 (1979).
65. Steward, F.C., "Plant Physiology", Vol. VB, Academic Press, New York (1969).

66. Thorpe, T.A., (ed.), "Frontiers of Plant Tissue Culture 1978", University of Calgary Press, Calgary, (1978).
67. Vasil, V., and Vasil, I.K., *Z. Pflanzenphysiol. 92,* 379 (1979).
68. Vasil, I.K., Ahuja, M., and Vasil, V., *Advances in Genetics, 20,* 127 (1979).
69. Von Arnold, S., and Eriksson, T., *Plant Sci. Lett. 15,* 363 (1979).
70. Wakasa, K., Koga, Y., and Kudo, M., *Japan J. Breed. 28,* 113 (1978).
71. Walkey, D.G.A., and Woolfitt, J.M.G., *J. Hort. Sci. 45,* 205 (1970).
72. Yie, S.-T., and Liau, S.I., *In Vitro 13,* 564 (1977).
73. Zanir, D., Jouer, R.A., and Kedar, N., *Plant Sci. Lett.* (in press).

GROWTH AND BEHAVIOR OF CELL CULTURES: EMBRYOGENESIS AND ORGANOGENESIS

David A. Evans
William R. Sharp

Campbell Institute for Research and Technology
Cinnaminson, New Jersey

Christopher E. Flick

Center for Somatic Cell Genetics and Biochemistry Research
State University of New York
Binghamton, New York

I. INTRODUCTION

A. Rationale for Plant Regeneration

The primary goal of plant tissue culture research is crop improvement. Before cellular genetic techniques can be applied to crop improvement, though, efficient protocols for plant regeneration must be implemented. Plant regeneration can be utilized to recover unique variants induced *in vitro*. For example, mutagenesis, protoplast fusion, or DNA uptake may be used to genetically modify a plant cell with clonal propagation utilized to recover genetically altered regenerated plants. Alternately, evidence exists which suggests that populations of regenerated plants may contain tremendous variability. In this case tissue culture is used to induce genetic variability or to recover pre-existing natural genetic variability. In either case, regenerated variant plants can then be used to complement existing breeding programs. Unfortunately, as most agriculturally important crop species cannot be regenerated with the ease of

ISBN 0-12-690680-7

model systems such as carrot and tobacco, it is necessary to establish usable protocols for somatic embryogenesis and organogenesis in crop species.

Numerous reviews of plant regeneration and *in vitro* propagation have been published in the past few years. Unfortunately, the summary tables of plant regeneration in most of these reviews contain only the names of species and sample references (e.g., 173, 174, 184, 291, 292). From these reviews it is impossible to directly ascertain the necessary information to reproduce these plant regeneration experiments. Usually no information on explant source, media, or hormone concentrations has been included. In this review we have not only listed all crop species for which plant regeneration has been documented, but have also emphasized the techniques of plant regeneration. Consequently we have consciously eliminated some citations with insufficient details or in which complete plants were not obtained. In addition this review emphasized crop species. This chapter is limited to somatic organogenesis and embryogenesis. *In vitro* cultures of protoplasts, meristems and anthers have been covered in detail in other chapters of this volume (see Chapters 4, 6, 8). For recent complete lists of species that will undergo plant regeneration we suggest Murashige (174), Vasil and Vasil (291) and Vasil *et al.* (292).

II. LABORATORY PROTOCOL

A. *General Methodology*

1. Explants. Establishment of callus growth with subsequent organogenesis or embryogenesis has been obtained from many species of plant cells cultured *in vitro*. Most viable plant cells can be induced to undergo mitosis *in vitro*. Plant regeneration has been successfully accomplished from explants of cotyledon (107), hypocotyl (114), stem (44), leaf (92), shoot apex (119), root (86), young inflorescences (155), flower petals (100), petioles (203), ovular tissue (130) or embryos (182). For any given species or variety, a particular explant may be necessary for successful plant regeneration, e.g., embryonic tissue is required for certain cereals. Explants consisting of shoot tips or isolated meristems, which contain mitotically active cells, have been especially successful for callus initiation and subsequent plantlet regeneration (173, 175). Explants from both mature and immature organs can be regenerated directly or can be induced to proliferate and form callus on the appropriate culture medium prior to plant regeneration. Size and shape of the explant may be critical. The increased cell number present in explants of greater mass increases the probability of obtaining a viable culture. Yeoman (308) has shown that differences in the critical size of carrot root and artichoke tuber explants reflects cell size. In each case the optimum explant contains 20-25,000 cells. Table I contains a sample of explants introduced *in vitro*, using surface sterilization procedures for various explants. In each case commercial hypochlorite in varying concentrations, alone or in combination with ethanol, is proposed. In each case field-grown material is more difficult to sterilize than material obtained from greenhouse grown plants (see also Chapter 2).

2. *Callus Proliferation.* Only a small percentage of the cells in a given explant contribute to the formation of callus. The site for initiation of callus proliferation is generally situated at the surface of the inoculum or at the excised surface. Yeoman has suggested maximizing the cut surface exposed to the culture medium (308). Callus may also be commonly formed from parenchyma, as in *Nicotiana* (58) or from the procambium of the leaf sheath nodes as in rice (304). Detailed discussions of callus formation have been made by Gautheret (75) and Aitchison *et al.* (5).

Numerous cytological studies have been made on callus and suspension cells, notably with carrot (89, 185, 186), and sycamore (32). General information pertaining to the cytology of cells *in vitro* suggests that cell size ranges between 200-475 μm in length versus 45-90 μm in width for elongated cells; 60-130 μm in diameter for spherical cells; and 120-175 μm in length versus 60-95 μm width for oval cells (115). Average cell size increases with increased time in culture (269). Growth regulator concentrations (94) and nutritional factors (281) also regulate cell size. Friability, or the tendency for cells to separate from each other is a property of the cell wall which can be increased at higher concentrations of auxin in the culture medium (283). It can also be increased by lowering the cytokinin concentration and/or the addition of gibberellin to the medium (147). Friable callus is essential for establishment of liquid suspension cell cultures.

The season of the year can influence the success of obtaining callus from explants, especially when the donor plant is field grown in temperate climates. Seasonal variations have been observed in the concentration of endogenous auxins (303) and have been reported to influence the establishment of potato meristem cultures (167) and

TABLE I. Preparation of Explants for Plant Regeneration Experiments

Explant	Plant	Size	Sterilization protocol	Remarks	Reference
Leaf blade	tomato	6 x 8 mm	wash in detergent, 10 min in 7% Clorox[a], wash twice in water	young leaves near shoot apex	198
Stem	rape	5 mm	rinse in 70% ethanol, 6 min in 10% Clorox, wash four times in water	basal end in contact with medium	263
Embryo	cacao	2.5-25 mm intact	15 min in 10% Clorox with 0.1% Tween 20	use pods larger than 12 cm	200
Storage organ	artichoke	2.4 x 2 mm	30 min in 20% Clorox, wash repeatedly	use storage parenchyma region	308
Seed (root)	petunia	3 mm	30 min 50% Clorox, wash three times in water	use 4-6 days after germination	40
Seed (hypocotyl)	flax	4-8 mm	1 min in 70% ethanol, 20 min in 20% Clorox, wash in water	use 5 days after germination	68

[a] Clorox[R], a commercial bleach, is a 5% solution of sodium hypochlorite.

conifer cultures (96). Spring and summer were found to be the best seasons for starting cultures. The developmental stage and physiological state of the plant at the time of culture also must be considered, because such factors as dormancy of the cambium or lateral buds, induction of flowering, etc., may affect the response and/or the success of initiating cultures.

Callus has been successfully induced from numerous explants. Within a given plant, success of callus initiation is dependent on the explant source. For example, growth regulator requirements in the culture medium are different for mesocarp versus cotyledon tissue for *Persea* (22). In *Persea*, the cotyledonary tissue is autonomous for cytokinin synthesis while the mesocarp tissue is cytokinin requiring. Different concentrations of 2,4-D in the culture medium are necessary for obtaining callus from root, scutellum, cotyledonary nodes, coleoptile, and leaf sheath nodes of rice (304). These variations in culture medium requirements reflect phenotypic differences in the physiology of the different explants. Differences also occur in the callus morphology from *Sinapis* root and hypocotyl (11) and in cell size differences in adult and juvenile tissues of *Hedera* (259). However, Barker did not observe gross morphological differences between callus derived from various organs of *Tilia* or *Triticum* (16). Ploidy differences have also been reported to differ in callus obtained from different tissue explants of rice (305). Physiological differences between root and hypocotyl callus cultures can be seen in the inability of soybean hypocotyl callus cultures to support symbiotic growth of *Rhizobium*, while nodules develop in callus cultures from root explants of the same plant (103).

Such variations *in vitro* may reflect the differences in the phenotypic physiological expression of the cells in the original explant. On the other hand, the phenotypic expression of a cell may be modified by isolation and culture. Explants of different tissues may achieve the same state of cytodifferentiation on a given medium or express epigenetic differences which reflect tissue origin. Obviously, the *in vitro* response for a given cell in culture is determined by environmental and genetic interactions.

The size of the explant and, in a few cases, the mode of culture or polarity of the explant in the culture medium effect callus development. Smaller explants are more likely to form callus while the larger explants maintain greater morphogenetic potential (195). Small tissue explants frequently require a more complex medium than do large ones, especially in the case of small organ cultures. The effect of explant tissue placement on solid medium can be observed by the improved growth of pine embryos with their cotyledons placed in direct contact with the agar medium as compared to when nutrients are absorbed through the root (25). The polarity of the tissue explant especially in the case of stem and petiole explants affects the location of callus formed.

3. *Culture Medium* (see also Chapter 2). The essential components of plant tissue culture medium have been summarized (72). Formulations of the four media most often used for plant regeneration are summarized in Table II. It

TABLE II. Mineral Salt Concentrations of Culture Media Most Often Used for Plant Regeneration

	Medium (reference)			
	MS (177)	B5 (70)	SH (234)	W (300)
Macronutrients (mM)				
NH_4NO_3	20.6	-	-	-
KNO_3	18.8	25	25	0.8
$CaCl_2 \cdot 2H_2O$	3.0	1.0	1.4	-
$MgSO_4 \cdot 7H_2O$	1.5	1.0	1.6	2.9
KH_2PO_4	1.25	-	-	-
$(NH_4)_2SO_4$	-	1.0	-	-
$(NH_4)H_2PO_4$	-	-	2.6	-
$NaH_2PO_4 \cdot H_2O$	-	1.1	-	0.12
$Ca(NO_3)_2 \cdot 4H_2O$	-	-	-	0.12
KCl	-	-	-	0.9
Na_2SO_4	-	-	-	1.4
Micronutrients (μM)				
KI	5	4.5	6	4.5
H_3BO_3	100	50	80	25
$MnSO_4 \cdot 4H_2O$	100	-	-	31.4
$MnSO_4 \cdot H_2O$	-	60	60	-
$ZnSO_4 \cdot 7H_2O$	30	7	3.5	10.5
$Na_2MoO_4 \cdot 2H_2O$	1.0	1.0	0.4	-
MoO_3	-	-	-	trace
$CuSO_4 \cdot 5H_2O$	0.1	0.1	0.8	0.001
$CoCl_2 \cdot 6H_2O$	0.1	0.1	0.4	-
$Fe_2(SO_4)_3$	-	-	-	6.3
Fe-EDTA chelate	100	100	55	-

is evident that MS medium (177) contains a high concentration of nitrogen as ammonium (20 mM), while White's medium (300) is ammonium-free. As high ammonium concentrations reduce cell growth in some plant systems (67), it may be difficult to transfer cells from White's (W), B5 (70), or SH (234) medium onto MS medium. White's medium is a low salt medium while the other three media are classified as high salt media. SH medium resembles B5 medium with some salts present in slightly higher concentrations. LS medium (148) has also been used for plant regeneration. The macro and micronutrients of LS and MS media are identical. As differences between these two media are found only in concentrations of vitamins and organic supplements, we will refer only to MS medium. Other modifications to the organic additives of MS medium have been reported (69). The organic supplements required in plant culture media include a carbon source and vitamins. Sucrose is used as a carbon source but may be substituted with glucose. Other sugars are used less often. Vitamins most often added to culture medium include inositol, nicotinic acid, pyridoxine, thiamine, calcium pantothenate and biotin. Thiamine is required for plant growth, while the remainder enhance growth in some systems. Other vitamins have been used in plant systems. Vitamin E has been shown to regulate cell aggregation (197) while vitamin D may enhance root formation (27). Kao and Michayluk use a number of vitamins in the culture of plant protoplasts and isolated single cells (116). Various plant extracts or undefined additives are sometimes added to the culture medium to increase cell growth.

4. *Modes of Culture.* Two modes of cell culture are generally used: (1) the cultivation of clusters of cells on a solid medium (e.g., agar, gelatin, filter paper, millipore filters) and (2) the cultivation of cell suspensions in liquid medium. It is best to initiate new cultures on solid medium as some essential nutrients may readily leak from small explants placed in large volumes of liquid medium. A suspension cell culture is usually initiated by placing friable callus into liquid culture medium. The suspension usually consists of free cells and aggregates of 2-100 cells. Suspension cultures should be subcultured at least once a week while callus cultures should be subcultured less frequently. The rate of subculture of each should increase when cultures are established to achieve optimal growth rate and genetic stability (60).

5. *Plant Regeneration.* Growth regulator concentrations in the culture medium are critical to the control of growth and morphogenesis, as first indicated by Skoog and Miller

(251). Generally a high concentration of auxin and a low concentration of cytokinin in the medium promotes abundant cell proliferation with the formation of callus. Often, 2,4-D is used alone to initiate callus. On the other hand, low auxin and high cytokinin concentrations in the medium result in the induction of shoot morphogenesis. Auxin alone or with a very low concentration of cytokinin is important in the induction of root primordia.

In tissue culture systems where the objective is vegetative propagation, investigators generally categorize the developmental sequence of events in three stages (173). Stage I occurs following the transfer of an explant onto a culture medium. Enlargement of the explant and/or callus proliferation occurs during this stage. Stage II is characterized by a rapid growth increase of organs and may require subculture onto medium with or without altered growth regulator concentrations. Subculture may result in the induction of adventitious organs or embryos. In stage III regenerated plants are removed from *in vitro* culture. This requires conversion of the plant to the autotrophic state. Factors to be considered in stage III, include root formation and moisture stress.

6. *Abbreviations*

auxins:

IAA = indole acetic acid;
IBA = indole butyric acid;
2,4-D = 2,4-dichlorophenoxyacetic acid;
NAA = naphthaleneacetic acid;
pCPA = para-chlorophenoxyacetic acid;
NOA = β-napthoxyacetic acid;
BTOA = 2-benzothiazole acetic acid;
PIC = picloram;
2,4,5-T = 2,4,5-trichlorophenoxyacetic acid;

cytokinins:

KIN = kinetin;
6BA = 6 benzyladenine (benzyl amino purine)
2iP = 2 isopentenyl adenine;
ZEA = zeatin;

other growth regulators:

ADE = adenine;
CW = coconut water;
CH = casein hydrolysate;
ABA = abscisic acid;
GA_3 = gibberellic acid.

B. *Species Capable of Embryogenesis*

Somatic embryogenesis *can* proceed directly from either a population or populations of sporophytic or gametophytic cells of a cultured explant, e.g., hypocotyl in *Daucus carota* and *Ranunculus sceleratus*. However, in the majority of cases, asexual somatic embryogenesis results from indirectly determined embryogenic cells. Here, specific growth regulator concentrations and/or cultural conditions are required for initiation of callus and the epigenetic redetermination of cells to the embryogenic pattern of development, e.g., secondary phloem of carrot; inner hypocotyl tissues of wild carrot; leaf tissue of coffee; and pollen for rice and certain other Gramineae.

An understanding of these two different patterns of development depends upon consideration of the determinative events of cytodifferentiation during the mitotic cell cycle. It is well known that the fate of determined daughter cells following mitosis occurs at least one mitotic cell cycle prior to differentiation (308). Cells with an epigenetic commitment for direct asexual embryogenesis are the daughters of a prior determinative cell division. Such cells may undergo a post-mitotic arrest until environmental conditions are favorable for commencement of the developmental sequence of embryogenesis.

1. *Somatic Embryogenesis in Carrot*. Most of the available cytological and histological information on the origin and development of somatic embryos *in vitro* is based on studies of callus and suspension cultures of carrot (10, 91, 113, 123, 124, 164, 223, 262). The protocol for achieving somatic embryogenesis in carrot suspension cultures is presented in Fig. 1. The principal steps in this protocol are briefly discussed below:

(1) Place a 0.5-1 cm petiole explant or 0.5 cm^2 storage root explant on MS agar medium + 4.5 μM 2,4-D using aseptic techniques.

(2) Sufficient callus proliferation should occur following a four week growth period for initiation of cell suspension cultures.

(3) Subculture 2.5-3.5 gm of callus into 50 ml of MS liquid medium + 4.5 μM 2,4-D contained in a 250 ml Erlenmeyer flask. Flasks are placed on a gyrotary shaker and maintained at 25°C in the presence of illumination (ca. 1000 lux) and agitated at 160 rpm.

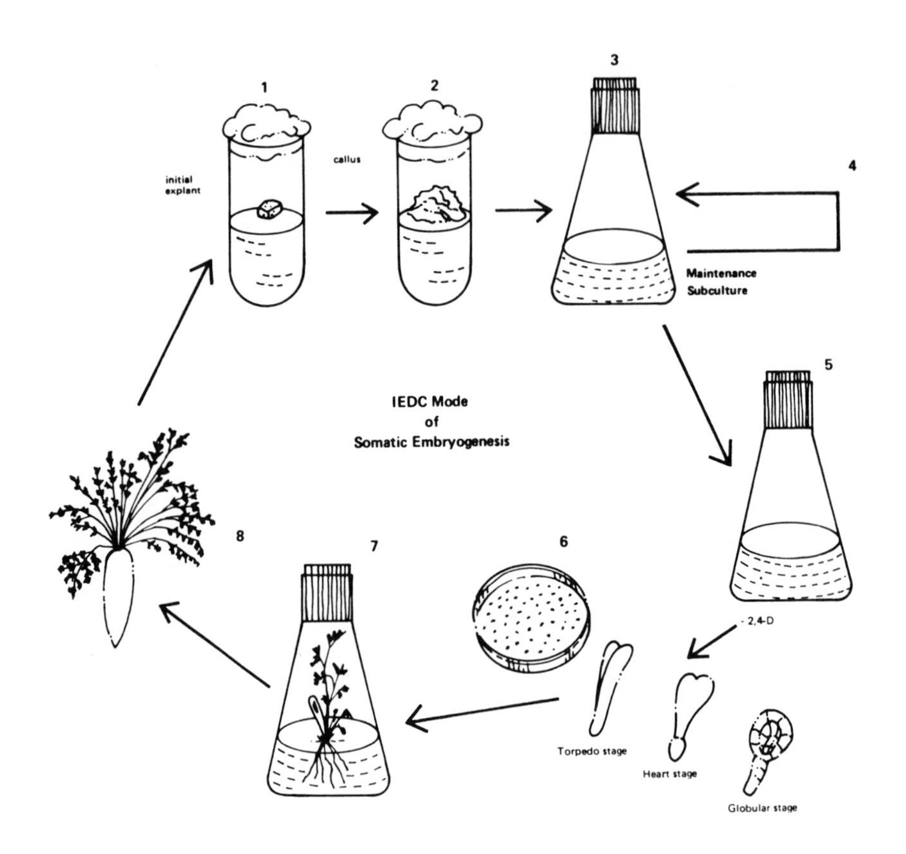

FIGURE 1. Protocol for somatic embryogenesis in carrot. For IEDC (induced embryogenic determined cells) mode of embryogenesis, explants are excised from petiole or roots and placed onto MS medium with 2,4-D for callus induction (1). Callus formation is obtained in 3-4 weeks (2). Suspension cultures can be established by subculturing callus into medium without agar (3). Suspension cultures can be maintained with 2,4-D in the medium (4), or somatic embryos induced when 2,4-D is removed (5). Somatic embryos develop when plated onto growth regulator-free MS medium (6). Root formation is obtained on growth regulator-free half-strength MS medium (7), with recovered plants transferred to the greenhouse (8).

(4) Suspension cultures are subcultured at 14-18 day intervals. Rapidly proliferating cultures can be placed on a 10 day subculture schedule.

(5) Aseptically aspirate and decant the medium from step 4 and subculture cells onto culture medium devoid of 2,4-D for initiation of embryo development.

(6) Aliquots of cells grown in step 5 for 10-18 days are plated out on 1/2 strength agar medium devoid of 2,4-D for plantlet development.

(7) Plantlets are transferred to Jiffy pots^R or vermiculite for subsequent development.

The frequency of somatic embryogenesis can be increased by altering components of the culture medium. These factors include concentrations of growth regulators, particularly auxins, source and concentration of nitrogen, and the concentration of other growth additives (in references 62, 243).

2. *Somatic Embryogenesis in Citrus*. A thorough review of tissue culture of *Citrus* has been published by Button and Kochba (30). They report that most *Citrus* species are polyembryonic and produce from one to 40 adventitive embryos in the nucellus (66). Polyembryony is believed to be a recessive, hereditary character controlled by multiple genes (154). These genes may regulate the synthesis of a potent inhibitor of embryogenesis in nucellar cells of monoembryonic *Citrus* varieties (59). In polyembryonic varieties, the induction of nucellar embryos *in vivo* appears to be dependent upon a stimulus. This is provided by pollination, and possibly by fertilization and the early development of a zygotic embryo (65). The frequency of polyembryony is influenced by the nutritional status of the fruit (287), the pollen parent, and external environmental factors (65).

The mineral constituents of MS medium are generally adequate, although not necessarily optimal, for induction of nucellar embryos in cultures of all polyembryonic or monoembryonic varieties (30). The major difference between MS medium and other formulations used is in the concentration and source of nitrogen (Table II). MS medium contains a relatively high concentration of ammonium in addition to a high nitrate level. It appears that embryogenesis in unorganized *Citrus* tissues may be enhanced by NH_4NO_3 as reported by Reinert and Tazawa for *Daucus* cell cultures (224). The calcium concentration is also somewhat higher in the MS formulation than in other culture media.

A sucrose concentration approaching 5% (w/v) is generally recommended. Glycine has been found to increase slightly the yield of lemon albedo callus (178) and is generally included at concentrations of 26.7-53.4 μM. Supplementation with complex substances rich in amino acids, e.g., CH, and extracts of salt and yeast have proved beneficial. These responses may have been due either to the supply of particular amino acids or to the provision of reduced nitrogen not included in some basal media (30). Malt extract (ME) has been found generally to be either essential (21) or at least beneficial to embryogenesis in nucellar cultures (28, 130, 131, 169, 215). In some systems ME can be replaced by combinations of orange juice, NAA, and ADE (215, 216) or by ADE and KIN (169), and to some extent by ADE alone (130). Button and Botha found that ME stimulated the development of pseudobulbils from single cells derived from embryogenic "Shamouti" callus (29). Ovular callus proliferation was enhanced by 500 mg/l ME (217) but similar concentrations were detrimental to the development of embryos within ovules, and to the growth of plantlets in culture (218).

In general it appears that auxins are not essential for *Citrus* proliferation particularly in explants from immature fruit tissues. Auxin used together with a cytokinin can stimulate embryogenesis (131, 169).

3. *General protocol for Citrus tissue culture.*

(1) Immature fruits are harvested at 100-120 days following pollination.

(2) Wash fruits with tap water, then surface sterilize by immersing for 10 minutes in 10% commercial hypochlorite. A few drops of Tween 20^R emulsifier are included in the disinfectant as a surfactant.

(3) Without rinsing, the fruits are cut open under a laminar flow hood and the ovules excised and transferred to a sterile petri dish.

(4) With the aid of a microscope, using 30-60 X magnification, the zygotic embryo and nucellus are removed from each ovule. The nucellus is placed into culture and the zygotic embryo is discarded.

The protocol for *Citrus* tissue culture is summarized in Figure 2.

Callus formation from the nucellar explant occurs in ca. 2-3 weeks with a polyembryonal mass formed in 6 weeks. Thereafter individual plantlets can be separated for reculture and eventual transfer to pots. Indirect induction of adventitive embryos (30) occurs in the polyembryonic varieties (131, 154, 218, 230).

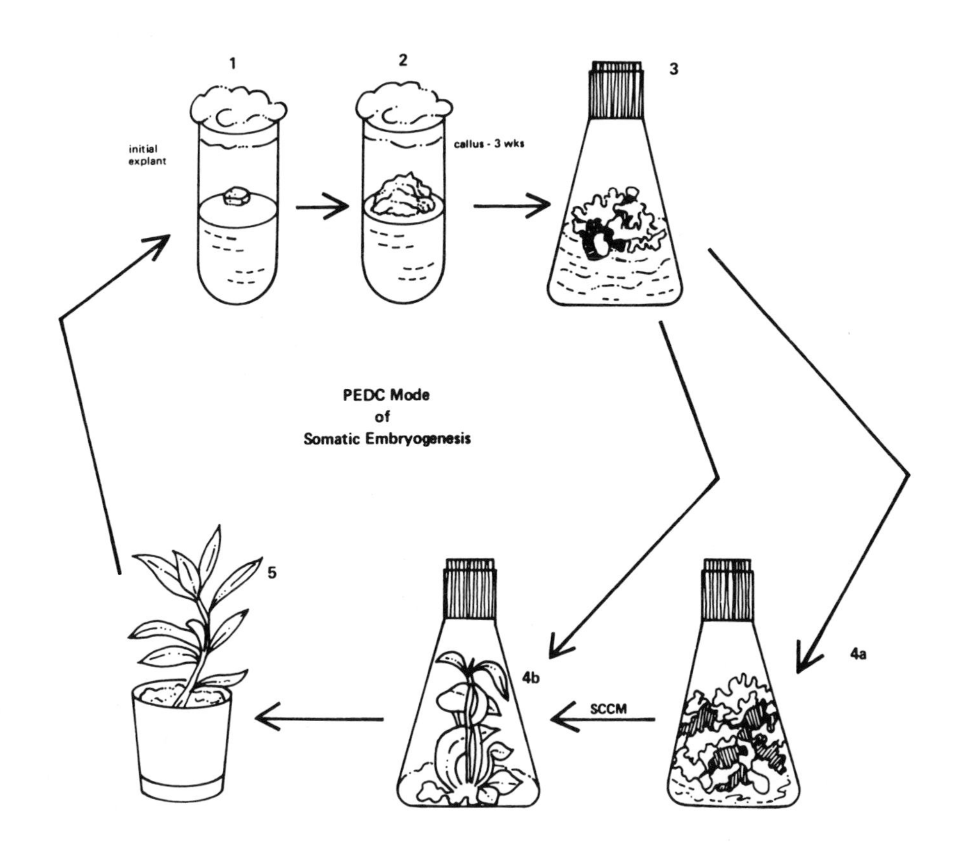

FIGURE 2. Protocol for somatic embryogenesis in Citrus. For PEDC (pre-embryogenic determined cells) mode of embryogenesis, nucellar explants are excised from immature fruits and placed onto MS medium with KIN and IAA (1). Callus formation is obtained in 3 weeks (2). Upon subculture to a medium devoid of growth regulators a polyembryonal mass is produced (3). Explants of the polyembryonal mass are subcultured onto medium with malt extract and ADE to achieve stem cell clonal multiplication (SCCM), resulting in an increase of cells capable of embryogenesis (4a). Alternately, embryos can be recovered directly by subculturing onto growth regulator-free medium (4b). Plantlets can be transferred to the greenhouse (5).

Kochba *et al.* obtained an embryonic callus from entire, unfertilized ovules and isolated nucelli of "Shamouti" orange (*C. sinensis*) (131). A similar callus was derived from sliced, unfertilized ovules of *Citrus aurantifolia*. Mitra and Chaturvedi noted a decrease in embryo formation after repeatedly subculturing their callus on media supplemented with ADE, KIN, and NAA (169). In the "Shamouti" callus however, Kochba and Spiegel-Roy observed continued embryogenesis or rather stem cell clonal multiplication (SCCM) which was promoted by ME and ADE (130).

After repeated subculturing, a number of "Shamouti" callus lines differing greatly in their embryogenic potential were selected. These lines all became independent of exogenous auxin and cytokinin (132). Embryo formation in callus lines of low embryogenic potential was strongly stimulated by aging the callus prior to subculture. A similar response was observed when callus was maintained on a medium devoid of sucrose prior to subculture (129). Stimulation of embryo differentiation was also observed when the callus was subjected to fairly high doses of gamma irradiation (256). Maximum stimulation was obtained by 16 kRad, while doses of 28-32 kRad were lethal. Irradiation of medium alone, when containing malt extract caused as much embryo stimulation as did irradiation of the callus alone (130).

4. *General Procedures*. General and specific information pertaining to the culture regimes for initiation of somatic embryogenesis is summarized in Tables III, IV, V, VI and VII.

It is important to note that ca. 40.0% of crop species undergoing somatic embryogenesis originate from either embryo or hypocotyl explants during primary culture. Furthermore, ca. 70.0% of the explants are cultured on MS medium or a modification of MS (Table III). These media regimes include 2,4-D in the primary medium for 57.7% of the crop species and 57.1% of the non-crop species. KIN is used in the primary medium for 50.0% of the crop species and 14.3% of the non-crop species. NAA has been used in the primary medium to a lesser extent in crop and non-crop species at frequencies of 26.9% and 21.4% respectively. Effective concentration ranges for these growth regulators are 0.5-27.6 μM for 2,4-D, 0.5-10.7 μM for NAA, and 0.5-5.0 μM for KIN (Table VII). It should be noted that the primary culture medium for date palm contains 452 μM 2,4-D. The growth regulators pCPA, 6BA, IBA, 2iP, NOA, BTOA are used to a lesser extent (Tables III, IV, V, VI, VII). The additive CW is used at concentrations of 10-15% per liter in the primary medium of 3.8% of the crop plants and 21.4% of the non-crop plants.

TABLE III. *Crop Species Capable of Somatic Embryogenesis Following Primary and Secondary Culture*

Crop	Species	Growth regulators: 1° Medium	Growth regulators: 2° Medium	Medium	Explant	Reference
Anise	*Pimpinella anisum*	5 μM 2,4-D	none	B5	hypocotyl	108
Asparagus	*Asparagus officinalis*	5.4 μM NAA 4.7 μM KIN	0.5-5.7 μM IAA 0.4-17.7 μM 6BA	LS or MS	cladodes shoots	226,227
Cacao	*Theobroma cacao*[a]	none 10% CW	6.4 μM NAA 10% CW	MS	imm. embryo cotyledon	200
Cauliflower	*Brassica oleracea*	5.7 μM IAA 2.0 μM KIN	5.7 μM IAA	MS	leaf	199
Caraway	*Carum carvi*	10.7 μM NAA	none	MS	petiole	6
Carrot	*Daucus carota*	4.5 μM 2,4-D	none	MS	storage root	90
Celery	*Apium graveolens*	2.2 μM 2,4-D 2.7 μM KIN	2.7 μM KIN	MS	petiole	312
Coffee	*Coffea arabica*	18.4 μM KIN 4.5 μM 2,4-D	2.3 μM KIN 0.27 μM NAA	MS	leaf	253
Coffee	*Coffea canephora*	2 μM KIN 10 μM 2,4-D	2.5 μM KIN 0.5 μM NAA	MS	leaf	254
Coriander	*Coriandrum sativum*	10.7 μM NAA	none	MS	embryo	258
Cotton	*Gossypium klotzschianum*	0.5 μM 2,4-D	11.4 μM IAA 4.7 μM KIN	MS	hypocotyl	207

TABLE III. (continued)

Crop	Species	Growth regulators		Medium	Explant	Reference
		1° Medium	2° Medium			
Date Palm	*Phoenix dactylifera*	452 µM 2,4-D 4.9 µM 2iP	none	MS	ovule	228
Dill	*Anethum graveolens*	10.7 µM NAA	none	MS	embryo	237, 258
		2.3 µM 2,4-D 2.3 µM KIN	none	White	inflorescence	
Eggplant	*Solanum melongena*	5 µM NOA	4.7 µM KIN	MS	hypocotyl	114
Fennel	*Foeniculum vulgare*	27.6 µM 2,4-D 1 µM KIN	none	Nitsch	stem	153
Garlic	*Allium sativa*	10 µM CPA 2 µM 2,4-D 0.5 µM KIN	10 µM IAA 20 µM KIN	AZ	stem	1
Ginseng	*Panax ginseng*	2.2 µM 2,4-D 0.8 µM KIN	0.4 µM 2,4-D	MS	pith	34
Grapes	*Vitis* spp.	4.5 µM 2,4-D 0.4 µM 6BA	10.7 µM NAA 0.4 µM 6BA	MS	flower, leaf	144
Oil Palm	*Elaeis guineensis*	4.5 µM 2,4-D 2.3 µM KIN	5.7 µM IAA	Heller	embryo	209
Orange	*Citrus sinensis*	0.5 µM KIN 5.7 µM IAA	none	Murashige, Tucker	ovule	130
Parsley	*Petroselinum hortense*	27 µM 2,4-D	none	Hildebrandt C	petiole	290
Pumpkin	*Cucurbita pepo*	4.9 µM IBA	1.4 µM 2,4-D	MS	cotyledon hypocotyl	110, 111

TABLE III. (continued)

Crop	Species	Growth regulators		Medium	Explant	Reference
		1° Medium	2° Medium			
Sandalwood	*Santalum album*	9.1 μM 2,4-D 23.2 μM KIN	none	White	embryo	219
	S. album	4.5 μM 2,4-D 0.9-2.3 μM KIN	1.5-5.8 μM GA_3	MS White	stem	145
Sweet Gum	*Liquidambar styraciflua*	5.3 μM NAA 8.8 μM 6BA	none	Blaydes	hypocotyl	252
Water Parsnip	*Sium suave*	10.7 μM NAA	none	MS	embryo	8
Papaya	*Carica papaya*	1 μM NAA 10 μM 2iP	0.1 μM NAA 0.01 μM 6BA	White	petiole	51

[a] Requires a subculture on a maintenance medium prior to 2° culture.

TABLE IV. *Other Species Capable of Somatic Embryogenesis After Primary and Secondary Culture*

Species	Growth regulators		Medium	Explant	Reference
	1° Medium	2° Medium			
Ammi majus	28.5 μM IAA	11.4 μM IAA	MS	hypocotyl	83, 236
Antirrhinum majus (Snapdragon)	4.5 μM 2,4-D	1.2 μM NOA 10% CW	MS	leaf, stem	233
Atropa belladonna	10.7 μM NAA 2.3 μM KIN	2.7 μM NAA 12.4 μM NOA	White WB	root	273
Biota orientalis	0.6 μM IAA	0.6 μM IAA	White	cotyledon	139
Cheiranthus cheiri	4.5 μM 2,4-D	0.5 μM 2,4-D	MS	seedling	126
Conium maculatum (Poison Hemlock)	10.7 μM NAA	none	MS	embryo	258
Iris spp.	4.5 μM 2,4-D	0.6 μM IAA 0.4 μM 6BA	MS	shoot apex	226
Macleaya cordata (Plume-poppy)	5 μM 2,4-D 5 μM KIN	none	basal	leaf	137
Nigella damascema (Fennell flower)	9.1 μM 2,4-D 10% CW	9.1 μM 2,4-D	MS	flower pedicel	212
Nigella sativa	2.7 μM NAA 15% CW	2.9 μM IAA	MS	leaf, root, stem	14
Pergularia minor	9.1 μM 2,4-D	0.6 μM IAA	MS	stem	205
Petunia hybrida	4.5 μM 2,4-D 0.9 μM 6BA	0.5 μM ZEA	MS	stem, leaf	221
Ranunculus sceleratus	5.7 μM IAA 10% CW	5.7 μM IAA 10% CW	White	stem	140
Tylophora indica	4.5 μM 2,4-D 5.2 μM BTOA	0-0.5 μM 2,4-D	White	stem	220

TABLE V. Frequency of Individual Growth Regulator Supplements in Primary and Secondary Culture Medium in the Initiation of de facto Somatic Embryogenesis

Growth regulators	*1° culture*		*2° culture*	
	Crops	*Non-crops*	*Crops*	*Non-crops*
Auxin				
IAA	11.5%	21.4%	19.2%	42.9%
IBA	3.8%	-	-	-
2,4-D	57.7%	57.1%	7.7%	21.4%
NAA	26.9%	21.4%	19.2%	7.1%
NOA	3.8%	-	-	14.3%
BTOA	-	7.1%	-	-
pCPA	3.8%	-	-	-
Cytokinin				
KIN	50%	14.3%	23.1%	-
6-BA	7.7%	7.1%	11.5%	7.1%
2iP	7.7%	-	-	-
ZEA	-	-	-	7.1%
Other				
GA_3	-	-	3.8%	-
CW	3.8%	21.4%	3.8%	14.3%
No regulators	-	-	46.2%	14.3%

The culture medium used for induction of embryo development during secondary culture is devoid of growth regulators for 46.2% of the crop species and 14.3% of the non-crop species. Examination of Table VI reveals that the growth regulator requirement in the culture medium of crop species consists of 2,4-D alone in 19.2% of the primary medium and 2,4-D plus a cytokinin in 41.3% of the primary medium. The non-crop species require either 2,4-D alone in the primary medium (28.6%) or 2,4-D plus a cytokinin (14.2%). During secondary culture, 46.2% of the crop species require no growth regulator while 38.3% require either IAA or NAA plus cytokinin. The non-crop species require either no growth regulators (14.3%) or one of the auxins, IAA or NAA, plus a cytokinin in 42.8% of the cases. Concentration ranges of NAA (1.0-16.7 μM) and Kin (2.3-20 μM) are used in these media. Information pertaining to the other growth regulators used during secondary culture is summarized in Tables V, VI, and VII.

TABLE VI. *Frequency of Specific Growth Regulator Supplements or Combinations thereof in Primary and Secondary Culture Medium Used for de facto Somatic Embryogenesis*

Growth regulator	1° medium		2° medium	
	Crops	Non-crops	Crops	Non-crops
No regulator(s)	-	-	46.2%	14.3%
2,4-D	19.2%	28.6%	7.7%	21.4%
2,4-D + pCPA + KIN	3.8%	-	-	-
2,4-D + 6BA	3.8%	7.1%	-	-
2,4-D + KIN	26.1%	7.1%	-	-
2,4-D + NAA + KIN	3.8%	-	-	-
2,4-D + 2iP	3.8%	-	-	-
2,4-D + BTOA	-	7.1%	-	-
2,4-D + CW	-	7.1%		
IAA	-	14.3%	3.8%	28.6%
IAA + KIN	11.5%	-	11.5%	-
IAA + 6BA	-	-	3.8%	7.1%
IAA + CW	-	7.1%	-	7.1%
IBA	3.8%	-	-	-
NAA	11.5%	7.1%	-	-
NAA + KIN	3.8%	7.1%	7.7%	-
NAA + CW	-	7.1%	3.8%	-
NAA + NOA	-	-	-	7.1%
NAA + 6BA	3.8%	-	7.7%	-
NAA + 2iP	3.8%	-	-	-
NOA	3.8%	-	-	-
NOA + CW	-	-	-	7.1%
KIN	-	-	7.7%	-
ZEA	-	-	-	7.1%
GA_3	-	-	3.8%	-
CW	3.8%	-	-	-

It is instructive to examine the general growth regulator regimes which have been successfully used for embryogenic cell determination during primary culture and permission of embryogenic development during secondary culture. It can be seen that the growth regulator regime for 11 crop species require a primary medium with growth regulator(s) and a secondary medium devoid of growth regulators. These crop species include anise, carrot, caraway, celery, citrus, date palm, dill, parsley, sandalwood,

TABLE VII. Growth Regulator Concentration Ranges in 1° and 2° Culture Media used for Induction of de facto Somatic Embryogenesis

Growth regulator	Effective Concentration (μM) 1° Culture medium Crops	Non-crops	2° Culture medium Crops	Non-crops
IAA	5.7	0.6-28.5	0.05-11.4	0.6-11.4
IBA	4.9			
2,4-D	0.5-27.0	4.5-27.6	1.4	0.5-9.1
NAA	1.0-5.4	2.7-10.7	0.5-10.7	2.7
NOA	5.0			1.2-12.4
BTOA		5.2		
CPA	10.0			
KIN	0.5-2.7	1.0-5.0	2.3-20	
6BA	0.4-8.8	0.9	0.4-17.7	0.4
2iP	4.9-10.0			
ZEA				0.5
GA_3			1.5-5.8	
CW	10%	10-15%	10%	10%

sweet gum, and water parsnip (Table I). Non-crop species adhering to this scheme are poison hemlock, plume-poppy and fennel (Table II). Asparagus, cauliflower, coffee, garlic, grape and papaya follow a growth regulator regime requiring both an auxin and cytokinin during both primary and secondary culture. The auxin during secondary culture may be the same auxin at a reduced concentration or a weaker auxin at the same or higher concentration. The same is true for *Atropa belladonna* (Table IV). Cotton, eggplant, and oil palm require an auxin for primary culture and a weak auxin during secondary culture at the same or higher concentration and the inclusion of cytokinin. Cacao requires an auxin during secondary culture.

C. Species Capable of Somatic Organogenesis

1. Organogenesis in Tobacco. Tobacco has been used as a model system for *in vitro* studies on regeneration, since the classical studies of Skoog and Miller (251). Totipotency was first demonstrated with *Nicotiana tabacum* by regeneration of mature plants from single cells (288). It is also one of the few systems for which some basic correlates for organogenesis have been elucidated (278). Production of haploid plants by the *in vitro* culture of excised anthers is easily achieved with tobacco (41). Plant regeneration from isolated protoplasts was first accomplished with *N. tabacum* (272) and the first somatic hybrid plant was obtained between two *Nicotiana* species, *N. glauca* and *N. langsdorffii* (33). The most frequently used pathways for plant regeneration in tobacco are summarized in Figure 3.

The following is the protocol for tobacco leaf explant:

(1) Plants can be grown under normal greenhouse conditions but must be used prior to flowering. The youngest fully expanded leaves are collected from tobacco plants either 1-2 hours after sunrise or from dark-pretreated tobacco plants.

(2) Leaves are washed with mild detergent in water and rinsed with first 70% EtOH, then distilled water.

(3) Leaves are sterilized in 20% commercial hypochlorite solution for 20 minutes. It is necessary to expose all leaf surfaces to the sterilizing solution. This is often done by placing leaves in a small beaker on a gyrotary shaker.

(4) Leaves are rinsed in sterile distilled water three times and air dried for approximately 5 minutes.

(5) Leaves are cut into 1 x 1 cm sections and transferred to culture medium.

(6) Leaf explants should be placed onto agar-based MS medium with 1-5 μM 6BA for shoot formation. Shoots should appear in 3-4 weeks.

(7) Alternately, if callus formation is desired, leaf explants should be placed onto MS medium with 4.5 μM 2,4-D and 1-2 grams per liter of casein hydrolysate. Callus formation should be visible in 4 weeks.

(8) Cultures should be placed in the light (ca. 1000 lux) with 12-24 hours of daylight each day. Both shoot and callus formation can be obtained when cultures are grown at 22-28°C., i.e., ambient temperature should be adequate in most cases.

(9) When young shoots appear they should be transplanted to rooting medium. Root formation can be observed in either hormone-free MS medium or in half-strength MS medium.

(10) Rooted plants can be transplanted to Jiffy^{R} pots. Humidity must be high for the first few days following transplantation to Jiffy^{R} pots.

Varieties of cultivated tobacco, *Nicotiana tabacum*, represent the easiest material to manipulate *in vitro*. The nutrient solution most often used for *in vitro* cultivation of plant species, Murashige and Skoog culture medium (MS), was formulated from results of growth experiments with *N. tabacum* (177). The effects of growth regulating auxins and cytokinins

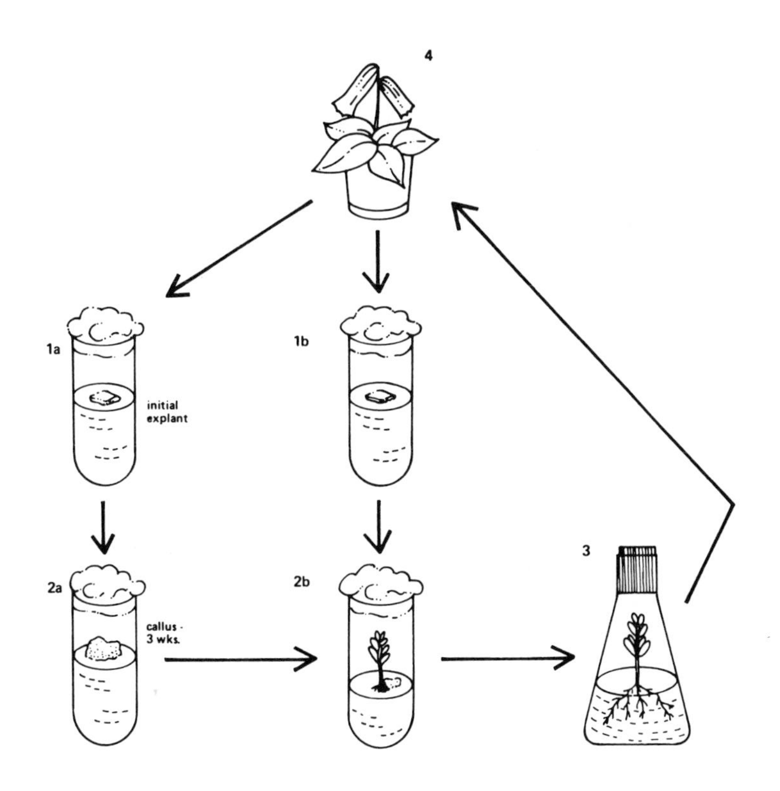

FIGURE 3. Protocol for plant regeneration of tobacco. Explants are excised from pith or leaf explants and placed onto MS medium with 2,4-D for callus initiation (1a). Callus formation is obtained in 3 weeks (2a). Callus is then transferred to MS medium with 6BA for shoot formation (2b). Alternately explants may be placed onto MS medium with 6BA to obtain direct somatic organogenesis (1b). Shoots are obtained in 3 weeks (2b). Shoots are transferred to a rooting medium (e.g., growth regulator free MS medium) for root formation (3). Rooted plantlets can be transferred to the greenhouse (4).

are quite specific and reproducible. Callus has been initiated from leaf or stem explants of *N. tabacum* and all *Nicotiana* spp. examined using the MS medium with the addition of 4.5 μM 2,4-D and 2 g/l CH (Evans, unpublished). Callus can be initiated on MS medium with other hormone concentrations, e.g., 11.4 μM IAA and 2.3 μM KIN (177), but in all cases an auxin is necessary, and in the case of 2,4-D, may be sufficient for the induction of callus. Once formed, callus can be maintained on MS medium with 4.5 μM 2,4-D, while the CH becomes unnecessary, or the callus may be shifted to an alternate medium, e.g., B5 medium with 4.5 μM 2,4-D (74).

In addition to direct plant regeneration, presented in the above protocol, it is possible to obtain either callus-mediated plant regeneration or plant regeneration from cell suspension cultures in tobacco. If callus growth is initiated and maintained in medium with 2,4-D, the 2,4-D must be replaced with a cytokinin to achieve shoot formation. In this manner plant regeneration has been obtained from long term suspension cultures of tobacco (74). Numerous combinations of growth regulating auxins and cytokinins have been varied by researchers attempting to achieve plant regeneration in tobacco. A summary of hormone treatments sufficient for plant regeneration in tobacco is listed in Table VIII.

2. *Organogenesis in Grasses and Cereals.* Species of numerous families of dicotyledonous plants can readily be induced to undergo plant regeneration *in vitro*, but in general monocotyledons have been more difficult to culture. This is unfortunate as graminaceous species are extremely important sources of nutrition. Because of the large number of agriculturally important species, most investigators using graminaceous species for studies on plant regeneration have restricted themselves to cultivated crops.

The two subfamilies, Poacoideae (grasses) and Panicoideae (cereals) have each been cultivated *in vitro* with limited success. The primary limitation has been explant source, as meristematic tissues have been used to initiate callus cultures capable of plant regeneration from each cereal species and most grass species cultured *in vitro*.

Sugar cane is certainly the most malleable of the grass species examined *in vitro*. Although the immature inflorescence is the most successful explant, a variety of explants may be used for plant regeneration including apical meristems, young leaves, and pith parenchyma (149). Plants have been regenerated from long term callus cultures in sugar cane (179). Though inconclusive, some evidence exists which suggests the plants obtained *in vitro* may originate from single cells via somatic embryogenesis (180). MS medium, when supplemented with 2 - 13.6 μM 2,4-D, has

TABLE VIII. Growth Regulator Concentrations used for Shoot Formation in Tobacco

Growth regulator concentrations	*Medium*	*Explant*	*Reference*
11.4 μM IAA 9.3 μM KIN	*MS*	*stem*	*176*
1 μM ZEA	*MS*	*stem*	*190*
5.7 μM IAA 1 μM SD-8339	*MS*	*stem*	*190*
0.1 μM IAA 1 μM KIN 10% CW	*White*	*leaf*	*87*
5.7 μM IAA 1 μM 6BA	*MS*	*stem*	*190*
1 μM IAA 1 μM KIN	*MS*	*flower*	*286*
0.6 μM IAA 10 μM benzo(a)pyrene	*MS*	*stem*	*133*
0.6 μM IAA 8.9 μM 6BA	*White*	*petiole*	*205*
9.3 μM KIN chlorflurenolmethylester	*MS*	*leaf, callus*	*191*
1 μM IAA 10 μM 6BA	*MS*	*flower*	*284*
1 μM IBA 10 μM KIN	*MS*	*flower*	*285*
1-5 μM 6BA	*MS*	*leaf, suspension culture*	*74*

been used for callus initiation. Plants can be obtained from callus upon removal of 2,4-D from the medium. Cytokinin (9.3 μM KIN) is present in some regeneration media used for sugar cane (e.g., 98), but may be unnecessary (150). Shoots, when obtained in sugar cane, have been reportedly difficult to root (98). Young shoots, though, when separated from growing callus and placed on medium with IAA form roots in a high percentage of cultures (see also Chapter 11).

MS medium has been used to culture 7 of 9 grass species (Table IX), while B5 and SH medium were each used for one species. For *Dactylis glomerata*, growth in SH medium, a medium devised for monocots, was greater than on either B5 or

TABLE IX. *Graminaceous Species Capable of Somatic Organogenesis after Primary and Secondary Culture*

Crop	Species	Growth regulators		Medium	Explant	Reference
		1°	2°			
Grasses						
Big bluestem	*Andropogon gerardii*	2.3 μM 2,4-D 0.9 μM KIN	less than 4.5 μM 2,4-D	MS	inflor-escence	35
Brome grass	*Bromus inermus*	4.5 μM 2,4-D	none	B5	mesocotyl	71
Fescue	*Festuca arundinacea*	40.7 μM 2,4-D	2.3 μM 2,4-D	MS	embryo	152
Indian grass	*Phargmites communis*	4.5 μM 2,4-D	none	MS	stem	232
Indian grass	*Sorghastrum nutans*	22.6 μM 2,4-D 0.9 μM KIN	less than 4.5 μM 2,4-D	MS	inflor-escence	36
Orchard grass	*Dactylis glomerata*	67.8 μM 2,4-D 1.0 μM KIN	4.5 μM 2,4-D 1.0 μM KIN	SH	caryopsis	42
Ryegrass	*Lolium multiflorum* *Lolium perenne*	6.8 μM 2,4-D 37.1 μM IAA 1.0 μM KIN	3.4 μM 2,4-D 18.6 μM IAA 0.5 μM KIN	MS	embryo	3
Sugar cane	*Saccharum spp.*	13.6 μM 2,4-D	0 or 26.9 μM NAA	MS	leaf inflor-escence	97 179
Hybrid	*Lolium multiflorum* X *Festuca arundinacea*	9-18 μM 2,4-D	1.1 μM 2,4-D	MS	stem peduncle	122

TABLE IX. (continued)

Crop	Species	Growth regulators 1°	2°	Medium	Explant	Reference
Cereals						
Barley	*Hordeum japonicum*	22.6 μM 2,4-D	1.4 μM KIN 2.6 μM GA_3	MS	embryo	196
Barley	*Hordeum vulgare* cv. Himalaya	15 μM 2,4-D 10 μM IAA 1.5 μM 2iP	none	MS	meristem	37
Barley	*H. vulgare* var. Akka	4.5 μM 2,4-D	none	MS	embryo	47
Corn	*Zea mays* cvs. A188, R-Navajo	9.0 μM 2,4-D	none	MS	embryo	81
Corn	*Zea mays* var. Prior, Inrakorn lines M9473, M0003	67.8 μM 2,4-D	none	MS	mesocotyl	95
Millet, bullrush	*Pennisetum typhoideum*	45.2 μM 2,4-D	1.1 μM IAA	MS	mesocotyl	214
Millet, common	*Panicum miliaceum*	45.2 μM 2,4-D	none	MS	mesocotyl	213
Millet, finger	*Eleusine coracana*	45.2 μM 2,4-D 10-15% CW	1.1 μM NAA or 1.1 μM IAA	MS	mesocotyl	214
Millet, Koda	*Paspalum scrobiculatum*	45.2 μM 2,4-D	1.1 μM IAA	MS	mesocotyl	214

TABLE IX. (continued)

Crop	Species	Growth regulators		Medium	Explant	Reference
		1°	2°			
Oats	*Avena sativa*	2.3-13.6 µM 2,4-D	none	B5	embryo	45
Oats	*A. sativa* cv. Tiger	9.1 µM 2,4-D 10.7 µM pCPA	30.0 µM NAA 10.0 µM IAA 1.5 µM 6BA	SH	hypocotyl	151
Rice	*Oryza sativa* var. Kyote Asahi	10.0 µM 2,4-D	none	MS	root	187
Rice	*O. sativa* CI 8970-S	9-27.1 µM 2,4-D	0-9.8 µM 2iP	MS	root, leaf	99
Rice	*O. sativa* var. Krasnodarskii	9-18.1 µM 2,4-D	79.9 µM IAA 9.3 µM KIN	MS	embryo, stem	50
Sorghum	*Sorghum bicolor* var. N. Dakota	22.6-67.8 µM 2,4-D	26.9 µM NAA	MS	meristem	160
Sorghum	*S. bicolor* line X4004	5.0 µM 2,4-D 10.0 µM ZEA	0-5 µM IAA	MS	embryo	73
Wheat	*Triticum aestivum* cv. Mengavi	22.6 µM 2,4-D 5.4 µM pCPA	5.4 µM NAA 23.3 µM KIN	basal	embryo	38
Wheat	*T. aestivum* cv. Chinese Spring	4.5-9 µM 2,4-D	5.7 µM IAA 4.6 µM ZEA	T,B5,MS	rachis and seed	78
Wild wheat	*T. longissimum*	0.1 µM 2,4-D	5.7 µM IAA 4.6 µM ZEA	MS	embryo	78
Hybrid barley	*Hordeum vulgare* X *H. japonicum*	22.6 µM 2,4-D	1.4 µM KIN 2.6 µM GA_3	MS	embryo	196
Hybrid wheat Triticale	*Triticum* sp. X *Secale cereale*	27.1 µM 2,4-D	none	MS	embryo	242

MS medium (42). The three media differ primarily in source of inorganic nitrogen (Table II). Thus an investigation of culture media which should include consideration of the SH medium, may be useful for initiation of new grass species *in vitro*. The frequency of plant regeneration was not reported for most grass species, nonetheless reported species varied from 3.1% (*Festuca*) to 66% (*Phragmites*) plant regeneration. Large numbers of shoots per explant, suggesting usefulness in clonal multiplication, have been reported for callus of both ryegrass with 20-50 plantlets per culture (3) and Indian grass with 5-20 plantlets per culture (36). Root formation has been relatively easy when shoots are separated to fresh medium with no hormones (36, 122), with reduced mineral concentration (3, 42, 152), or with a high auxin to cytokinin ratio (232). Plants regenerated from tissue cultures of grasses may be quite useful for crop improvement.

Emphasis has been placed exclusively on agriculturally important species in studies of plant regeneration in the cereals. This is perhaps unfortunate as a number of examples exist in which wild related species are more prolific *in vitro* than cultivated crops as in protoplasts of *Lycopersicon* species (311). Nonetheless, plant regeneration has been reported in most cultivated cereal species. As success has been limited to organized tissue of juvenile origin, it has been emphasized that plant regeneration may arise from pre-existing meristematic centers (127, 229, 276). Extensive research has been carried out with corn, wheat, and rice. While plants can be obtained *in vitro* from each of these species, corn is certainly the most recalcitrant of these species, as plants have been obtained only from organized explants.

Callus can be initiated from a number of explants but plant regeneration is limited (80). Immature embryos have been used most often for formation of a callus capable of plant regeneration (81). MS medium has been used in all cases resulting in a totipotent callus. While 2,4-D is the sole hormone used for callus initiation, the concentration has been varied from 2.3 (229) to 67.8 μM (95). Callus proliferation can be enhanced by the addition of 21.5 μM NAA and 0.25 μM 2iP to 4.5 μM 2,4-D (81). The orientation of the immature embryo is important as the scutellum side must be oriented upwards when immature embryos are cultured (80) and downwards when young seedlings are cultured (95) for maximum callus induction. In all cases reported to date, plant regeneration is obtained by removing the 2,4-D (63, 81, 95, 229). These results have led King *et al.* to conclude that reports of plant regeneration from cultured corn explants and most other cereals represents repression by 2,4-D of shoot primordia during callus initiation, followed by derepression of pre-existing primordia when the 2,4-D is removed (127). Nonetheless, the capacity for plant regeneration has been maintained following subculture of scutellar-derived callus for 19 to 20 months (63, 81). At least 16 *Zea mays* varieties, stocks, and genetic lines have been tested for *in vitro* regeneration capacity with varied results. Under the culture conditions used only three of these 16 genetic lines have resulted in consistent plant regeneration; the lines A188 and A188 X R-navajo (80) and the variety 'Prior' (95).

Cultivated wheat has been investigated in great detail. Of 15 cultivars and varieties examined in six different laboratories, only five resulted in plant regeneration. Callus proliferation has been classified extensively for both wild and cultivated species of *Triticum*. Gosch-Wackerle *et al.* induced callus from the rachis and embryos of seven species of *Triticum*, including diploid (*T. monococcum*, *T. speltoides*, and *T. tauschii*, 2n=14), tetraploid (*T. timepheevii* and *T. turgidum*, 2n=28) and hexaploid (*T. aestivum*, 2n=42) species (78). They used T-medium, after Dudits *et al.* (57) with 4.5 μM 2,4-D to initiate rachis callus and MS medium with 9 μM 2,4-D or B5 medium with 4.5 μM 2,4-D to initiate callus from embryos. Callus could be obtained for each species, but *T. tauschii* was certainly the least prolific species on these media. These results confirm the earlier finding that it was possible to initiate viable callus from these seven species (249). Of all the species and cultivars initiated *in vitro*, only two wheat species have undergone plant regeneration (Table IX). Unlike corn, wheat has a requirement for an auxin for plant regeneration. In most cases, callus has been placed onto appropriate medium with IAA (57, 78), NAA (38) or pCPA (193) rather than 2,4-D, but apparently shoots can be obtained when the concentration of 2,4-D is greatly reduced (248). The cultivar most often used is Chinese Spring, but regeneration has been achieved with the cultivars Salmon, Maris Ranger, and Mengavi and the variety Tobari 66. Plant regeneration has been observed from a number of explants including callus derived from rachises, shoots, seeds and embryos (Table IX). The greatest frequency of regeneration (45-68%) was obtained from embryo-derived callus initiated on MS medium with 9 μM 2,4-D and after the first or second transfer subcultured onto regeneration medium with 4.6 μM ZEA and 5.7 μM IAA (78). In most cases, callus rapidly lost its organogenic capability.

Rice is perhaps the easiest cereal species to regenerate *in vitro*. Callus can be obtained from numerous young explants of rice using MS medium with 9-45.2 μM 2,4-D (50).

Shoot regeneration has been obtained from callus derived from seeds (187), endosperm (183), 3-4 day old roots and leaves (99), 2 week old roots and shoots (188), immature and mature embryos, root tips, scutellum, plumule, stem and panicle (50). In each case shoots are obtained following removal of the 2,4-D from the medium. Although a cytokinin or auxin is unnecessary, shoot formation can be enhanced by addition of 0.3-19.7 μM 2iP (99) or 79.9 μM IAA and 9.3 μM KIN (50). The frequency of plant regeneration may reach 100% in certain conditions, e.g., 4 month old callus cultures derived from young root or leaf explants. As in most other cereal species, the age of rice callus is inversely proportional to regenerative capacity (see also Chapter 10).

Plants can be regenerated from at least eleven additional cereal species or hybrids (Table IX). In oats regenerative capacity can be retained in callus cultures maintained for 12-18 months, but in most species the ability to regenerate is lost after only a few subcultures (1-4 months). In oats, genotype represents an important factor in plant regeneration. Of 25 genotypes initiated *in vitro* (45), 9 could not be regenerated, while an additional 5 have *very* low frequency of plant regeneration.

Barley plants have been obtained from callus derived from shoot apices (37, 128), mature embryos (118) and more recently immature embryos (47, 196). Greater shoot regeneration was obtained for immature embryos on B5 medium (50%), than MS medium (10%) when callus was transferred from medium with 2,4-D to hormone-free medium (47). In *Sorghum bicolor* (sorghum) plants have been regenerated from callus derived from both seedling shoots (160), immature embryos (73), and mature embryos (277). While organized tissue was used in each case, multiple shoots were obtained upon plant regeneration in each case. MS has been shown to be better than B5 medium for plant regeneration in sorghum (73). Callus can be induced with 5 to 67.8 μM 2,4-D (Table IX). Plants can be regenerated if subcultured to regeneration medium within one to two months. The frequency of regeneration for immature embryos is 20-50%. An auxin requirement exists for plant regeneration in *Sorghum*. Shoots could be obtained from immature embryos in the presence of 2,4-D as an auxin, but only if a high concentration of cytokinin were added to the medium (e.g., 10-50 μM ZEA). Maximum shoot and plantlet production was obtained though, if callus was subcultured on medium with IAA (73), NAA (160) or a combination of 6BA, NAA and GA_3 (277).

Four species of millets have been regenerated *in vitro* (213, 214). In each case mesocotyl tissue was cultured on MS medium. Callus was induced with 45.2 μM 2,4-D with or without 10-15% coconut water. Regeneration was obtained when the 2,4-D was either completely eliminated, resulting in hormone-free medium (*P. miliaceum*) or replaced with IAA or NAA. The frequency of regeneration for all four species ranged from 36-80% as tested 3-4 months after callus initiation.

Secale cereale (rye) is one major cereal species that has not been tested for *in vitro* regeneration, although anther cultures of rye have been regenerated (274). Nonetheless, hybrids of wheat and rye (*Triticale* AD-20) have been regenerated *in vitro* and respond in the same manner as wheat (208).

3. *Organogenesis in Other Crop Species*. Additional crop species capable of plant regeneration are listed in Tables X and XI. These crop species have been grouped according to the reported protocols for plant regeneration. Species capable of direct somatic organogenesis do not require a callus intermediate and are regenerated directly from explants placed in culture. The majority of crop species have been grown in a callus stage prior to plant regeneration (Tables IX and XI). Regeneration may proceed via a direct or a callus-mediated route. The method most frequently used for each crop species has been presented. All graminaceous protocols include a callus stage and are comparable to species capable of somatic embryogenesis. Graminaceous species have been summarized separately, due to the unique features of this family and the required use of primary and secondary media. In most crop species, leaf, cotyledon and hypocotyl explants have been used for somatic embryogenesis (Table XII). Leaf explants have been used most frequently for direct regeneration (44%), while hypocotyl and cotyledon (30 and 26% respectively), have been used for callus mediated regeneration. Each of these tissues is a non-meristematic explant. In contrast, primarily meristematic explants have been used to obtain regeneration in graminaceous species. Embryo, inflorescence and mesocotyl explants (48, 24, and 29% respectively) have been used most frequently. In 18 of 23 (78.3%) graminaceous species examined regeneration has been obtained exclusively from meristematic tissue. MS medium is used most often for somatic organogenesis (74.6% of species). MS is used in 87.5% of species capable of direct shoot formation, 58% of crop species capable of callus-mediated regeneration and 85% of graminaceous species. White's (9.5%), Gamborg's B5 (7.9%), and Schenk-Hildebrandt's (4.8%) media have each been used with more than one species.

TABLE X. Crop Species Capable of Direct Somatic Organogenesis

Crop	Species	Growth regulators used for shoot formation	Medium	Explant	Reference
Blackberry	*Rubus* sp.	0.4 μM 6BA 0.3 μM GA_3 4.9 μM IBA	MS	shoot tip	24
Broccoli	*Brassica oleracea*	46-51 μM IAA 14-28 μM KIN	MS	leaf	112
Cabbage, leaf-mustard	*Brassica juncea*	2.7 μM NAA 8.9 μM 6BA	MS	leaf, cotyledon	109
Chinese kale	*Brassica alboglabra*	21.4 μM NAA 2.3 μM KIN	MS	cotyledon	312
Ginger	*Zingiber officinale*	4.4 μM 6BA	MS	rhizome	106
Horseradish	*Amoracia lapathifolia*	5.4 μM NAA 0.5-2.3 μM KIN	MS	leaf	168
Lettuce	*Lactuca sativa*	2.9 μM IAA 2.3 μM KIN	Miller	leaf	134
Linseed	*Linum usitatissimum*	1 μM ZEA	MS	hypocotyl	68
Paper mulberry	*Broussonetia kazinoki*	0.5 μM 6BA	MS	hypocotyl	194
Pepper	*Capsicum annuum*	4.4-8.9 μM 6BA 0-5.7 μM IAA	MS	hypocotyl, cotyledon	85
Pimpernel	*Anagallis arvensis*	2.9 μM KIN 0.06 μM IAA 400 mg/l CH	MS	stem, leaf, hypocotyl	12

TABLE X. (continued)

Crop	Species	Growth regulators used for shoot formation	Medium	Explant	Reference
Pineapple	*Ananas sativus*	9.7 µM NAA 9.8 µM IBA 9.8 µM KIN	MS	axillary bud	162
Rape	*Brassica napus*	5 µM 6BA 0.1 µM IAA	MS	stem	120
Sweet potato	*Ipomoea batatus*	3.8 µM ABA 0.1 µM KIN 0.2 µM 2,4-D	White	tuber	306
Tobacco	*Nicotiana tabacum*	1 µM 6BA	MS	leaf	74
Tomato	*Lycopersicon esculentum*	22.8 µM IAA 18.6 µM KIN	MS	leaf	198

TABLE XI. *Crop Species Capable of Callus-Mediated Somatic Organogenesis*

Crop	Species	Growth regulators: Callus	Growth regulators: Shoots	Medium	Explant	Reference
Alfalfa	*Medicago sativa*	9 μM 2,4-D 9.3 μM KIN 10.7 μM NAA	none	Blaydes	hypocotyl	20
Almond	*Prunus amygdalis*	26.9 μM NAA 10% CW	26.9 μM NAA 2.3-4.7 μM KIN	MS	leaf, embryo cotyledon	165
Artichoke, globe	*Cynara scolymus*	1 μM ZEA 1 μM 2,4-D	1 μM 6BA 0.1 μM 2,4-D	MS	cotyledon, petiole	54
Bean, pinto	*Phaseolus vulgaris*	18.6 μM KIN 5.7 μM IAA 1/4 bean seed/ml	11.4 μM IAA 5.4 μM NAA 0.9 μM KIN 1/4 bean seed/ml	67-V	leaf	43
Brussels sprouts	*Brassica oleracea*	11.4 μM IAA 2.3 μM KIN 4.9 μM IBA	none	MS	petiole, stem	39
Buckwheat	*Fagopyrum esculentum*	22.6-45.2 μM 2,4-D	none	MS	cotyledon, hypocotyl	307
Cabbage, red	*Brassica oleracea*	4.5 μM 2,4-D 0.5 μM KIN	11.4 μM IAA 9.3 μM KIN or 5.7 μM IAA 2.3 μM KIN	MS	cotyledon, seed hypocotyl	13
Cassava	*Manihot esculenta*	0.4 μM 6BA 1.1 μM NAA 1.9 μM GA_3	0.4 μM 6BA 1.1 μM NAA	MS	stem	279

TABLE XI. (continued)

Crop	Species	Growth regulators		Medium	Explant	Reference
		Callus	Shoots			
Clover, Berseem	*Trifolium alexandrinum*	5.4 μM NAA 7 μM KIN	2.7 μM NAA 2.3 μM KIN	MS	hypocotyl, suspension culture	171
Clover, crimson	*Trifolium incarnatum*	11.0 μM NAA 10.0 μM 2,4-D 10.0 μM KIN	11.0 μM NAA 15.0 μM ADE	B5, SH	hypocotyl	105
Clover, ladino	*Trifolium repens*	19.6 μM 2,4,5-T 0.5 μM KIN	2.3 μM 2,4-D 0.5 μM KIN	BM	seed	197
Clover, red	*Trifolium pratense*	0.25 μM picloram 0.5 μM 6BA	0.03 μM picloram 0.05-44.4 μM 6BA	PC-L2	cotyledon	202
Endive	*Cichorium endiva*	27 μM 2,4-D or 0.5 μM NAA	0.19 μM KIN	MS	embryo	289
Foxglove	*Digitalis purpurea*	4.5 μM 2,4-D 0.5 μM KIN	0.6 μM IAA 4.7 μM KIN	MS	seedling	102
Geranium	*Pelargonium* spp.	5.4 μM NAA 23.2 μM KIN	0.5 μM NAA 46.5 μM KIN	MS	petiole	250
Japanese persimmon	*Diospyros kaki*	16.1 μM NAA 0.5 μM KIN	5.4 μM NAA 0.5-4.7 μM KIN	MS	immature embryo	310
Kale	*Brassica oleracea*	10.7 μM NAA 2.3 μM KIN	2.3 μM KIN	MS	stem	104
Onion	*Allium cepa*	5 μM 2,4-D	5 μM 2iP	B5	bulb	64

TABLE XI. (continued)

Crop	Species	Growth regulators		Medium	Explant	Reference
		Callus	Shoots			
Pea	*Pisum sativum*	10.7 µM NAA 4.7 µM 6BA	22.2 µM 6BA 1.1 µM IAA	MS	epicotyl	159
Pigeon pea	*Cajanus cajan*	not reported	4.7 µM KIN 0.06 µM IAA 400 mg/l CH	White	hypocotyl	241
Potato	*Solanum tuberosum*	2.3 µM IAA 1.0 µM GA_3 3.7 µM KIN	1.8 µM 6BA	MS	tuber	146
Stone crop	*Sedum telephium*	4.5 µM 2,4-D	44.4 µM 6BA	B5	leaf	23
Stylo, Carribean	*Stylosanthes hamata*	9 µM 2,4-D 0.2 µM KIN	14 µM KIN	SH	radicle, cotyledon	235
Sugar beet	*Beta vulgaris*	5.7 µM IAA 0.5 µM KIN	4.7 µM KIN 0.5 µM GA_3	PBO	leaf	52
Taro	*Colocasia esculenta*	10 µM CPA 2 µM 2,4-D 0.5 µM KIN	0.5 µM NAA 9 µM KIN	AZ	corm	2
Winged bean	*Psophocarpus tetragnolobus*	0.5 µM KIN 5.4-26.9 µM NAA	none	not reported	hypocotyl, cotyledon	293
Yam	*Dioscorea deltoidea*	4.5 µM 2,4-D	1.2 µM IBA 2.2 µM 6BA	MS	hypocotyl	82

TABLE XII. Frequency of Explants Used for Somatic Organogenesis

Explant	*Direct regeneration*	*Callus-mediated regeneration*	*Graminaceous species*
Non-meristematic			
Leaf	44%	15%	8.7%
Cotyledon	19%	26%	-
Hypocotyl	25%	30%	4.3%
Stem	12.5%	11%	13.0%
Petiole	-	11%	-
Root	6%	-	4.3%
Epicotyl	-	4%	-
Meristematic			
Bud, shoot tip	12.5%	-	8.7%
Seed	-	11%	4.3%
Mesocotyl	-	-	26.1%
Inflorescence	-	-	21.7%
Embryo	-	15%	52.2%
Underground stem	6%	11%	-
Caryopsis	-	-	4.3%

The auxin of choice for callus formation prior to plant regeneration was 2,4-D. Culture medium containing 2,4-D has been used to initiate a callus capable of plant regeneration in 35 of 50 (70%) crop species examined. This includes 100% of the 23 graminaceous species reported. NAA has been used for callus induction with 40.7% of non-graminaceous species, but has not been used for callus induction of any graminaceous species. KIN has been used in combination with an auxin for callus induction of 59.3% of crop species, but only 14.7% of the Gramineae. Other cytokinins have been used less frequently (Table XIII). 2,4-D has been used alone for callus initiation in 78.3% of the graminaceous species. For the five remaining species, 2,4-D is combined with another auxin, or more often, with a cytokinin (Table XIV). For other crop species, combinations of NAA-KIN (18.5%), 2,4-D-KIN (11.1%), and IAA-KIN (11.1%) are frequently used for callus initiation. 2,4-D has also been used alone to initiate callus (18.5%).

A cytokinin is required in the culture medium of each species capable of direct somatic organogenesis. The cytokinin may be KIN (50%) or 6BA (43.8%) or less frequently

TABLE XIII. *Frequency of Individual Growth Regulators Used for Plant Regeneration in Species Capable of Somatic Organogenesis*

Growth regulators	Direct regeneration	Callus-mediated regeneration		Gramineae	
		Callus	Shoot formation	1° medium	2° medium
Auxin					
IAA	37.5%	14.8%	18.5%	8.7%	34.8%
IBA	12.5%	3.7%	3.7%	-	-
2,4-D	6.3%	44.4%	7.4%	100%	26.1%
NAA	25%	40.7%	29.6%	-	26.1%
pCPA	-	3.7%	-	8.7%	-
picloram	-	3.7%	3.7%	-	-
Cytokinin					
KIN	50%	59.3%	51.9%	17.4%	26.1%
6BA	43.8%	11.1%	25.9%	-	4.3%
2iP	-	-	3.7%	4.3%	4.3%
ZEA	6.3%	3.7%	-	4.3%	8.7%
Other Growth Regulators					
GA_3	6.3%	7.4%	3.7%	-	8.7%
ABA	6.3%	-	-	-	-
2,4,5-T	-	3.7%	-	-	-
CW	-	3.7%	-	4.3%	-
CH	6.3%	-	3.7%	-	-
ADE	-	-	3.7%	-	-
No Growth Regulator	-	-	14.8%	-	34.8%

TABLE XIV. *Auxin and Cytokinin Combinations Used for Plant Regeneration in Species Capable of Somatic Organogenesis*

Growth regulators	Direct shoot formation	Callus-mediated shoot formation		Graminaceous species	
		Callus	Shoot	1° medium	2° medium
No regulators	-	-	14.8%	-	39.1%
2,4-D	-	18.5%	-	78.3%	17.4%
2,4-D + KIN	6.3%	11.1%	3.7%	13.0%	4.3%
2,4-D + 6BA	-	-	3.7%	-	-
2,4-D + ZEA	-	3.7%	-	4.3%	-
2,4-D + CW	-	-	-	4.3%	-
IAA	-	-	-	-	13.0%
IAA + IBA + KIN	-	3.7%	-	-	-
IAA + NAA + KIN	-	-	3.7%	-	-
IAA + 2,4-D + KIN	-	-	-	4.3%	4.3%
IAA + 2,4-D + 2iP	-	-	-	4.3%	-
IAA + NAA + 6BA	-	-	-	-	4.3%
IAA + KIN	25%	11.1%	11.1%	-	4.3%
IAA + 6BA	12.5%	-	3.7%	-	-
IAA + ZEA	-	-	-	-	8.7%
IBA + 6BA	6.3%	-	3.7%	-	-
IBA + NAA + KIN	6.3%	-	-	-	-
NAA	-	7.4%	3.7%	-	17.4%
NAA + 2,4-D + KIN	-	7.4%	-	-	-
NAA + KIN	12.5%	18.5%	18.5%	-	4.3%
NAA + 6BA	6.3%	7.4%	3.7%	-	-
pCPA + 2,4-D	-	-	-	8.7%	-
pCPA + 2,4-D + KIN	-	3.7%	-	-	-

TABLE XIV. (continued)

Growth regulators	Direct shoot formation	Callus-mediated shoot formation		Graminaceous species	
		Callus	Shoot	1° medium	2° medium
picloram + 6BA	-	3.7%	3.7%	-	-
2,4,5-T + KIN	-	3.7%	-	-	-
KIN	-	-	14.8%	-	8.7%
6BA	18.8%	-	7.4%	-	-
2iP	-	-	3.7%	-	4.3%
ZEA	6.3%	-	-	-	-
KIN + GA_3	-	-	-	-	8.7%

ZEA (Table XIII). In most cases, the cytokinin is used in combination with an auxin. Combinations of IAA-KIN (25%), IAA-6BA (12.5%) and NAA-KIN (12.5%) have been used most frequently (Table XIV). In some species regeneration can be achieved with only 6BA (18.1%) or ZEA (6.3%). These cytokinins, though used less frequently than KIN, are considered stronger than KIN.

For species where plant regeneration is achieved from induced callus, shoot formation is obtained by transferring callus to medium with growth regulators present in different concentrations. Some species (14.8%) can be regenerated by transferring callus from callus induction medium to hormone-free medium (Table XI). In each of these cases, callus was induced in a medium with a high auxin concentration. In all other cases of plant regeneration from callus, a cytokinin or cytokinin precursor (ADE) is present in the shoot formation medium. KIN (51.9%) and to a lesser extent 6BA (25.9%) have been used (Table XIII). KIN may be used alone (14.8%) or in combination with NAA (18.5%) or IAA (11.1%). Graminaceous species have required use of cytokinin less frequently than other crop species. KIN (26.1%) has been used most frequently, while 6BA (4.3%) was rarely used. This may simply reflect the prejudice of grass and cereal researchers as there are no data showing that 6BA cannot be used with graminaceous species. The effective concentration ranges used successfully for somatic organogenesis are summarized in Table XV.

D. *Generalizations for Embryogenesis and Organogenesis*

It is evident that a growth regulator requirement exists for the expression of organogenesis or embryogenesis. In general, high auxin in combination with low cytokinin concentrations promote callus proliferation. Alternately, low auxin in combination with high cytokinin concentrations promote shoot or plantlet formation. It is evident that during embryogenesis the primary culture medium is involved in the modification of genetic expression. Histological evidence supports this theory. Either an auxin or in at least one case (coffee), a cytokinin may be required to achieve this genetic modification.

The primary culture medium of graminaceous species (Table IX) and of species capable of somatic embryogenesis (Tables III and IV) are similar. In each case MS medium is used most frequently. Also callus is induced with a high auxin concentration. All 23 graminaceous species use 2,4-D in the primary culture medium while 2,4-D (57.7%) or another auxin is used in the primary culture medium of embryogenic

TABLE XV. *Growth Regulator Concentration Ranges in Callus and Shoot Media for Induction of Somatic Organogenesis*[a]

Growth regulators	Direct shoot formation	Callus-mediated regeneration		Graminaceous species	
		Callus	Shoot	1° medium	2° medium
Auxin					
IAA	0.06-51.0	2.3-11.4	0.06-11.4	10.0-37.1	1.1-79.9
NAA	2.7-21.4	0.5-26.9	0.5-26.9	-	1.1-30.0
2,4-D	0.2	1.0-45.2	0.1-2.3	2.3-67.8	1.1-4.5
IBA	4.9-9.8	4.9	1.2	-	-
picloram	-	0.25	0.03	-	-
pCPA	-	10.0	-	5.4-10.7	-
Cytokinin					
KIN	0.1-28.0	0.2-23.2	0.2-46.5	0.9-1.0	0.5-23.3
6BA	0.4-8.9	0.4-4.7	0.05-44.4	-	1.5
2iP	-	-	5.0	1.5	9.8
ZEA	1.0	1.0	-	10.0	4.6
Other Growth Regulators					
GA_3	0.3	1.0-1.9	0.5	-	2.6
ABA	3.8	-	-	-	-
2,4,5-T	-	19.0	-	-	-
CW	-	10%	-	10-15%	-
CH	400 mg/l	-	400 mg/l	-	-

[a]Concentration in μM.

species. Of the embryogenic crop species with 2,4-D in the primary culture medium, 12.5% are transferred to a secondary medium with reduced 2,4-D. The remaining species are transferred to a secondary culture medium devoid of 2,4-D. This secondary medium may contain a weaker auxin (31.3%), only a cytokinin (6.3%), only GA_3 (6.3%) or no growth regulators (50%). In both the embryogenic and graminaceous species, regeneration is achieved by subculturing callus to a medium for induction of the embryo or shoot pattern of development. Although definitive experiments have not been completed for all taxa, it appears that commitment for embryo or shoot morphogenesis occurs in the primary medium. For embryogenesis, the secondary culture medium requires either elimination or reduction of the auxin present in the primary culture medium. The secondary culture medium of 46% of crop species has no growth regulator requirement. The remaining species contain (1) a reduced concentration of the same auxin or (2) the same or a higher concentration of a weaker auxin in the secondary culture medium. A similar protocol is followed for organogenesis in graminaceous species. In every graminaceous species, the 2,4-D concentration is either reduced (26%) or eliminated (73.9%). In those species in which 2,4-D is eliminated from the culture medium, callus has been transferred to hormone-free medium (34.8%), to medium with no auxin but with a cytokinin (13.0%), or to a medium with a weaker auxin (39.1%).

Only 12 of 27 crop species capable of two step organogenesis contain 2,4-D in the primary culture medium. In each of these species the pattern of development is similar to that reported for the Gramineae. In all but one species (artichoke) the 2,4-D must be eliminated prior to shoot development. The shoot medium may contain a weaker auxin (41.6%), only a cytokinin (33.3%), or no growth regulators (16.6%).

Graminaceous species have been regenerated primarily from meristematic explants (78.3%). On the other hand, embryogenic species or species capable of callus-mediated regeneration have used mostly non-meristematic explants (e.g., Table XII).

It should be noted that some plant families undergo embryogenesis more readily than other plant families (62). The plant family with the highest frequency of somatic embryogenesis is Umbelliferae, while the Gramineae contains *no* embryogenic species to date. However, in most cases the essential histological work has not been performed to determine the mode of morphogenesis.

III. APPLICATION

A. *Mass Clonal Plant Propagation*

Propagation via *in vitro* techniques may have several advantages over conventional methods. Most of these advantages are dependent on the speed of tissue culture propagation. (1) Tissue culture can be used for plants that are normally propagated slowly. Clonal propagation of plants via tissue culture was first commercially used with orchids (173). Orchids are propagated slowly and tissue culture techniques could be used to speed propagation. (2) Tissue culture can be used to reduce propagation time of species in high demand. *In vitro* propagation has proved useful in propagation of ornamentals (e.g., houseplants, lilies) and crop species (*Asparagus*) that are propagated rapidly, but inadequately to satisfy market demands. (3) In addition tissue culture is extremely useful in the establishment of new varieties when only a few plants are initially available (174). In this case, a virus resistant or horticultural variant can be multiplied rapidly for commercial planting. Tissue culture propagation should prove useful for multiplication of hybrid seed. A basis for propagation of hybrid *Asparagus* has been explored (56). All of these potential advantages reflect the speed of *in vitro* propagation. Additional advantages relate to the variability observed among propagules. Plants arising from tissue culture may be free of insects or viruses (175). In addition, in some cases greater vigor, earlier flowering, higher yields and other desirable qualities have been observed in plants recovered from tissue culture (175). For example, the Boston fern develops into a plant with a single growth center when propagated by runners, but tissue culture propagated plants have several growth centers, therefore, appearing fuller and more attractive (174). Some of these improved qualities may relate to elimination of undefined pathogens.

Clonal propagation may proceed via direct or callus-mediated somatic organogenesis or somatic embryogenesis. As genetic variability has been associated with callus-mediated plant regeneration, this method is not recommended for clonal propagation, but may be ideal for recovery of useful variant lines. In general, multiplication *in vitro* proceeds through (1) shoot tip culture resulting in proliferation of axillary buds or (2) somatic embryogenesis. In the first method, shoots are rooted or used to establish new shoot cultures. As this method represents maintenance of the shoot apex by culturing meristematic explants, genetic uniformity is favored. Somatic embryogenesis is limited to a few species (Tables III, IV) but results in the most rapid mode of plant regeneration. Clonal propagation can be used in commercial greenhouse operations to reduce the number of stock plants maintained and avoid the seasonal variation of propagation procedures. A more complete list of techniques for regeneration of horticulturally valuable species has been published (62).

B. *Genetic Variability of Regenerated Plants*

Phenotypic variation has been observed in plants regenerated *in vitro*. Such genetic variability may be agriculturally useful when applied to existing breeding programs. Variability obtained in callus clones regenerated from *Saccharum* spp. has been used in breeding programs (98). This variability recovered in regenerated plants reflects either pre-existing cellular genetic differences or tissue culture-induced variability. Unfortunately the crucial experiments necessary to statistically distinguish between these two factors have not been devised. Skirvin and Janick systematically compared plants regenerated from callus of five cultivars of *Pelargonium* spp. (250). Plants obtained from geranium stem cuttings *in vivo* were uniform, whereas plants from *in vivo* root and petiole cuttings and plants regenerated from callus were quite variable. Changes were observed in plant and organ size, leaf and flower morphology, essential oil constituents, fasciation, pubescence, and anthocyanin pigmentation. It is likely that both genetic and tissue culture-induced variability has been recovered in these geranium lines. It has been observed that plants regenerated *in vitro* from tomato stem sections are quite uniform (53), while plants regenerated from leaf sections contain tremendous variability (Sharp and Lipschutz, unpublished). Variability, resulting in the isolation of disease resistant clones, has also been obtained in protoplast-derived plants of *Solanum tuberosum* (161, 245). To distinguish epigenetic and genetic variability, the F_1 generation must be analyzed.

Syóno and Furuya recovered phenotypic variability in plants regenerated from stem-derived callus, particularly abnormal floral morphology, but only in plants regenerated from long-term cultures (270). Long-term cultures do result in tissue culture-induced variability in chromosome number in both callus (176) and in plants regenerated from callus (231). Phenotypic changes associated with change of chromosome number probably represents genetic changes. In addition to chromosome aberrations, phenotypic variability manifested in plants regenerated from somatic cells could be the result of mitotic crossing over (61) or somatic mutation (201). Both processes are continuous ongoing genetic events occurring in somatic cells and subject to environmental modification.

C. Virus Elimination

During recent years, meristem-tip culture has been used to produce virus free stock. The technique is discussed by Kartha in this volume (Chapter 6). Callus culture, coupled with a plant regeneration protocol, has also been used for obtaining healthy plants from mosaic virus (TMV) infected tobacco callus (93). Infectivity tests with single cells isolated from TMV infected callus indicate that only 30-40% of the cells are infected. The virus elimination apparently occurs because cell proliferation takes place at a greater rate than viral replication. Viral infection may be completely eliminated via repeated subculture (268). Temperature can also be used for viral elimination (172).

The uneven distribution of the virus within an infected leaf may be used to obtain virus-free plants (172). Culture of 1 mm explants, on a regeneration medium, from the dark-green island areas of a TMV infected leaf results in plants of which ca. 50% are virus-free or contain virus particles in low concentration. This may explain the ease of regenerating virus-free plants from protoplasts liberated from infected leaf material. Shepard has reported that 7.5% of 4140 plants regenerated from potato virus X (PVX) infected tobacco leaves were virus free (244).

D. Cloning and Artificial Seed

High frequency somatic embryogenesis offers the opportunity to develop an artificial seed operation. Unfortunately substantial research must be accomplished pertaining to synchronization of embryo development and perfection of large scale separation procedures for collecting synchronous embryos before an artificial seed operation can be established. Sieves or glass beads may be used for the separation and collection of uniform embryos (295). Thereafter, the embryos should be maintained in an arrested state of development as an artificial seed until the time of sowing. The arrested embryo development could be achieved using cold storage or mitotic inhibitors.

The most promising artificial seed technology suitable for use with somatic embryos is fluid drilling. A wide range of crops have been tested for fluid drilling of pre-germinated seed with satisfactory results (26). These crops include tomato, lettuce, cucumber, cabbage, okra, squash, broccoli, turnip, carrot, corn and celery. Somatic embryos are available in carrot and celery (Table III). Fluid drilling of pre-germinated seeds involves: (1) seed germination in controlled optimum environmental conditions; (2) mixing selected germinated seeds with a gel-type medium having the consistency of wallpaper paste to protect and suspend the seeds for uniform distribution, and (3) sowing the seeds and gel through a fluid drill seeder (26). Fluid drilled seeds which are suspended in liquid may be pre-germinated under ideal conditions before placing in the soil. Research continues pertaining to the refinement of fluid drilling technology with pre-germinated seed. This technology allows for early uniform plant development which can be reflected in an earlier harvest (46). Work accomplished in tomato illustrates some of the advantages of fluid drilling (206). Most tomato varieties will not germinate at soil temperatures less than 10°C and even at 11.5°C three weeks are required for emergence. Pre-germinated seed placed in 11.5°C soil, however, emerges in 6.6 days (206).

The fluid drilling technology appears to be ideal for a commercial operation dependent upon asexual propagation and artificial seed for sowing. The gel is comparable to agar as it can be fortified with growth regulators, plant nutrients, pesticides, and other plant protection agents. Somatic embryos planted using this procedure in temperate environments should probably be treated with a cryoprotectant prior to the fluid drilling operation as a safeguard in the event of frost.

Bryan *et al.* have listed some of the gels and fluid drilling machines which have been successfully used for sowing germinated seeds (26). These include: Metylan clear wallpaper paste (Henkel Inc.); Laponite (Laporte Industries [North America] Ltd.); Viterra 2 (Union Carbide); Super Slurper (General Mills) and Fluid Carrier Powder No. 1 (Fluid Drilling Ltd.). Hand-operated fluid drill seeding machines are manufactured by Flowsow Engineering Co. in Sussex, England, and by Fluid Drilling Ltd. in Stratford-on-Avon, England. The Flowsow seeder is used to deposit a gel-seed mixture in clumps about 7 cm apart and delivers 22 ml gel/m of ground travel. The Fluid Drilling Ltd. Hand Drill sows a continuous bead of gel and is adjustable for several gel rates.

IV. SUMMARY AND FUTURE PROSPECTS

A. *Genetic Aspects of Plant Regeneration*

1. Chromosome Variation: Regeneration from Established Cultures. An altered chromosome number, either polyploid or aneuploid, in general reflects underlying genetic alterations. Variations of chromosome number of plants in long-term cell suspension cultures has been well documented (117). Chromosomal variation in cell suspension and callus cultures have been reviewed extensively (49, 247, 265). It has been suggested that the progressive increase in variation of cultured cells is proportional to a progressive loss in organ forming capacity (282). Murashige and Nakano have shown that shoot forming capacity of aneuploid cells is severely reduced (176), but on the other hand, Sacristan and Melchers were able to regenerate numerous aneuploids with relative ease from tumorous cell lines (231). Unfortunately the chromosome instability of cells in suspension culture has been compared with cells of regenerated plants in only a few cases (48). In no case has greater chromosome variability been found in regenerated plants than in the callus from which plants were derived. On the other hand, despite the occurrence of a wide range of chromosome numbers in callus cultures, only diploid plants have been regenerated from *Daucus carota* (170), *Oryza sativa* (187), *Prunus amygdalus* (165), and *Triticum aestivum* (249). These species range in somatic chromosome number from 2n=2x=16, *P. amygdalus*, to 2n=6x=42, *T. aestivum*. Consequently, in these species, under the plant regeneration protocols used, diploids are selectively favored. Orton has compared chromosome number of suspension and callus cultures of *Hordeum vulgare*, *Hordeum jubatum*, and their interspecific hybrid with plants regenerated from these cultures (196). In each case it is evident that no polyploid and a greatly reduced number of aneuploid callus cells are capable of plant regeneration. Identical chromosome number, though, is insufficient to conclude genetic stability. Only in long term cultures of *Lilium longiflorum* have regenerated plants been shown to have the normal karyotype (247). Polyploid plants have been recovered from a number of cultures (48), including *Asparagus officinalis* (271), *Nicotiana tabacum* (121), *Lilium longiflorum* (247) and haploid *Pelargonium* (18).

Despite selective pressure against plant regeneration, a wide range of aneuploid plants has been recovered from tissue cultures of numerous species. Predominantly reductions in the chromosome number have been observed in plants regenerated from callus cultures of triploid (2n=3x=21) ryegrass hybrids

(4), while a wide range of aneuploids, both additions and reductions in chromosome number, were obtained in both sugar cane (149) and *N. tabacum* (231). Each of these chromosomal variants is associated with phenotypic variation including agriculturally useful traits such as disease resistance (143). Variability associated with aneuploidy, if not accompanied by a concomitant depression of yield is particularly valuable in vegetatively propagated agricultural crops. Aneuploidy and morphological variability have been observed in plants regenerated from protoplasts (161) and in most somatic hybrid plants (e.g., 166).

Chromosome number mosaicism of regenerated plants has been reported in *Nicotiana* (191), *Hordeum* (196), *Triticum durum* (19), sugar cane (149), and *Lycopersicon peruvianum* (257). The common occurrence of chromosome number mosaicism in regenerated plants suggests that plantlets originate from two or more initial cells (19) or that new chromosome variability is generated *in vivo* after plant regeneration (196). Somatic mosaicism has also been reported in somatic hybrid plants derived from protoplast fusion where plants are presumably derived from single cells (157). There have been suggestions that the chromosome number mosaicism of regenerated plantlets is reduced during the subsequent development of regenerants to mature plants. Nonetheless, this mosaicism may be established as periclinal or mericlinal 'chimeras' (257) or transmitted to subsequent generations (191). The maintenance of chromosome mosaicism *in vivo* in regenerated plants may be under genetic control (192).

2. *Genetic Selection to Increase the Frequency of Regeneration.* Attempts to extend established plant regeneration protocols to closely related species have often been unsuccessful. For example, despite using growth regulator concentrations sufficient for plant regeneration in nearly all closely related species, plants could not be regenerated from *Nicotiana knightiana* (156). Differences in regenerative capacity observed between two donor plants may reflect underlying genetic differences. If plant regeneration *in vitro* is viewed as a phenotype it is not surprising that differences exist between species, cultivars, or genotypes that are otherwise closely related. The value of screening diverse genetic lines for plant regeneration has been emphasized (159). Experiments with reported differences in regenerative capacity between donor plants have been summarized in Table XVI. In each example listed, at least one genetic line was incapable of plant regeneration. The value of screening genetic lines is more evident in attempts to regenerate recalcitrant species, such

TABLE XVI. Variation in Frequency of Plant Regeneration Observed Between Plants

Species examined	*Plants examined were different*	*Number of plants screened*	*Number of plants capable of regeneration*	*Reference*
Triticum	*Species Varieties*	12	1	208
Triticum	*Species Varieties*	4	1	249
Avena sativa	*Genotypes*	25	16	45
Medicago sativa	*Cultivars*	9	8	20
Pisum sativum	*Genotypes*	16	6	159
Saccharum	*Cultivars*	11	10	97
Saccharum	*Cultivars*	4	3	15
Trifolium pratense	*Cultivars*	5	4	202
Zea mays	*Genotypes*	5	4	81
Lycopersicon esculentum	*Genotypes (mutants)*	15	12	17
Nicotiana tabacum	*Genotypes (mutants)*	5	3	294

as peas (6 of 16 lines regenerate), oats (1 of 3 lines regenerate), corn (4 of 5 lines regenerate), and wheat (2 of 16 lines regenerate). In species presently incapable of plant regeneration, screening genotypes may prove extremely useful, especially when hormone variations are simultaneously tested. Plant regeneration frequency can also be increased by genetic selection. Initial regeneration frequencies of 12% were obtained from hypocotyl explants of alfalfa (20), but this frequency was increased by selecting for plant regeneration. After two cycles of selection, the frequency of regeneration was increased in one genetic line of alfalfa from 12 to 67%.

Mutant cell lines have proven quite useful in recent years in studies of somatic cell genetics. Chlorophyll deficient and analogue resistant lines have been used in protoplast fusion (292). In most cases attempts to regenerate plants from induced mutations or from fusion products have proven unsuccessful (301, 299). In some cases, the loss of regenerative capacity is dependent on the specific mutant. This is particularly true for chlorophyll deficient mutants. Of five chlorophyll mutants of *N. tabacum* examined, plants were regenerated from three mutants (*Su*, *Ws*, *Ys*), but could not be regenerated from the white plastid and spontaneous chimera lines despite using all genotypes of each mutant on three regeneration media (294). Similarly, of eight chlorophyll deficient mutants of tomato, two mutants (*Xa*-2 and *yg*-4) were incapable of shoot formation despite being tested on 13 media for plant regeneration (17). Similarly, dwarf mutants, associated with reduced viability, do not regenerate as well as normal lines. When five dwarf mutants of *Arabidopsis thaliana* were cultured in 8 combinations of NAA and 6BA sufficient for plant regeneration of normal wild type donor plants, two mutants (*le* and *g*) were incapable of plant regeneration. Of the three dwarf mutants capable of regeneration, *t* and St_1 could be regenerated on only one or two of the eight growth regulator combinations, while the *f* mutant regenerated on all eight combinations (79).

Mutants induced *in vitro* may have further reduced regeneration capabilities due both to pleiotropic effects of the mutants isolated and to tissue culture induced variation. For example, the W001 and W002 5-methyl tryptophan (5MT) resistant mutants of *Daucus carota* form less than 0.01% embryos than normal *D. carota* in optimal media for each (267). The reduction in frequency is the result of tryptophan overproduction in the 5MT resistant lines. Tryptophan is readily converted to IAA and high auxin concentrations inhibit somatic embryogenesis. Other regulatory or auxotrophic mutants with altered metabolism may respond in a similar manner. Kanamycin resistant cells of *N. sylvestris* are incapable of plant regeneration (55), yet restoration of regenerative capacity can be achieved by fusing mutant protoplasts with *N. knightiana* protoplasts (156). In some cases mutant isolation has resulted in aneuploid clonal lines. As discussed earlier, aneuploid lines have reduced regenerative capacity. Examples include analogue-resistant *N. sylvestris* cells (299) and temperature sensitive *N. tabacum* cells (158). Consequently it is not surprising that mutants induced *in vitro* have reduced capacity for plant regeneration.

B. An Understanding of the Developmental Process

1. Introduction. The two general patterns of *in vitro* embryogenic development, direct and indirect may be further characterized by their relative times of determination and differentiation into embryogenic cells (243). Direct embryogenesis proceeds from pre-embryogenic determined cells (PEDC) (125, 138), while indirect embryogenesis requires the redetermination of differentiated cells, callus proliferation, and differentiation of embryogenic determined cells (IEDC).

Apparently, PEDC's await either synthesis of an inducer substance or removal of an inhibitory substance, requisite to resumption of mitotic activity and embryogenic development. Conversely, cells undergoing IEDC differentiation require a mitogenic substance to re-enter the mitotic cell cycle and/or exposure to specific concentrations of growth regulators. Cyto-differentiation and the emergence of multicellular organization are multi-step processes in which each step leads to the establishment of a particular pattern of gene activation, allowing transition to the next essential state of development (260). Arrestment may occur at any step in this process.

The view that external applications of growth regulators can be permissive or inhibitive of differentiation but not determinative has been evolved (260). Tisserat *et al.* concur that the explant and certain of its associated physiological qualities are the most significant determinants of embryo initiation, while the "*in vitro*" environment acts primarily to enhance or repress the embryogenic process (280). That is, the cells that undergo embryo initiation are predetermined, and their subsequent exposure to exogenous growth regulators simply allows embryogenesis to occur (280). Street believes that growth regulators may be regarded best as activating agents toward previously induced cells which are preconditioned to respond in specific ways (260). The processes of *de novo* organogenesis are similar in many respects to that of somatic embryogenesis (278).

2. Pre-Embryogenic Determined Cells (PEDC).

a. Citrus nucellus. The natural occurrence of polyembryony in many species of *Citrus* (9, 297) and its economic importance have long been realized (296). Thus, not only a zygotic embryo, but also several adventive embryos may be found within a single seed. These adventive embryos have been shown to originate in single cells of the nucellus, near the micropyle of the ovule, and appear to be initiated after fertilization, soon before or after the first zygotic division (9).

In addition to the natural occurrence of nucellar polyembryony, *in vitro* cultures of nucellar explants may give rise to embryos and eventually to fully developed plants (217, 131). Nucellar tissues from both unfertilized as well as fertilized ovules may undergo embryogenesis *in vitro* (28, 169). Cultures of monoembryonic cultivars like 'Shamouti' also have been shown to be embryogenic (31, 215, 216). Early characterization of embryogenesis in nucellar cultures consists of an initial proliferation of callus and subsequent development of "pseudobulbils" in the absence of exogenous plant growth regulators (217). Some of these pseudobulbils continue embryogenic development and eventually become entire plantlets. However, in other experiments with several cultivars of normally monoembryonic species, callus and pseudobulbil formation were prerequisite to embryo development, and embryos arose directly from nucellar tissue (28). Mitra and Chaturvedi reported that embryos may arise directly from the nucellus, or indirectly from nucellar callus (169). Many of these experiments include growth regulators such as NAA and KIN, as well as complex addenda like malt or yeast extract and coconut water. These may be beneficial in increasing the frequency of observed embryos. However, it must be stressed that embryogenesis may be observed in the absence of these components.

Button *et al.* characterized embryogenesis in habituated nucellar callus cultures (31). This callus was found to be composed not of unorganized parenchymatous tissue, but solely of numerous proembryos. Embryogenesis has been observed to occur from single cells in the periphery of the callus, as well as from existing proembryos. Some of these developing embryos may enlarge only to a globular stage, commonly referred to as "pseudobulbils". These rarely develop into plants. Other proembryos follow the developmental sequence characteristic of zygotic embryogenesis and eventually develop into plants. The fact that this callus was habituated, i.e., autonomous for exogenous growth regulators, in no way decreased its embryogenic potential. That the presence of exogenous growth regulators actually depressed embryogenesis lends further support to the concept that exogenous growth regulators may be viewed best as inductive agents for determination.

We are in agreement with Street (260) and Tisserat *et al.* (280) with regard to PEDC. However, these investigators have not recognized the occurrence of two distinct patterns of development, i.e., PEDC and IEDC. Their concept of embryogenesis is limited to predetermined embryogenic cells, where growth regulators serve only to initiate embryo development from PEDCs and/or the cloning of these PEDCs. However, an alternative concept must be developed to explain how induced embryogenic determined cells (IEDCs) become committed to embryogenic development, since the occurrence of such cells requires redetermination and commitment to embryogenesis.

It is our view that embryogenesis from nucellar tissues, both *in vitro* and *in vivo*, may be considered best as cases of PEDC-mediated embryogenesis. The cells of the nucellus are actually pre-embryogenic determined cells (PEDCs), and their proliferation as a callus mass and subsequent embryogenesis may be viewed best as simply the cloning of the PEDCs. Thus, it appears that embryogenesis in the nucellus is autonomous and may be observed even in the absence of exogenous growth regulators. Although low concentrations of kinetin and NAA have been shown to be beneficial, they are not absolute requirements of embryogenic determination. Exogenous growth regulators probably contribute to the cloning of these PEDCs, thus increasing the relative number of embryogenic cells. Of course, it is also possible that an additional population of induced cells (PEDCs) further contribute to the relative number of embryos.

b. Other examples of PEDC. Spontaneous origin of viable embryos from the superficial cells of plantlets arising in culture via embryogenesis has been reported where somatic embryos originate *in vitro* without an intervening callus stage. This now has been observed in *Atropa belladonna* (222), carrot (164), *Datura innoxia* (76), *Ranunculus sceleratus* (140). In *Ranunculus sceleratus*, it has been established that the embryos arise from single cells of the shoot axis epidermis. Here, as in the natural or rare origin of embryos from synergids and antipodal cells of the embryo sac, the competence for embryogenesis is retained in cells of specialized function. This raises the possibility that differentiation - or at least some pathways of differentiation - and competence to embark upon morphogenesis are not necessarily incompatible (261).

3. Induced-Embryogenic Determined Cells (IEDC). The concept of IEDC explains the redetermination of differentiated cells to the embryogenic pattern of development. Evidence exists that the auxin or auxin/cytokinin concentration in the primary culture medium or conditioning medium is not only critical to the onset of mitotic activity in non-mitotic differentiated cells but to the epigenetic redetermination of these cells to the embryogenic state of development.

A statement characterizing the role of growth regulators in gene expression as direct or indirect cannot be made. Regardless of how growth regulators control gene expression, evidence exists that the auxin, 2,4-D, elicits a response at the transcriptional and translational levels during primary culture. Subsequently, an additional response at the transcriptional and translation levels occurs shortly after subculture onto a secondary or induction medium (238, 239, 240).

Numerous examples of IEDC have been reported in the literature in which embryogenic cells, resulting from cellular redetermination proceed through the various stages of embryo development and form plants. Development of the embryogenic determined cells is usually restricted during culture on the primary or conditioning medium. Thereafter cells need to be subcultured onto a secondary medium for induction or continuation of development in the embryogenic determined cells. The latter medium is usually referred to as an induction medium. Documented examples of IEDC have been reported in the literature for the following taxa: *Macleaya cordata* (135, 136), *Cichorium endiva* (289), *Daucus carota* (88, 113, 225), *Atropa belladonna* (141), *Petunia hybrida* (221), *Cucurbita pepo* (111), *Corylus avellana* (210), *Apium graveolens* (302), *Nigella sativa* (14), *Vitis vinifera* (144), *Asparagus officinalis* (226), *Iris* spp. (226), *Carum carvi* (7), *Apium graveolens* (312), *Gossypium klotzschianum* (207), *Theobroma cacao* (200), *Solanum melongena* (163), *Phoenix dactylifera* (228), and *Coffea arabica* (255).

Androgenesis and gynogenesis involve the development of IEDCs, although it is beyond the scope of this review to present a detailed account of embryogenesis in haploid cells. In these instances, either the microspore or megaspore must undergo a quantal mitotic division, i.e., a division in which the developmental fate of the sister cells are different (84). This results in an embryogenic determined cell. In the former a quantal mitotic division results in a regenerative cell and a vegetative cell. The vegetative cell, or in some instances the generative cell (211), or a fusion cell is determined as an embryogenic mother cell (189, 264, 266).

REFERENCES

1. Abo El-Nil, M.M., *Plant Sci. Lett. 9*, 259 (1977).
2. Abo El-Nil, M.M., and Zettler, F.W., *Plant Sci. Lett. 6*, 401 (1976).
3. Ahloowalia, B.S., *Crop Sci. 15*, 449 (1975).
4. Ahloowalia, B.S., *In* "Current Chromosome Research", (K. Jones and P.E. Brandham, eds.), p. 115. Elsevier/North Holland, (1976).

5. Aitchison, P.A., MacLeod, A.J., and Yeoman, M.M., *In* "Plant Cell and Tissue Culture", (H.E. Street, ed.), p. 276. Blackwell, Oxford, (1977).
6. Ammirato, P.V., *Bot. Gaz. 135,* 328 (1974).
7. Ammirato, P.V., *Plant Physiol. 59,* 579 (1977).
8. Ammirato, P.V., and Steward, F.C., *Bot. Gaz. 133,* 149 (1971).
9. Bacchi, O., *Bot. Gaz. 105,* 221 (1943).
10. Backs-Husemann, D., and Reinert, J., *Protoplasma 70,* 49 (1970).
11. Bajaj, Y.P.S., and Bopp, M., *Z. Pflanzenphysiol. 66,* 378 (1972).
12. Bajaj, Y.P.S., and Mader, M., *Physiol. Plant. 32,* 43 (1974).
13. Bajaj, Y.P.S., and Nietsch, P., *J. Expt. Bot. 26,* 883 (1975).
14. Banerjee, S., and Gupta, S., *Physiol. Plant. 38,* 115 (1976).
15. Barba, R., and Nickell, L.G., *Planta 89*, 299 (1969).
16. Barker, W.G., *Can. J. Bot. 47,* 1334 (1969).
17. Behki, R.M., and Lesley, S.M., *Can. J. Bot. 54,* 2409 (1976).
18. Bennici, A., *Z. Pflanzenzuchtg. 72,* 199 (1974).
19. Bennici, A., and D'Amato, F., *Z. Pflanzenzuchtg. 81,* 305 (1978).
20. Bingham, E.T., Hurley, L.V., Kaatz, D.M., and Saunders, J.W., *Crop Sci. 15,* 719 (1975).
21. Bitters, W.P., Murashige, T., Rangan, T.S., and Nauer, E., *Calif. Citrus Nurserymen's Soc. 9,* 27 (1970).
22. Blumenfeld, A., and Gazit, S., *Physiol. Plant. 25,* 369 (1971).
23. Brandao, I., and Salema, R., *Z. Pflanzenphysiol. 85,* 1 (1977).
24. Broome, O.C., and Zimmerman, R.H., *Hort. Sci. 13,* 151 (1978).
25. Brown, C.L., and Gifford, E.M., *Plant Physiol. 33,* 57 (1958).
26. Bryan, H.H., Stall, W.M., Gray, D., and Richmond, N.S., *Proc. Fla. State Hort. Soc. 91,* 88 (1978).
27. Buchala, A.J., and Schmid, A., *Nature 280,* 230 (1979).
28. Button, J., and Bornman, C.H., *South Afr. J. Bot. 37,* 127 (1971).
29. Button, J., and Botha, C.E.J., *J. Expt. Bot. 26,* 723 (1975).
30. Button, J., and Kochba, J., *In* "Plant, Cell, Tissue and Organ Culture", (J. Reinert and Y.P.S. Bajaj, eds.), p. 70. Springer-Verlag, Berlin (1977).

31. Button, J., Kochba, J., and Bornman, C.H., *J. Expt. Bot. 25*, 446 (1974).
32. Carceller, M., Davey, M.R., Fowler, M.W., and Street, H.E., *Protoplasma 73*, 367 (1971).
33. Carlson, P.S., Smith, H.H., and Dearing, R.D., *Proc. Nat. Acad. Sci., USA. 69*, 2292 (1972).
34. Chang, W.C., and Hsing, Y.I., *Nature 284*, 341 (1980).
35. Chen, C.H., Stenberg, N.E., and Ross, J.G., *Crop Sci. 17*, 847 (1977).
36. Chen, C.H., Lo, P.F., and Ross, J.G., *Crop Sci. 19*, 117 (1979).
37. Cheng, T.Y., and Smith, H.H., *Planta 123*, 307 (1975).
38. Chin, J.C., and Scott, K.J., *Ann. Bot. 41*, 473 (1977).
39. Clare, M.V., and Collin, H.A., *Ann. Bot. 38*, 1067 (1974).
40. Colijn, C.M., Kool, A.J., and Nijkamp, H.J.J., *Protoplasma 99*, 335 (1979).
41. Collins, G.B., and Legg, P.D., *In* "Plant Cell and Tissue Culture", (W.R. Sharp *et al.*, eds.), p. 585. Ohio State Univ. Press, Columbus, (1978).
42. Conger, B.V., and Carabia, J.V., *Crop Sci. 18*, 157 (1978).
43. Crocomo, O.J., Sharp, W.R., and Peters, J.E., *Z. Pflanzenphysiol. 78*, 456 (1976).
44. Crocomo, O.J., Sharp, W.R., and de Carvalho, M.T.V., *Proc. 1st Congresso Alagoas, Brasil, 21-22 Jan. 1979* (In press), (1980).
45. Cummings, D.P., Green, C.E., and Stuthman, D.D., *Crop Sci. 16*, 465 (1976).
46. Currah, I.E., Fluid Drilling Research, National Veg. Res. Station Rept., July 1977. Wellesbourne, Warwick, England, (1977).
47. Dale, P.J., and Deambrogio, E., *Z. Pflanzenphysiol. 94*, 65 (1979).
48. D'Amato, F., *In* "Plant Cell, Tissue and Organ Culture", (J. Reinert and Y.P.S. Bajaj, eds.), p. 343. Springer-Verlag, Berlin, (1977).
49. D'Amato, F., *In* "Frontiers of Plant Tissue Culture 1978", (T.A. Thorpe, ed.), p. 287. Univ. Calgary Press, Calgary, Canada, (1978).
50. Davoyan, E.I., and Smetanin, A.P., *Fiziologiya Rastenii. 26*, 323 (1979).
51. de Bruijne, E., de Langhe, E., and van Rijk, R., *Meded. Fak. Landbouwwet., Gent. 39*, 637 (1974).
52. de Greef, W., and Jacobs, M., *Plant Sci. Lett. 17*, 55 (1979).
53. de Langhe, E., and de Bruijne, E., *Scientia Hort. 4*, 221 (1976).

54. Devos, P., de Langhe, E., and de Bruijne, E., *Meded. Fak. Landbouwwet., Gent. 27,* 829 (1974).
55. Dix, P.J., Joo, F., and Maliga, P., *Molec. Gen. Genet. 157,* 285 (1977).
56. Dore, C., *Ann. Amelio. Plantes 25,* 201 (1975).
57. Dudits, D., Nemet, G., and Haydu, Z., *Can. J. Bot. 53,* 957 (1975).
58. Ellis, R.P., and Bornman, C.H., *J. S. Afr. Bot. 37,* 109 (1971).
59. Esan, E.B., Ph.D. Thesis, University of California, Riverside, (1973).
60. Evans, D.A., and Gamborg, O.L., *Plant Sci. Lett.* (in press), (1980).
61. Evans, D.A., and Paddock, E.F., *In* "Plant Cell and Tissue Culture: Principles and Applications", (W.R. Sharp *et al.*, eds.), p. 315. Columbus, (1978).
62. Evans, D.A., Sharp, W.R., and Flick, C.E., *In* "Horticultural Reviews", Vol. 3. (J. Janick, ed.), AVI Publishing Co., Westport, Conn., (in press), (1980).
63. Freeling, M., Woodman, J.C., and Cheng, D.S.K., *Maydica 21,* 97 (1976).
64. Fridborg, G., *Physiol. Plant. 24,* 436 (1971).
65. Frost, H.B., and Soost, R.K., *In* "The Citrus Industry", Vol. II. (W. Reuther, L.D. Batchelor and H.J. Webber, eds.), p. 290. Univ. of California Press, Berkeley, (1968).
66. Furusato, K., *Rep. Kikara Inst. Biol. Res. 8,* 40 (1957).
67. Gamborg, O.L., and Shyluk, J.P., *Plant Physiol. 45,* 598 (1970).
68. Gamborg, O.L., and Shyluk, J.P., *Bot. Gaz. 137,* 301 (1976).
69. Gamborg, O.L., and Wetter, L.R., *In* "Plant Tissue Culture Methods", National Research Council, Saskatoon, Canada, (1975).
70. Gamborg, O.L., Miller, R.A., and Ojima, K., *Exp. Cell Res. 50,* 148 (1968).
71. Gamborg, O.L., Constabel, F., and Miller, R.A., *Planta 95,* 355 (1970).
72. Gamborg, O.L., Murashige, T., Thorpe, T.A., and Vasil, I.K., *In Vitro 12,* 473 (1976).
73. Gamborg, O.L., Shyluk, J.P., Brar, D.S., and Constabel, F., *Plant Sci. Lett. 10,* 67 (1977).
74. Gamborg, O.L., Shyluk, J.P., Fowke, L.C., Wetter, L.R., and Evans, D.A., *Z. Pflanzenphysiol. 95,* 225 (1979).
75. Gautheret, R.J., La Culture des Tissues Végétaux, Masson, Paris, (1959).
76. Geier, T., and Kohlenbach, H.W., *Protoplasma 78,* 381 (1973).

77. Gengenbach, B.G., Green, C.E., and Donovan, C.M., *Proc. Natl. Acad. Sci., USA 74,* 5113 (1977).
78. Gosch-Wackerle, G., Avivi, L., and Galun, E., *Z. Pflanzenphysiol. 91,* 267 (1979).
79. Goto, N., *Japan. J. Genet. 54,* 303 (1979).
80. Green, C.E., *Hort. Sci. 12,* 131 (1977).
81. Green, C.E., and Phillips, R.L., *Crop Sci. 15,* 417 (1975).
82. Grewal, S., and Atal, C.K., *Ind. J. Expt. Biol. 14,* 352 (1976).
83. Grewal, S., Sachdeva, U., and Atal, C.K., *Ind. J. Expt. Biol. 14,* 716 (1976).
84. Grobstein, C., *Science 143,* 643 (1964).
85. Gunay, A.L., and Rao, P.S., *Plant Sci. Lett. 11,* 365 (1978).
86. Gunckel, J.E., Sharp, W.R., Williams, B.W., West, W.C., and Drinkwater, W.O., *Bot. Gaz. 133,* 254 (1972).
87. Gupta, G.R.P., Guha, S., and Maheshwari, S.C., *Phytomorphology 16,* 175 (1966).
88. Halperin, W., *Planta 88,* 91 (1969).
89. Halperin, W., and Jensen, W.A., *J. Ultrastruct. Res. 18,* 428 (1967).
90. Halperin, W., and Wetherell, D.F., *Amer. J. Bot. 51,* 274 (1964).
91. Halperin, W., and Wetherell, D.F., *Nature 105,* 519 (1965).
92. Handro, W., *Ann. Bot. 41,* 303 (1977).
93. Hansen, A.J., and Hildebrandt, A.C., *Virology 28,* 15 (1966).
94. Harada, H., Ohyama, K., and Cherruel, J., *Z. Pflanzenphysiol. 66,* 307 (1972).
95. Harms, C.T., Lorz, H., and Potrykus, I., *Z. Pflanzenzuchtg. 77,* 347 (1976).
96. Harvey, A.E., and Grasham, J.L., *Can. J. Bot. 48,* 663 (1969).
97. Heinz, D.J., and Mee, G.W.P., *Crop Sci. 9,* 346 (1969).
98. Heinz, D.J., Krishnamurthi, M., Nickell, L.G., and Maretzki, A., *In* "Plant Cell, Tissue and Organ Culture", (J. Reinert and Y.P.S. Bajaj, eds.), p. 3. Springer-Verlag, Berlin, (1977).
99. Henke, R.R., Mansur, M.A., and Constantin, M.J., *Physiol. Plant. 44,* 11 (1978).
100. Heuser, C.W., and Apps, D.A., *Can. J. Bot. 54,* 616 (1976).
101. Hill, G.P., *Ann. Bot. 31,* 437 (1967).
102. Hirotani, M., and Furuya, T., *Phytochem. 16,* 610 (1977).
103. Holsten, R.D., Burns, R.C., Hardy, R.W.F., and Herbert, R.R., *Nature 232,* 173 (1971).

104. Horak, J., Lustinec, J., Mesicek, J., Kaminek, M., and Polacokova, D., *Ann. Bot. 39,* 571 (1975).
105. Horvath Beach, K., and Smith, R.R., *Plant Sci. Lett. 16,* 231 (1979).
106. Hosoki, T., and Sagawa, Y., *Hort. Sci. 12,* 451 (1977).
107. Hu, C.Y., and Sussex, I.M., *Phytomorphology 21,* 103 (1971).
108. Huber, J., Constabel, F., and Gamborg, O.L., *Plant Sci. Lett. 12,* 209 (1978).
109. Hui, I.H., and Zee, S.Y., *Z. Pflanzenphysiol. 89,* 77 (1978).
110. Jelaska, S., *Planta 103,* 278 (1972).
111. Jelaska, S., *Physiol. Plant. 31,* 257 (1974).
112. Johnson, B.B., and Mitchell, E.D., *Hort. Sci. 13,* 246 (1978).
113. Kamada, H., and Harada, H., *J. Expt. Bot. 30,* 27 (1979).
114. Kamat, M.G., and Rao, P.S., *Plant Sci. Lett. 13,* 57 (1978).
115. Kant, U., and Hildebrandt, A.C., *Can. J. Bot. 47,* 849 (1969).
116. Kao, K.N., and Michayluk, M.R., *Planta 126,* 105 (1975).
117. Kao, K.N., Miller, R.A., Gamborg, O.L., and Harvey, B.L., *Can. J. Genet. Cytol. 12,* 297 (1970).
118. Kartel, N.A., and Maneshin, T.V., *Soviet Plant Physiol. 25,* 223 (1978).
119. Kartha, K.K., Gamborg, O.L., Constabel, F., and Shyluk, J.P., *Plant Sci. Lett. 2,* 107 (1974).
120. Kartha, K.K., Michayluk, M.R., Kao, K.N., Gamborg, O.L., and Constabel, F., *Plant Sci. Lett. 3,* 265 (1974).
121. Kasperbauer, M.J., and Collins, G.B., *Crop Sci. 12,* 98 (1972).
122. Kasperbauer, M.J., Buckner, R.C., and Bush, L.P., *Crop Sci. 19,* 457 (1979).
123. Kato, H., *Sci. Papers College Gen. Educ., Univ. of Tokyo 18,* 191 (1968).
124. Kato, H., and Takeuchi, M., *Plant Cell Physiol. 4,* 243 (1963).
125. Kato, H., and Takeuchi, M., *Sci. Papers College Gen. Educ. Univ. of Tokyo 16,* 245 (1966).
126. Khanna, P., and Staba, J., *Bot. Gaz. 131,* 1 (1970).
127. King, P.J., Potrykus, I., and Thomas, E., *Physiol. Vég. 16,* 381 (1978).
128. Koblitz, H., and Saalbach, G., *Biochem. Physiol. Pflanzen 170,* 97 (1976).
129. Kochba, J., and Button, J., *Z. Pflanzenphysiol. 73,* 415 (1974).
130. Kochba, J., and Spiegel-Roy, P., *Z. Pflanzenzuchtg. 69,* 156 (1973).

131. Kochba, J., Spiegel-Roy, P., and Safran, H., *Planta 106*, 237 (1972).
132. Kochba, J., Spiegel-Roy, P., Saad, S., and Neumann, H., *In* "Production of Natural Compounds by Cell Culture Methods", (A.W. Alferman and E. Reinhard, eds.), p. 223. BPT-Report 1/78, München (1978).
133. Kochhar, T.S., and Shabhawal, P.S., *Physiol. Plant. 40*, 169 (1977).
134. Koevary, K., Rappaport, L., and Morris, L.L., *Hort. Sci. 13*, 39 (1978).
135. Kohlenbach, H.W., *Planta 64*, 37 (1965).
136. Kohlenbach, H.W., *Z. Pflanzenphysiol. 55*, 142 (1966).
137. Kohlenbach, H.W., *In* "Plant Tissue Culture and Bio-technological Application", (W. Barz *et al.*, eds.), p. 355. Springer-Verlag, Berlin, (1978).
138. Konar, R.M., and Nataraja, K., *Phytomorphology 15*, 132 (1965).
139. Konar, R.M., and Oberoi, Y.P., *Phytomorphology 15*, 137 (1965).
140. Konar, R.M., Thomas, E., and Street, H.E., *J. Cell Sci. 11*, 77 (1972).
141. Konar, R.M., Thomas, E., and Street, H.E., *Ann. Bot. 36*, 249 (1972).
142. Kostoff, D., "Cytogenetics of the Genus *Nicotiana*". States Printing House, Sofia, (1943).
143. Krishnamurthi, M., and Tlaskal, J., *Proc. Int. Soc. Sugar Cane Technol. 15*, 130 (1974).
144. Krul, W.R., and Worley, J.F., *J. Amer. Soc. Hort. Sci. 102*, 360 (1977).
145. Lakshmi Sita, G., Raghava Ram, N.V., and Vaidyanathan, C.S., *Plant Sci. Lett. 15*, 265 (1979).
146. Lam, S.L., *Amer. Potato J. 52*, 103 (1975).
147. Lance, B., Reid, D.M., and Thorpe, T.A., *Physiol. Plant. 36*, 287 (1976).
148. Linsmaier, E.M., and Skoog, F., *Physiol. Plant. 18*, 100 (1965).
149. Liu, M.C., and Chen, W.H., *Euphytica 25*, 393 (1976).
150. Liu, M.C., Huang, Y.J., and Shih, S.C., *J. Agric. Assoc. China, 77*, 52 (1972).
151. Lorz, H., Harms, C.T., and Potrykus, I., *Z. Pflanzenzuchtg. 77*, 257 (1976).
152. Lowe, K.W., and Conger, B.V., *Crop Sci. 19*, 397 (1979).
153. Maheshwari, S.C., and Gupta, G.R.P., *Planta 67*, 384 (1965).
154. Maheshwari, P., and Rangaswamy, N.S., *Indian J. Hort. 15*, 275 (1958).
155. Majumdar, S.K., *South Afr. J. Bot. 36*, 63 (1970).

156. Maliga, P., Lazar, G., Joo, F., Nagy, A.H., and Menczel, L., *Molec. Gen. Genet. 157*, 291 (1977).
157. Maliga, P., Kiss, Z.R., Nagy, A.H., and Lazar, G., *Molec. Gen. Genet. 163*, 145 (1978).
158. Malmberg, R.L., *Abstr. Int. Congress Plant Tissue and Cell Culture, 4th, Calgary, Canada*, p. 137 (1978).
159. Malmberg, R.L., *Planta 146*, 243 (1979).
160. Masteller, V.J., and Holden, D.J., *Plant Physiol. 45*, 362 (1970).
161. Matern, U., Strobel, G., and Shepard, J., *Proc. Nat. Acad. Sci., (USA) 75*, 4935 (1978).
162. Mathews, V.H., Rangan, T.S., Narayanaswamy, S., *Z. Pflanzenphysiol. 79*, 450 (1976).
163. Matsuoka, H., and Hinata, K., *J. Expt. Bot. 30*, 363 (1979).
164. McWilliam, A.A., Smith, S.M., and Street, H.E., *Ann. Bot. 38*, 243 (1974).
165. Mehra, A., and Mehra, P.N., *Bot. Gaz. 135*, 61 (1974).
166. Melchers, G., and Sacristan, M.D., *In* "La Culture des Tissus et des Cellules des Végétaux", (R.J. Gautheret, ed.), p. 169. Masson, Paris, (1977).
167. Mellor, F.C., and Stace-Smith, R., *Can. J. Bot. 47*, 1617 (1969).
168. Meyer, M.M., and Milbrath, G.M., *Hort. Sci. 12*, 544 (1977).
169. Mitra, G.C., and Chaturvedi, H.C., *Bull. Torrey Bot. Club 99*, 184 (1972).
170. Mitra, J., Mapes, M.O., and Steward, F.C., *Amer. J. Bot. 47*, 357 (1960).
171. Mokhtarzadeh, A., and Constantin, M.J., *Crop Sci. 18*, 567 (1978).
172. Murakishi, H.H., and Carlson, P.S., *Phytopathology 66*, 931 (1976).
173. Murashige, T., *Ann. Rev. Plant Physiol. 25*, 135 (1974).
174. Murashige, T., *In* "Frontiers of Plant Tissue Culture 1978", (T.A. Thorpe, ed.), p. 15. Univ. Calgary Press, Calgary, (1978).
175. Murashige, T., *In* "Propagation of Higher Plants through Tissue Culture", (K.W. Hughes, R. Henke, and M. Constantin, eds.), p. 14. U.S. Dept. of Energy, Knoxville, Tenn., (1979).
176. Murashige, T., and Nakano, R., *Amer. J. Bot. 54*, 963 (1967).
177. Murashige, T., and Skoog, F., *Physiol. Plant. 15*, 473 (1962).
178. Murashige, T., and Tucker, D.P.H., *In* "Proc. Ist Int. Citrus Symp.", Vol. 3. (H.D. Chapman, ed.), p. 1155. Univ. of Calif. Press, Berkeley, (1969).

179. Nadar, H.M., and Heinz, D.J., *Crop Sci. 17*, 814 (1977).
180. Nadar, H.M., Soepraptoto, S., Heinz, D.J., and Ladd, S.L., *Crop Sci. 18*, 210 (1978).
181. Naf, U., *Growth 22*, 167 (1958).
182. Nag, K.K., and Johri, B.M., *Phytomorphology 19*, 405 (1969).
183. Nakano, H., Tashiro, T., and Maeda, E., *Z. Pflanzenphysiol. 76*, 444 (1975).
184. Narayanaswamy, S., *In* "Plant Cell, Tissue and Organ Culture", (J. Reinert and Y.P.S. Bajaj, eds.), p. 179. Springer-Verlag, Berlin, (1977).
185. Neumann, K.H., *Mikroskopie 25*, 261 (1969).
186. Neumann, K.H., Cirelli, E., and Cirelli, B., *Physiol. Plant. 22*, 787 (1969).
187. Nishi, T., Yamada, Y., and Takahashi, E., *Nature 219*, 508 (1968).
188. Nishi, T., Yamada, Y., and Takahashi, E., *Bot. Mag. (Tokyo) 86*, 183 (1973).
189. Nitsch, C., *In* "Genetic Manipulations with Plant Material", (L. Ledoux, ed.), p. 197. Plenum Press, New York, (1974).
190. Nitsch, J.P., Nitsch, C., Rossini, L.M.E., and Bui Dang Ha, D., *Phytomorphology 17*, 446 (1967).
191. Ogura, H., *Jap. J. Genet. 51*, 161 (1976).
192. Ogura, H., *Jap. J. Genet. 53*, 77 (1978).
193. O'Hara, J.F., and Street, H.E., *Ann. Bot. 42*, 1029 (1978).
194. Ohyama, K., and Oka, S., *Abstr. Int. Congress Plant Tissue and Cell Culture, 4th, Calgary, Canada*, p. 33 (1978).
195. Okazawa, Y., Katsura, N., and Tagawa, T., *Physiol. Plant. 20*, 862 (1967).
196. Orton, T.J., *Theor. Appl. Genet. 56*, 101 (1980).
197. Oswald, T.H., Smith, A.E., and Phillips, D.V., *Physiol. Plant. 38*, 129 (1977).
198. Padmanabhan, V., Paddock, E.F., and Sharp, W.R., *Can. J. Bot. 52*, 1429 (1974).
199. Pareek, L.K., and Chandra, N., *Plant Sci. Lett. 11*, 311 (1978).
200. Pence, V.C., Hasegawa, P.M., and Janick, J., *J. Amer. Soc. Hort. Sci. 104*, 145 (1979).
201. Peterson, P.A., *Brookhaven Symp. in Biol. 25*, 244 (1974).
202. Phillips, G.C., and Collins, G.B., *Crop Sci. 19*, 59 (1979).
203. Pierik, R.L.M., *Z. Pflanzenphysiol. 60*, 343 (1972).
204. Prabhudesai, V.R., and Narayanaswamy, S., *Phytomorphology 23*, 133 (1973).

205. Prabhudesai, V.R., and Narayanaswamy, S., *Z. Pflanzenphysiol. 71*, 181 (1974).
206. Price, H.C., *Am. Vegetable Grower 26*, 18 (1978).
207. Price, H.J., and Smith, R.H., *Planta 145*, 305 (1979).
208. Prokhorov, M.N., Chernova, L.K., and Filin-Koldakov, B.V., *Doklady Akad. Nauk. 214*, 472 (1974).
209. Rabechault, H., Ahee, J., and Guenin, G., *C.R. Acad. Sci., Paris 270*, 3067 (1970).
210. Radojevic, L., Vujicic, R., and Neskovic, M., *Z. Pflanzenphysiol. 77*, 33 (1975).
211. Raghavan, V., *Science 191*, 388 (1976).
212. Raman, K., and Greyson, R.I., *Can. J. Bot. 52*, 1988 (1974).
213. Rangan, T.S., *Z. Pflanzenphysiol. 72*, 456 (1974).
214. Rangan, T.S., *Z. Pflanzenphysiol. 78*, 208 (1976).
215. Rangan, T.S., Murashige, T., and Bitters, W.P., *Hort. Sci. 3*, 226 (1968).
216. Rangan, T.S., Murashige, T., and Bitters, W.P., *In* "Proc. First Int. Citrus Symp." Vol. 1, (H.D. Chapman, ed.), p. 225. Univ. of Calif. Press, Berkeley, (1969).
217. Rangaswamy, N.S., *Experientia 14*, 111 (1958).
218. Rangaswamy, N.S., *Phytomorphology 11*, 101 (1961).
219. Rao, P.S., *Phytomorphology 15*, 175 (1965).
220. Rao, P.S., and Narayanaswamy, S., *Physiol. Plant. 27*, 271 (1972).
221. Rao, P.S., Handro, W., and Harada, H., *Physiol. Plant. 28*, 458 (1973).
222. Rashid, A., and Street, H.E., *Planta 113*, 263 (1973).
223. Reinert, J., *Ber. Deutsch. Bot. Ges. 71*, 15 (1959).
224. Reinert, J., and Tazawa, M., *Planta 87*, 239 (1969).
225. Reinert, J., Bajaj, Y.P.S., and Zbell, B., *In* "Plant Tissue and Cell Culture", (H.E. Street, ed.), p. 389. Blackwell, Oxford, (1977).
226. Reuther, G., *Ber. Deutsch. Bot. Ges. 90*, 417 (1977).
227. Reuther, G., *Acta Horticulturae 78*, 217 (1977).
228. Reynolds, J.F., and Murashige, T., *In Vitro 15*, 383 (1979).
229. Rice, T.B., Reid, R.K., and Gordon, P.N., *In* "Propagation of Higher Plants through Tissue Culture", (K.W. Hughes, R. Henke and M. Constantin, eds.), p. 262. U.S. Dept. of Energy Conf. 7804111, (1978).
230. Sabharwal, P.S., *In* "Plant Tissue and Organ Culture - a Symposium", (P. Maheshwari and N.S. Rangaswamy, eds.), p. 332. Delhi Intern. Soc. Plant Morphologists, (1963).
231. Sacristan, M.D., and Melchers, G., *Molec. Gen. Genet. 105*, 317 (1969).

232. Sangwan, R.S., and Gorenflot, R., *Z. Pflanzenphysiol. 75*, 256 (1975).
233. Sangwan, R.S., and Harada, H., *J. Expt. Bot. 26*, 868 (1975).
234. Schenk, R.V., and Hildebrandt, A.C., *Can. J. Bot. 50*, 199 (1972).
235. Scowcroft, W.R., and Adamson, J.A., *Plant Sci. Lett. 7*, 39 (1976).
236. Sehgal, C.B., *Current Sci. 41*, 263 (1972).
237. Sehgal, C.B., *Phytomorphology 28*, 291 (1978).
238. Sengupta, C., and Raghavan, V., *In* "Plant Cell and Tissue Culture", (W.R. Sharp *et al.*, eds.), p. 841. Ohio State Univ. Press, Columbus, (1979).
239. Sengupta, C., and Raghavan, V., *J. Expt. Bot. 31*, 247 (1980).
240. Sengupta, C., and Raghavan, V., *J. Expt. Bot. 31*, 259 (1980).
241. Shana Rao, H.K., and Narayanaswamy, S., *Radiat. Bot. 15*, 301 (1975).
242. Sharma, G.C., Bello, L.L., and Sapra, V.T., *In* "Propagation in Higher Plants through Tissue Culture", (K.W. Hughes, R. Henke and M. Constantin, eds.), p. 258. U.S. Dept. of Energy Conf. 7804111, (1979).
243. Sharp, W.R., Sondahl, M.R., Caldas, L.S., and Maraffa, S.B., *In* "Horticultural Reviews", Vol. 2. (J. Janick, ed.), p. 268. AVI Publishing Co., Westport, Conn., (1980).
244. Shepard, J.F., *Virology 66*, 492 (1975).
245. Shepard, J.F., Bidney, D., and Shahin, E., *Science 208*, 17 (1980).
246. Sheridan, W.F., *Planta 82*, 189 (1968).
247. Sheridan, W.F., *In* "Genetic Manipulations with Plant Materials", (L. Ledoux, ed.), p. 263. Plenum Press, New York, (1975).
248. Shimada, T., *Jap. J. Genet. 53*, 371 (1978).
249. Shimada, T., Sasakuma, T., and Tsunewalei, K., *Can. J. Genet. Cytol. 11*, 294 (1969).
250. Skirvin, R.M., and Janick, J., *J. Amer. Soc. Hort. Sci. 101*, 281 (1976).
251. Skoog, F., and Miller, C.O., *Symp. Soc. Exp. Biol. 11*, 118 (1957).
252. Sommer, H.E., and Brown, C.L., *Forest Sci. 26*, 257 (1980).
253. Sondahl, M.R., and Sharp, W.R., *Z. Pflanzenphysiol. 81*, 395 (1977).
254. Sondahl, M.R., and Sharp, W.R., *In* "Plant Cell and Tissue Culture", (W.R. Sharp *et al.*, eds.), p. 527. Ohio State Univ. Press, Columbus, (1979).

255. Sondahl, M.R., Spahlinger, D.A., and Sharp, W.R., *Z. Pflanzenphysiol. 94,* 101 (1979).
256. Spiegel-Roy, P., and Kochba, J., *Radiat. Bot. 13,* 97 (1973).
257. Sree Ramulu, K., Devreux, M., Ancora, G., and Laneri, U., *Z. Pflanzenzuchtg. 76,* 299 (1976).
258. Steward, F.C., Ammirato, P.V., and Mapes, M.O., *Ann. Bot. 34,* 761 (1970).
259. Stoutemeyer, V.T., and Britt, O.K., *Amer. J. Bot. 52,* 805 (1965).
260. Street, H.E., *In* "Regulation of Developmental Processes in Plants", (H.R. Schutte and D. Gross, eds.), p. 192. Fisher, Jena, (1978).
261. Street, H.E., *In* "Plant Cell and Tissue Culture", (W.R. Sharp *et al.*, eds.), p. 123. Ohio State Univ. Press, Columbus, (1979).
262. Street, H.E., and Withers, L.A., *In* "Tissue Culture and Plant Science", (H.E. Street, ed.), p. 71. Academic Press, London, (1974).
263. Stringham, G.R., *Z. Pflanzenphysiol. 92,* 459 (1979).
264. Sunderland, N., *In* "Haploids in Higher Plants, Advances and Potential", (K.J. Kasha, ed.), p. 91. Univ. of Guelph, Guelph, Ontario, (1974).
265. Sunderland, N., *In* "Plant Tissue and Cell Culture" 2nd edition. (H.E. Street, ed.), p. 177. Blackwell, Oxford, (1977).
266. Sunderland, N., and Dunwell, J.M., *In* "Plant Tissue and Cell Culture", (H.E. Street, ed.), p. 223. Blackwell, Oxford, (1977).
267. Sung, Z.R., Smith, R., and Horowitz, J., *Planta 147,* 236 (1979).
268. Svobodova, J., *In* "Viruses in Plants", (A. Beemster and J. Dykstra, eds.), p. 48. North Holland, Amsterdam, (1965).
269. Syōno, K., *Plant Cell Physiol. 6,* 371 (1965).
270. Syōno, K., and Furuya, T., *Bot. Mag. (Tokyo) 85,* 273 (1972).
271. Takatori, F.H., Murashige, T., and Stillman, J.I., *Hort. Sci. 3,* 20 (1968).
272. Takebe, I., Labib, G., and Melchers, G., *Naturwissenschaften 58,* 318 (1971).
273. Thomas, E., and Street, H.E., *Ann. Bot. 34,* 657 (1970).
274. Thomas, E., and Wenzel, G., *Naturwissenschaften 62,* 40 (1975).
275. Thomas, E., and Wenzel, G., *Z. Pflanzenzucht. 74,* 77 (1975).

276. Thomas, E., and Wernicke, W., *In* "Frontiers of Plant Tissue Culture, 1978", (T.A. Thorpe, ed.), p. 403. Univ. of Calgary Press, Canada (1978).
277. Thomas, E., King, P.J., and Potrykus, I., *Naturwissenschaften 64*, 587 (1977).
278. Thorpe, T.A., *Int. Rev. Cytol.* Suppl. 11A (I.K. Vasil, ed.), p. 71. Academic Press, New York, (1980).
279. Tilquin, J.P., *Can. J. Bot. 57*, 1761 (1979).
280. Tisserat, B., Esan, E.B., and Murashige, T., *In* "Horticultural Reviews", Vol. I. (J. Janick, ed.), p. 1. AVI Publishing Co., Westport, Conn., (1979).
281. Toren, J., *Rev. Gen. Bot. 62*, 392 (1955).
282. Torrey, J.G., *Physiol. Plant. 20*, 265 (1967).
283. Torrey, J.G., and Reinert, J., *Plant Physiol. 36*, 483 (1961).
284. Tran Thanh Van, K., *In* "Plant Tissue Culture and Its Bio-technological Application", (W. Barz *et al.*, eds.), p. 367. Springer-Verlag, Berlin, (1977).
285. Tran Thanh Van, K., and Trinh, H., *In* "Frontiers of Plant Tissue Culture", (T.A. Thorpe, ed.), p. 37. Univ. of Calgary Press, Calgary, (1978).
286. Tran Thanh Van, K., Thi Dien, N., and Chlyah, A., *Planta 119*, 149 (1974).
287. Traub, H.P., *Science 83*, 165 (1936).
288. Vasil, I.K., and Hildebrandt, A.C., *Science 150*, 889 (1965).
289. Vasil, I.K., and Hildebrandt, A.C., *Amer. J. Bot. 53*, 860 (1966).
290. Vasil, I.K., and Hildebrandt, A.C., *Amer. J. Bot. 53*, 869 (1966).
291. Vasil, I.K., and Vasil, V., *Int. Rev. Cytol.*, Suppl. 11A (I.K. Vasil, ed.), p. 145. Academic Press, New York, (1980).
292. Vasil, I.K., Ahuja, M.R., and Vasil, V., *Adv. Genet. 20*, 127 (1979).
293. Venketeswaran, S., and Huhthinen, O., *In Vitro 14*, 355 (1978).
294. Vyskot, B., and Novak, F.J., *Z. Pflanzenphysiol. 81*, 34 (1977).
295. Warren, G.S., and Fowler, M.W., *Plant Sci. Lett. 9*, 71 (1977).
296. Webber, H.J., *Proc. Amer. Soc. Hort. Sci. 28*, 57 (1931).
297. Webber, H.J., *Bot. Rev. 6*, 575 (1940).
298. Webster, P.L., and Davidson, D., *J. Cell Biol. 39*, 332 (1968).
299. White, D.W.R., and Vasil, I.K., *Theoret. Appl. Genet. 55*, 107 (1979).

300. White, P.R., "The Cultivation of Animal and Plant Cells", 2nd Edition. Ronald Press, New York, (1963).
301. Widholm, J.M., *In* "Plant Tissue Culture and Its Bio-technological Application", (W. Barz *et al.*, eds.), p. 112. Springer-Verlag, Berlin, (1977).
302. Williams, L., and Collin, H.A., *Ann. Bot. 40*, 325 (1976).
303. Wodzicki, T.J., *Acta Soc. Bot. Polon. 47*, 225 (1978).
304. Wu, L., and Li, H.W., *Cytologia 36*, 411 (1971).
305. Yamada, Y., Tanaka, K., and Takahashi, E., *Proc. Japan Acad. 43*, 156 (1967).
306. Yamaguchi, T., and Nakajima, T., *In* "Plant Growth Substances 1973", (S. Tamura, ed.), p. 1121. Hirokawa-Shoten, Tokyo, Japan, (1974).
307. Yamane, Y., *Jap. J. Genet. 49*, 139 (1974).
308. Yeoman, M.M., *Int. Rev. Cytol. 29*, 383 (1970).
309. Yeoman, M.M., and MacLeod, A.J., *In* "Plant Cell and Tissue Culture", (H.E. Street, ed.), p. 31. Blackwell, Oxford, (1977).
310. Yokoyama, T., and Takeuchi, M., *Phytomorphology 26*, 273 (1976).
311. Zapata, F.J., Evans, P.K., Power, J.P., and Cocking, E.C., *Plant Sci. Lett. 8*, 119 (1977).
312. Zee, S.Y., and Wu, S.C., *Z. Pflanzenphysiol. 93*, 325 (1979).

ISOLATION, FUSION AND CULTURE OF PLANT PROTOPLASTS

Oluf L. Gamborg

International Plant Research Institute
San Carlos, California

Jerry P. Shyluk

Prairie Regional Laboratory
National Research Council of Canada
Saskatoon, Saskatchewan, Canada

Elias A. Shahin

International Plant Research Institute
San Carlos, California

I. INTRODUCTION

Somatic hybridization by protoplast fusion is a potential method for producing hybrids between closely related as well as unrelated plants. The cell fusion method is the probable means of producing desirable crosses in plant breeding which are not feasible by conventional methods (29, 30, 35).

The production of new hybrids is only one of a variety of applications of isolated protoplasts. By definition, protoplasts are plant cells with the plasma membrane but without the cell wall. The absence of the rigid cellulose wall and the complete exposure of the plasma membrane makes protoplasts a particularly useful system for investigating the uptake of macromolecules and transport phenomena. Protoplasts have also been employed extensively in studies on plant virus uptake.

ISBN 0-12-690680-7

New insights have been gained into the uptake and infection process as well as virus replication. Protoplasts are also becoming the material of choice in the elucidation of the specificity and mode of action of fungal and bacterial plant pathogens. Toxic compounds produced by some fungal and bacterial pathogens have a deleterious effect on protoplasts from susceptible plants but those from resistant plants are unaffected. These compounds can then be used in screening for protoplasts which are resistant to the plant pathogen.

Isolated protoplasts are capable of taking up plant organelles such as nuclei and chloroplasts. Uptake of protein, DNA, and other macromolecules has also been widely reported. DNA from bacteria as well as higher organisms is taken up by protoplasts and there are indications that transfer of genetic information may be achieved through this uptake of DNA. Protoplasts have even proven useful as sources of organelles (60). These topics have been covered in detail recently in Thorpe (73), Sharp *et al.* (67), Vasil (76) and Vasil *et al.* (78).

II. ISOLATION OF PLANT PROTOPLASTS

The introduction in the 1960's of methods for the isolation of viable protoplasts by enzyme treatment made it possible to obtain large quantities for experimental purposes. Since then, progress has been made in methods of handling and manipulating isolated protoplasts. In suitable culture medium and environmental conditions, the protoplasts regenerate a cell wall, undergo cell division and may grow into complete plants (see also, 25, 35).

Protoplasts are usually isolated by treating tissues with a mixture of cell wall degrading enzymes in solutions which contain osmotic stabilizers to preserve the structure and viability of the protoplasts. The relative ease with which protoplast isolation can be achieved depends upon a variety of factors. The most important of these are the physiological state of the tissues and cell materials, the choice of enzymes, the composition of the solutions, and the concentration and type of osmotic stabilizer.

Protoplasts have been isolated from tissues of a large number of plant species (Table I). A convenient and suitable source of protoplasts is leaf mesophyll tissue. The most satisfactory results have been obtained with fully expanded leaves from young plants or new shoots. Protoplast yield, viability, and quality are influenced substantially by the age of the leaf as well as the environmental conditions of

TABLE I. Some Crop Plants from which Protoplasts have been Isolated and Cultured

Crop	Species	References
Cereal, Grains, and Grasses		
Barley	Hordeum vulgare	46, 51
Maize	Zea mays	9, 58
Millet	Pennisetum americanum	77
Rice	Oryza sativa	72, 74
Sorghum	Sorghum bicolor	9, 10
Wheat	Triticum spp.	20, 72
Root, Tuber, Fiber, and Oil Seed Crops		
Cassava	Manihot esculentum	66
Cotton	Gossypium hirsutum	3
Flax	Linum usitatissimum	31
Potato	Solanum tuberosum	6, 55, 68
Rapeseed	Brassica napus	47
Rapeseed	Brassica campestris	37
Sugar beet	Beta vulgaris	56
Sugar cane	Saccharum spp.	52
Leguminous Plants		
Alfalfa	Medicago sativa	44
Bean	Phaseolus vulgaris	57
Broad bean	Vicia faba	4
Cow pea	Vigna sinensis	18
Pea	Pisum sativum	1, 14, 33
Soybean	Glycine max	16, 45
Sweet clover	Melilotus spp.	16
Vegetables and Fruits		
Asparagus	Asparagus officinalis	11
Cabbage	Brassica oleracea	36
Carrot	Daucus carota	22, 40
Cucumber	Cucumis sativus	17
Citrus	Citrus sinensis	75
Tomato	Lycopersicon esculentum	12, 55
Tomato	Lycopersicon peruvianum	79

Table I. (continued)

Crop	*Species*	*References*
Woody Species		
Caragana	*Caragana arborescens*	*16*
Douglas fir	*Pseudotsuga menziesii*	*50*
Pine	*Pinus pinaster*	*19*
Rose	*Rosa spp.*	*70*
Other Crops		
Tobacco	*Nicotiana sp.*	*8*

light, temperature, soil fertility, and humidity under which the plants are grown. Since the physiological state of the tissue is critical, the plants are usually grown in environmental growth chambers or greenhouses where light and temperature can be carefully controlled. With respect to light intensity, a range of 0.3 to 1.0 Wcm^{-2} has proven suitable, although in special cases much more intense illumination can be beneficial. A photoperiod of 18:6 or 16:8 hours of light:dark is the most common. In some cases the yield of protoplasts has been greatly improved by placing the tissue in the dark for 24-72 hours prior to isolation, during which time starch degradation occurs. The presence of starch grains has been reported to have adverse effects on protoplast viability. The disappearance of starch improves the potential yield of protoplasts. The choice of temperature and humidity is generally dictated by the requirements to ensure vigorous plant growth. Maintaining the plants at a high level of fertility, particularly using nitrogen fertilizers, can be beneficial.

In addition to leaf tissue other plant materials have been used. These include shoot tips, cotyledons, flower petals, and microspores. Callus tissue and cell suspension cultures are frequently used as sources of protoplasts. Since callus tissues grown on agar often have a slow growth rate and have a broad diversity in cell age and physiological state these may be less suitable. Cells grown in liquid suspension can be expected to have a generation time of 24-30 hours when fully established, and consist of smaller cell clusters and single cells. For best production of protoplasts such cell cultures are subcultured at 3-4 day intervals to maintain maximum growth rate and uniformity of the cell population. Cells taken at the early log phase are generally the most suitable. In established cultures this would correspond to ca. two days after subculturing. Cultures which have been newly established in liquid suspension often consist of mixtures of cell sizes and ages as well as a proportion of dead cells and would yield a lower percentage of protoplasts.

A. *Enzymes and Isolation Medium*

The enzymes used in the isolation of protoplasts comprise three general classes; cellulases, hemicellulases, and pectinases. Commercial preparation of the enzymes are usually used (Table II). The range of available sources is limited. The preparations are not pure and contain other enzymes such as proteases, lipases, and nucleases, some of which may have deleterious effects. Some purification and desalting of the preparations is achieved by column gel filtration using Sephadex G50^{R} (32).

Both the composition and the concentration of enzymes affect the yield of protoplasts from a given tissue. The incubation mixture used in protoplast isolation consists of the enzymes dissolved in a solution containing a few salts or a culture medium, a buffer, and an osmotic stabilizer. Calcium (2-6mM) is an essential ingredient and phosphate (0.5-2.0 mM) is beneficial in the preservation of protoplasts viability. The most common osmotic stabilizers are sorbitol, mannitol, glucose, and sucrose. They are often used in mixtures and glucose and/or sucrose are always included as the carbon source. The optimal osmolarity varies from 0.3 to 0.7 M.

The prevailing pH is critical. An acidic pH in the range of 5.0-6.0 is optimal. The solution often contains a buffer such as phosphate or MES [2(N-Morpholino)-ethanesulfonic acid] at 3 mM to minimize shifts in pH during digestion.

B. *Isolation Procedures*

Several procedures and strategies are employed in protoplast isolation. A two-step procedure in which the tissues are treated consecutively in different mixtures of enzymes has been used. Pretreatment of tissue in nutrient media with hormones at different temperature and light conditions is used in the isolation of protoplasts from potato and cassava. (See Lab Procedure below). The pretreatment provides a conditioning of the cells for more efficient isolation, greater viability, and consequently a higher percentage division.

TABLE II. Enzymes Commonly used for Protoplast Isolation

Enzymes	*Commercial Source*
Onozuka R10, cellulase	*Kinki Yakult Biochemical, Nishinomiya, Japan*
Cellulysin, cellulase	*Calbiochem, San Diego, Ca., U.S.A.*
Driselase, cellulase	*Kyowa Hakko Kogyo Co., Tokyo, Japan*
Macerozyme R10, pectinase	*Kinki Yakult Biochemical, Nishinomiya, Japan*
Pectinase	*Sigma Chemical Co., St. Louis, Mo., U.S.A.*
Rhozyme HP-150 hemicellulase	*Rohm and Haas Co., Philadelphia, Pa., U.S.A.*

Protoplast isolation is carried out in flasks or in petri dishes. Leaves or other tissues are cut into thin sections of ca. 1 mm and whenever possible the leaf epidermis is removed by forceps or by brushing. During incubation (23-28°C) a slow shaking of 30-50 rpm or occasional stirring facilitates isolation. Excessive shaking results in protoplast destruction.

After incubation the undigested tissue is removed by gravity filtration on a nylon or a stainless steel filter. The protoplasts are collected and washed by centrifugation at ca. 100x g.

Protoplasts are isolated under aseptic conditions. The use of laminar flow sterile transfer units and careful sterilization of plant materials and solutions eliminates the need for antibiotics.

III. PROTOPLAST CULTURE

Protoplasts are cultured in liquid media or placed on nutrient agar in petri dishes. Protoplast suspensions can be cultured in liquid droplets in petri dishes sealed with Parafilm^{R}. During culture the protoplasts are incubated in humidified chambers at 25-28°C with a light intensity of 100 to 500 lux or in the dark. After cell wall regeneration and division have been initiated in liquid cultures the cells can be transferred to agar media.

A. *Culture Media*

Protoplasts have nutritional requirements similar to those of cultured plant cells (28). The mineral salt compositions established for plant cell cultures have been modified to meet particular requirements by protoplasts. Table III lists examples of media employed for culturing protoplasts. The medium consists of: (1) osmotic stabilizers, (2) inorganic nutrients, (3) carbon sources, (4) vitamins, (5) organic nitrogen, and (6) growth hormones.

1. Osmotic Stabilizers. Mannitol and sorbitol are the compounds most frequently used to maintain osmolarity. They are used separately or in combination. Very little is known about their rate of absorption and metabolism by plant cells, but sorbitol can be utilized by some plant cell cultures as a carbon source. Glucose is a suitable alternative to hexitols. Mineral salts have also been employed.

2. Inorganic Nutrients. Mineral salts supply the inorganic nutrients required by plant cells. The inorganic nitrogen sources should consist of both nitrate and ammonium. It is generally necessary for maximum growth of cultured plant cells to keep the concentration of nitrate and potassium at 20-30 mM. To lessen any toxic effect and improve utilization of ammonium, it may be beneficial to add an organic acid such as succinate. Protoplasts have been cultured in a wide range of mineral salt media varying widely in total salt concentration. It has proven beneficial to use a calcium concentration of 4-6 mM. After cell regeneration and division have been initiated, regular plant cell culture media are satisfactory (see 34, and Chapter 2).

3. Carbon Sources. Glucose is perhaps the preferred and most reliable carbon source. Plant cells grow about equally well on a combination of glucose and sucrose, but sucrose alone may not always be satisfactory for plant protoplasts. Some media contain 1-3 mM of ribose or another pentose as supplementary carbon sources.

TABLE III. Plant Protoplast Culture Media

Ingredients (mg/l)	Reference (10)	(42)	(24)
KNO_3	2500	1900	1480
NH_4NO_3	250	600	270
$(NH_4)_2SO_4$	134	-	-
$MgSO_4 \cdot 7H_2O$	250	300	340
$CaCl_2 \cdot 2H_2O$	735	600	570
KH_2PO_4	-	170	80
$NaH_2PO_4 \cdot H_2O$	150	-	-
KCl	-	300	-
Ferric EDTA (Versenate or Sequestrene[R]	40	40	40
330 Fe 12.22 Fe_2O_3)[a]	(30)	(30)	(30)
KI	0.75	0.75	0.25
H_3BO_3	3	3	2
$MnSO_4 \cdot 4H_2O$	10	10	5
$ZnSO_4 \cdot 7H_2O$	2	2	1.5
$Na_2MoO_4 \cdot 2H_2O$	0.25	0.25	0.1
$CuSO_4 \cdot 5H_2O$	0.025	0.025	0.015
$CoCl_2 \cdot 6H_2O$	0.025	0.025	0.01
Sucrose	-	125	150
Glucose	81000	68400	-
Ribose	500	125	-
Xylose	500	125	-
Mannitol	-	125	0.3M
Other carbohydrates[b]	-		-
Myo-inositol	100	100	100
Thiamine·HCl	10	10	4
Nicotinic acid	1	1	4
Pyridoxine·HCl	1	1	0.7
Biotin	0.01	0.005	0.04
Folic acid	0.4	0.2	0.4
Calcium pantothenate	1	0.5	-
Other vitamins[c]			-
Casamino acids (Bacto[R], vitamin-free)[d]	250	125	-
L-Glutamine	580	-	-
L-Arginine	210	-	-
L-Asparagine	300	-	-
Organic acids[e]	-		-
Coconut water (ml)[f]	-	10	-
2,4-dichlorophenoxyacetic acid	1	0.2	1.4
2,4,5-trichlorophenoxyacetic acid	0.5	-	-

Table III. (continued)

Ingredients (mg/l)	*Reference* (10)	(42)	(24)
Naphthaleneacetic acid	-	1	-
Benzyladenine	-	-	0.4
Zeatin	-	0.5	-
pH	5.8	5.7	5.6

[a] *Geigy Agricultural Chemical Co., Ardsley, N.Y., U.S.A.*
[b] *Fructose 125 mg/l, mannose 125 mg/l, rhamnose 125 mg/l, cellulose 125 mg/l, sorbitol 125 mg/l (42).*
[c] *Choline chloride 1 mg/l, riboflavine 0.2 mg/l, p-amino bezoic acid 0.02 mg/l, ascorbic acid 2 mg/l, vitamin A 0.01 mg/l, vitamin D_3 0.01 mg/l, vitamin B_{12} 0.02 mg/l (10).*
Same compounds as c in the following concentrations: (mg/l) 0.5, 0.1, 0.01, 1, 0.005, and 0.01 respectively) (42).
[d] *Difco Laboratories, Detroit, Mi., U.S.A.*
[e] *Citrate 10 mg/l, fumarate 10 mg/l, malate 10 mg/l, pyruvate 5 mg/l (42).*
[f] *From ripe coconuts.*

4. *Vitamins.* The vitamins used for protoplast culture include those present in standard plant tissue culture media. Plant cells in culture require thiamine. Nicotinic acid and pyridoxine may enhance growth. If protoplasts are to be cultured at very low density in defined media there may be a requirement for additional vitamins.

5. *Organic Nitrogen Supplements.* Protoplast media frequently contain one or more amino acids. A convenient approach is to add 0.01 to 0.25% of vitamin-free casamino acids (Difco) or casein hydrolyzate (enzymic digest). Such mixtures appear to meet the needs of protoplasts if inorganic nitrogen is inadequate. The addition of 1 to 5 mM L-glutamine can improve growth and coconut milk at 1-5% (v/v) can have a beneficial effect on the survival and growth of protoplasts.

6. *Growth Hormones.* Auxins and cytokinins are required to induce cell division and plant regeneration from proto-

plasts. Auxins which have proven effective include 2,4-dichlorophenoxyacetic acid, and naphthaleneacetic acid. Cytokinins which have been shown to be effective include kinetin, benzyladenine, zeatin, and isopentenyladenosine.

B. *Culture Conditions and Environment*

Success in culturing depends on the correct choice of pH and osmolarity. A pH in the range of 5.5 to 5.8 appears to be the most suitable, although a pH above 6 can be beneficial. The prevailing osmolarity for culturing protoplasts varies from 0.35-0.70 M depending upon the source. Protoplasts are cultured at a temperature varying from 22 to 28°C. Light has not been reported to be essential but may be harmful if supplied at intensities greater than ca. 2000 lux. Another factor which influences the capacity of protoplasts for cell division is the initial concentration. Protoplasts are generally cultured at a concentration varying from 10^4 to 10^6 per ml but success has been achieved in culturing very low concentrations. A feeder technique significantly enhanced division rates in tobacco protoplasts plated at 5-50 protoplasts per ml (60-100 per 6 cm petri dish). Freshly isolated leaf protoplasts were embedded in agar, which was layered on a preformed agar plate containing tobacco protoplasts which had been exposed to X-irradiation. The irradiated protoplasts failed to divide. Between 10-40% of the untreated protoplasts divided and shoot formation occurred within 7 weeks (61, 62). Protoplasts from *Vicia* cell cultures were cultured successfully in enriched liquid media at a density of one in 4 ml.

IV. PROTOPLAST DEVELOPMENT DURING ISOLATION AND CULTURE

Freshly isolated protoplasts are devoid of cell walls. The absence of microfibrils has been demonstrated by the use of freeze-etching and platinum/palladium replica techniques. However, it is conceivable that wall removal may not be complete in all protoplasts of a given population if variation exists in cell age and type of the source material. Using the replica technique it has been shown that the resynthesis of cell walls begins immediately after removal of enzymes. Microfibrils are observed within 1 hour of culture. Indications of wall formation can be obtained by using a fluorescent dye (Calcofluor White[R], American Cyanamid, Bound Brook, N.J.) and viewing in ultraviolet light with appropriate filters. Protoplasts offer an unparalleled system in which to study the early stages of wall biosynthesis and to establish which organelles may be involved.

The exposed membranes contain large numbers of evenly distributed sugar molecules detected by a concanavalin + haemocyanin labelling technique. Exposure of soybean protoplasts to ultraviolet irradiation causes cessation of wall regeneration, and the inhibition of nucleic acid and protein synthesis. Protoplasts from cell cultures of carrot exposed to irradiation or chemical mutagens have the capacity to synthesize DNA and repair damage to DNA.

A. *Division and Differentiation*

Protoplasts of many plant species have been observed to undergo sustained division (Table I). Wall regeneration generally precedes division. There have been no reports on the production of wall-less mutants, although it may be possible to produce such mutants by the treatment of protoplasts with mutagenic agents.

Mitosis in protoplasts appears to be similar to the process in plant cells. The difficulties which have prevented division in various protoplast species are related to the induction process. A high rate of division is difficult to achieve in most protoplast cultures. The consistently high division frequency reported for some sources suggest that in due time the problem limiting division frequency will be resolved. After protoplasts have regenerated cell walls and sustained division has been achieved, the cells can be placed on a medium which is conducive to differentiation and morphogenesis. If protoplasts from carrot or other embryo-forming cell cultures are used, the newly formed cells then form embryos and regenerate complete plants (Table I). In most species shoot initiation requires the use of cytokinins at 1-50 μM, but even in the presence of these compounds the callus may only form roots. Plant regeneration from protoplasts has been reported for alfalfa, tobacco, carrot, brome grass, rapeseed, asparagus, orange, millet, potato, and tomato.

B. *Protoplast Fusion*

Protoplasts can fuse spontaneously during the isolation process or after induction by exposure to special conditions. Spontaneous fusion can occur during isolation between two or more adjacent protoplasts. The process appears to take place when the plasmodesmata expand rather than break. The resulting homokaryons may contain numerous nuclei. These multinucleated protoplasts reform a cell wall and the nuclei may enter mitosis synchronously. Protoplasts isolated from meiocytes of lily fuse spontaneously after isolation by contact without a fusing agent. The membrane surface structure apparently facilitates adhesion and fusion, which occur within 5 minutes. The meiotic process continues and occurs synchronously in the homokaryons.

C. *Induced Fusion*

Fusion of protoplasts from different sources must be induced. The protoplasts must be agglutinated to establish large areas of membrane contact. Keller and Melchers introduced a successful fusion procedure consisting of incubating protoplasts in high alkaline, high Ca-salt solutions at 37°C for ca. 30 minutes (49). A successful and reproducible procedure for protoplast fusion involves the use of polyethylene glycol (PEG) (13, 43). High molecular weight PEG added to protoplasts at a concentration of 28-30% (w/v) causes immediate extensive adhesion. Fusion follows when the PEG is eluted/diluted. PEG-induced fusion is non-specific and has been effective for fusion of protoplasts from a wide variety of plant sources and can occur at frequencies of up to 50%. The procedure is outlined in the laboratory section. The protoplasts of two sources are prepared separately, collected by centrifugation, and mixed in the desirable ratios of ca. 1:1 to 1:4 in a solution containing osmotic stabilizer(s) and a calcium salt. The protoplasts are dispensed in droplets and allowed to settle at the bottom of a petri dish (or on cover slip). When small droplets of PEG solution (5 g of PEG-1540 added into 10 ml solution) are added, the protoplasts form adhesion bodies. This condition is maintained for 20-45 minutes. Elution of PEG is initiated by adding droplets of an alkaline, high calcium solution which is then followed by washing with culture medium. Fusion takes place during elution and the protoplasts regain their spherical structure. The protoplasts and the fusion products adhere to the surface, a feature which facilitates washing. After washing, a thin layer of medium is retained and the petri dishes incubated as usual.

PEG may function as a molecular bridge perhaps aided by calcium, and may facilitate the dissociation of the protoplast plasmalemma. Both the adhesion process and the actual fusion appear to require critical but different concentrations of PEG. Adhesion requires a concentration of 25-28% and occurs rapidly. The membrane contact extends to large areas but may be discontinuous forming intervening spaces. Upon dilution of the PEG the opposing plasmalemmas appear to erupt at several points and cytoplasmic continuity between adjacent protoplasts becomes established. At the outer edges, the membranes of the fusing protoplasts join and the intervening sections form vesicles which gradually degrade. Initially the two cytoplasms remain separated, but mixing occurs within 12-24 hours. The discovery of PEG as an efficient agent for plant protoplast fusion has prompted studies on the use of the polymer in fusion of other cell types. Apparently the compound is non-specific and induces fusion of protoplasts of bacteria, fungi, yeast, and algae (see 30). Moreover, PEG also facilitates fusion of mammalian cells with plant protoplasts and of plant protoplasts with those of algae and microorganisms (see 27).

D. *Development of Fusion Products*

The fusion products can be observed and studied by various methods. A differential staining procedure for nuclei permits monitoring of the fusion products. Results have shown that the ratio of 1:1 nucleus of each species occurs most frequently (14%) followed by those with 2:1 ratios. The large multinucleated protoplasts appear to deteriorate. The fusion products recover from the PEG treatment and reform a cell wall within 24 hours (15).

Combinations of fusion partners often involve protoplasts from a cell culture and from leaf material. The capacity of the fusion product to recover and develop appears to be inherited from the protoplast of the cell culture. After 1-2 days in culture several developments can be observed. Nuclear fusions have been observed. The event can be detected one day after fusion and requires several hours to complete (15). The synkaryons contain a mixed chromatin. Nuclear fusion possibly occurs through the formation of nuclear membrane bridges. Such phenomena have been observed in multinucleated homokaryons and in heterokaryons (27).

The ability of synkaryons to enter mitosis has not been demonstrated and fusion of interphase nuclei may not be essential in the formation of hybrid cells. The occurrence of chromosomes of both parents in mitosis has been shown for several intergeneric and interfamily heterokaryons which may suggest fusion of mitotic nuclei (15, 42).

Mitosis can occur synchronously in multinucleated homokaryons as well as in heterokaryons. Division has been observed in heterokaryons arising from fusion of several species of di- and monocotyledonous plants (Table IV).

The production of cell hybrids between plant families of *N. glauca* + soybean, carrot + barley, soybean + *B. napus* suggest the absence of an apparent somatic cell incompatibility. The hybrid nature of the cell progeny has been established on the basis of ultrastructural examination, chromosome identification, isoenzymes, and polypeptide patterns of the Fraction 1 protein. In the initial stages, the cells contain chloroplasts originating from leaf protoplasts as well as leucoplasts contributed by the cultured cells. In the nuclei, the heterochromatin is that of the two parents and chromosomes of both parental species can be recognized.

V. HYBRID SELECTION AND PLANT DEVELOPMENT

Various approaches have been used to isolate hybrids. In most fusion experiments, the division rates are relatively low. Moreover, one or both parental protoplast species also may divide and within a short period the hybrid cells cannot be distinguished from parental cells. Several selection methods have been successful (Table V). A special plating procedure was used to isolate hybrids of *N.glauca* + soybean (42), and of *Arabidopsis thaliana* + *Brassica campestris* (37). After allowing the hybrid cells to undergo a few divisions, the cell mixture was diluted. Microdroplets (ca. 500 ml) were placed in Cuprak[R] petri dishes designed with numerous small wells. Each droplet contained a single or a few cell clusters. By scanning under a light microscope the wells containing single hybrid cell clusters could be identified. This procedure has been used to obtain several hybrid cell lines of *N. tabacum* + soybean and *A. thaliana* + *B. campestris*. In the latter example mature plants were also regenerated.

A number of other types of potential chemical selection procedures may involve herbicides, phytotoxins, or antibiotics. Plants differ in their capacity to metabolize and therefore tolerate herbicides. Phytotoxins produced by plant pathogens are metabolic analogues and have been shown to be species specific (71). The compounds exert their effects at relatively low concentrations and can be used in positive selection and screening for cells and plants which are resistant to the compound and thus to the pathogen.

The principle of genetic complementation is likely to be one of the most reliable and effective selection approaches for the recovery of somatic hybrids in plants. In the first

TABLE IV. Examples of Plant Species in which Protoplast Fusion and Division of the Fusion Products has been Achieved

Source of protoplast		Reference
Barley (*Hordeum vulgare*)	+ Soybean (*Glycine max*)	46
Corn (*Zea mays*)	+ Soybean (*Glycine max*)	46
Pea (*Pisum sativum*)	+ Soybean (*Glycine max*)	15
Sweet clover (*Melilotus alba*)	+ Soybean (*Glycine max*)	16
Alfalfa (*Medicago sativa*)	+ Soybean (*Glycine max*)	16
Caragana (*Caragana arborescens*)	+ Soybean (*Glycine max*)	16
Rapeseed (*Brassica napus*)	+ Soybean (*Glycine max*)	48
Barley (*Hordeum vulgare*)	+ Carrot (*Daucus carota*)	21
Nicotiana glauca	+ *Nicotiana tabacum*	26
Nicotiana glauca	+ Soybean (*Glycine max*)	42
Nicotiana rustica	+ Soybean (*Glycine max*)	41
Faba bean (*Vicia faba*)	+ Petunia (*Petunia hybrida*)	5
Arabidopsis thaliana	+ *Brassica campestris*	37
Tomato (*Lycopersicon esculentum*)	+ Potato (*Solanum tuberosum*)	55
Sorghum (*S. bicolor*)	+ Corn (*Zea mays*)	10

reported experiments on somatic hybridization in plants two tobacco species were used (69). It was known that sexual hybrids of these species (*Nicotiana glauca x N. langsdorffii*) were oncogenic and produced genetic tumors. Cells from plant tumors can grow in culture in the absence of growth hormones (auxins). When protoplasts from these species were fused a proportion of the regenerated cells grew in the absence of hormones and the plants obtained were hybrids and produced tumors.

Chlorophyll-deficient (non-allelic) mutants have been used widely in selection of somatic hybrids (Table V). Nutritional mutants would provide very desirable material. Until recently with the report of a panthothenate-requiring mutant in *Datura* (63), such mutants were not available in higher plants.

Nitrate reductase deficient mutants of *Nicotiana tabacum* have been used in selection of hybrids. The mutant cells had a requirement for reduced nitrogen. The hybrids obtained by

TABLE V. *Examples of Chromosome Numbers Observed in Somatic Hybrid Plants*

Parental species	Selection method	Chromosome numbers	Reference
Nicotiana glauca + N. langsdorffii	Selective medium	55-64	(69)
N. tabacum + N. knightiana	Albino organogenesis	44-126	(54)
D. innoxia + Atropa belladonna	Albino	84-175	(53)
D. innoxia + D. stramonium	Albino	48-72	(65)
Daucus carota + D. capillifolius	Albino	34-54	(22)
D. carota + Aegopodium podagraria	Albinism	18	(23)
Arabidopsis thaliana + Brassica campestris	Mechanical albino	55-60	(38)
N. tabacum + N. sylvestris	Selective medium	76-80	(80)

fusion regained the ability to grow on media in which nitrate was the sole nitrogen source (39).

VI. EVIDENCE FOR GENE EXPRESSION AND OBSERVED PLOIDY IN SOMATIC HYBRIDS

The majority of hybrid plants which have been produced by somatic hybridization can also be obtained by sexual crossing. Direct comparison was possible therefore between the two types of plant hybrids, and the observations have been used to confirm that indeed hybridization had occurred. As expected, the morphological features of foliage, floral structure and color were intermediate and distinct from either of the parent plants. Somatic hybridization has yielded progeny of greater variability than is possible by sexual means because of the merging of the cytoplasm of the parent cells.

The somatic hybrid plants vary in ploidy and may deviate from that expected by adding the parental chromosome numbers. Table V lists hybrids on which chromosome counts were performed. The *N. glauca* + *N. langsdorffii* hybrids analyzed by Smith *et al.* had chromosome numbers close to 60 (69). The authors account for this observation by suggesting that hybrids arose by fusion of two protoplasts of *N. langsdorffii* (2n=18) and one of *N. glauca* (2n=24). Relatively few observations have been made on intergeneric hybrids and no generalizations are possible.

VII. BIOCHEMICAL CHANGES

Zymograms of constitutive enzymes have proven a useful method to confirm that hybridization has occurred. Interpretations from zymograms of a number of enzymes from somatic hybrids have shown that one of the following can happen: (1) the isoenzyme bands are additive for the total of that of the parents; (2) some bands present in parent tissues are missing in the hybrid; or (3), new bands occur in the hybrids.

In somatic hybridization, as opposed to sexual crosses, the cytoplasms of the parental types become integrated. Because of this unique feature of somatic hybridization, certain heritable factors associated with the cytoplasm can be transmitted. Male sterility is a desirable feature of economic importance in plant breeding. It has been possible to transfer the male sterility property from one species to another by protoplast fusion (2, 80). Restriction enzyme degradation and "finger printing" on polyacrylamide gels has become a useful method to analyze DNA sequences. The method has been employed in analyzing and mapping chloroplast DNA. Belliard *et al.*, using a restriction enzyme digest, compared the plastid DNA patterns of parents and hybrids and observed that the plastid DNA was that of either parent and not a mixture or a recombinant DNA (2). On the basis of results with restriction enzyme analyses of mitochondrial DNA of somatic hybrids, it has been shown that recombination occurs in mitochondrial DNA during somatic hybridization. The DNA fragments were not a mixture or sum of the parental types, but new sequences were produced in the somatic hybrid plants (2).

VIII. OTHER GENE TRANSFER METHODS

Genetic modification in plants is also being considered through uptake of DNA and organelles and single cell uptake in protoplasts. Genetic transformation through DNA uptake implies that DNA from one source is taken up, incorporated into the recipient cell in a stable form and that genetic information encoded in the foreign DNA is expressed as new stable characteristics. The uptake of organelles such as chloroplasts or cells of bacteria and algae into protoplasts may provide new and effective approaches to study nuclear-organelle and nuclear-cytoplasmic interactions, as well as serving as a method of intergeneric transfer of such processes as nitrogen fixation. The status of research in these areas and the prospects of any of the recombinant DNA approaches have been discussed in recent reviews (7, 30, 59, 64).

IX. LABORATORY METHODS

The laboratory outlines cover procedures for several tissues. The required equipment and supplies are listed. The procedure for other plants will require modifications according to the peculiarities within each species.

A. *Equipment*

(1) Laminar air flow transfer cabinet
(2) Analytical and top-pan balances
(3) Steam autoclave
(4) Benchtop centrifuge with a swinging bucket adapted for 15 ml conical tubes
(5) Inverted microscope (Wild-Leitz[R] with fluotar objective lens 10, 40)
(6) Gyrotory shaker (e.g., New Brunswick Model G-2)
(7) Magnetic stirrer and bars
(8) Stainless steel sieves, 60-80 μm mesh, fixed on a wide mouth funnel
(9) Stainless steel watch maker curved forceps
(10) Razor blades
(11) pH meter
(12) Conical test tubes, 15 ml, capped with aluminum foil, 50 ml open[1]
(13) Sterile petri dishes, 60 x 15 mm, 10 x 15 mm
(14) Beakers[1] 50 ml, 100 ml, 250 ml, covered with aluminum foil
(15) Pasteur pipettes with rubber bulbs (wrapped)[1]
(16) Pipettes, 2 ml (wide mouth) in pipette box[1]
(17) Parafilm[R]
(18) Millex[R] TM disposable filter unit 0.45 μm membrane (Millipore) sterile

[1]*Autoclaved prior to starting experiment.*

(19) Disposable plastic syringe, 1 ml, 10 ml, sterile (Plastipak)[R]
(20) Nalgene filter unit PS, 0.20 μm (Nalgen Sybron Corporation, Rochester, N.Y. 14602, U.S.A.), or Millipore membrane filters (0.22 μm pore size) and holders fitted in 125 or 500 ml Pyrex side-arm flasks[1]
(21) Sterile distilled water (100 ml/bottle)[1]
(22) Cover slips, 22 mm^2 [1]
(23) Tubes with screw caps, 25-150 mm[1]
(24) Cuprak[R] dishes, 60 x 15 mm (Costar 3268) (not essential)
(25) Disposable pipettes (Drummond[R] Microcaps) 20 μl (not essential)
(26) Graduated cylinders, 100 ml
(27) Bottles, screw capped, 125 ml, and Babcock[1]
(28) Funnels

B. *Materials*

1. *Plant Materials.* Leaf materials may be taken from young plants grown in shaded areas or seedlings grown aseptically in jars in a growth chamber at photoperiod 16:8 and ca. 25°C. Cells from suspension cultures are grown on shakers (150 rpm) and subcultured at 3-4 day intervals with cell generation times of 36 hours or less. Growth at 26-28°C in continuous light or dark may be desirable.

2. *Solutions.* Formulas for the PS-1, PS-2, and wash solution PS-3 are presented in Table VI. Refrigerated glass distilled water should be used. It is preferable to filter sterilize these solutions.

Enzyme solutions, E-1 and E-2, are presented in Table VII. E-1 is prepared using PS-1 solution, while E-2 uses PS-2 solution. Sources of chemicals are found in Table II.

a. *Cell culture media.* For preparation of stock solutions see Gamborg and Wetter (32) and Chapter 2 (this volume). The IB5C medium is a modified IB5 medium, as presented in Chapter 2, but in addition contains 500 mg/l N-Z-Amine[R] type A.

[1]*Autoclaved prior to starting experiment.*

TABLE VI. Solutions needed for Protoplast Isolation and Culture

Chemical	Molarity	Solution/100 ml		
		PS-1	PS-2	PS-3
$CaCl_2 \cdot 2H_2O$	6.0 mM	90.0 mg	90.0 mg	90.0 mg
$NaH_2PO_4 \cdot 2H_2O$	0.7 mM	11.0 mg	10.0 mg	-
$KH_2PO_4 \cdot H_2O$	0.7 mM	-	-	10.0 mg
Glucose	0.7 M	12.6 g	-	-
Glucose	0.5 M	-	-	9.1 g
Sorbitol	0.35 M	-	6.4 g	-
Mannitol	0.35 M	-	6.4 g	-
MES buffer[a]	3.0 mM	60.0 mg	60.0 mg	-
pH		5.8	5.8	5.8

[a] 2,(N-Morpholino)-ethane sulfonic acid.

TABLE VII. Enzyme Solutions for Protoplast Isolation

Enzyme	E-1 mg/per 100 ml PS-1	E-2 mg/per 100 ml PS-2
Onozuka R10	200	200
Hemicellulose, Rhozyme HP150	100	100
Pectinase	100	100
pH	5.8	5.8

b. Protoplast culture medium.

	amount/100 ml
B5 salts and vitamins (no sucrose) (10 x conc.)	10.0 ml
Glucose (0.38M)	6.84 g
Ribose	25.0 mg
Sucrose	25.0 mg
L-glutamine (4 mM)	58.0 mg
Casamino acids	25.0 mg
NH_4NO_3 (3 mM)	25.0 mg
$CaCl_2 \cdot 2H_2O$ (4 mM)	60.0 mg
Coconut milk	2.0 ml
Multivitamins, Stock Solution A	0.2 ml
Multivitamins, Stock Solution B	0.1 ml
Naphthaleneacetic acid (NAA)	1 μM
2,4-dichlorophenoxyacetic acid (2,4-D)	1 μM
Zeatin (or benzyladenine)	1 μM

Adjust to pH 5.8 and filter sterilize.

c. Stock solutions. A and B multivitamins.

(1) Multivitamin Solution A	mg/100 ml water
Calcium pantothenate	50
Ascorbic acid	100
Choline chloride	50
p-aminobenzoic acid	1
Folic acid	20
Riboflavin	10
Biotin	1

(2) Multivitamin Solution B (in 70% ethanol)	mg/100 ml
Vitamin A	1
Vitamin D_3	1
Vitamin B_{12}	2

(3) Coconut milk. Drain mature nuts, heat to ca. 80°C, filter and store frozen.

3. Solutions for fusion.

(1) Polyethylene glycol (PEG)	In 50 ml
0.2 M glucose	1.8 g
10 mM $CaCl_2 \cdot 2H_2O$	73.5 mg
0.7 mM KH_2PO_4	4.8 mg

Dissolve 25 g pharmaceutical grade PEG (MW 1500) in 5 ml of the above solutions, adjust to pH 5.8 and filter sterilize (0.22 μm Nalgene filter).

(2) Eluting solution	In 100 ml
a. 100 mM glycine	750 mg
0.3 M glucose	5.4 g

Adjust to pH 10.5 with NaOH pellets, filter sterilize and store at 4°C in the dark.

b. 100 mM $CaCl_2 \cdot 2H_2O$	1.47 g
0.3 M glucose	5.40 g

Filter sterilize, mix (a) and (b), 1:1 at the time of fusion.

C. *Procedures for Protoplast Isolation and Culturing* (*Fig.1*)

IMPORTANT: All operations should be performed aseptically in a laminar air-flow cabinet.

1. *From Leaf Materials*

(1) Immerse leaves in 70% ethanol for ca. 0.5-1.0 minutes, transfer to a 15 x 100 mm petri dish containing 10-20% Chlorox^R bleach and Tween^R-80 (1 drop/25 ml) and leave for 10-20 minutes.

(2) Rinse leaves with distilled sterile water (3X) *or* with a sterile sorbitol solution at the same osmolality as the enzyme solution to be used.

(3) Remove the lower epidermis by peeling it off with a curved pair of fine watch-maker forceps, cut off epidermisless sections of leaf (ca. 1 cm^2) and incubate in a 1:1 mixture of enzyme solution (E-1) and protoplast culture medium in a 100 x 15 mm petri dish. (For cereals: cut leaves into fine strips, parallel to veins. Use basal protion of leaf).

(4) Seal the dish with Parafilm^R and incubate for 4-16 hours at 22-24°C with slow shaking (50 rpm) on a gyrotory shaker or agitate gently by hand at hourly intervals.

(5) Observe the release of protoplasts on the stage of an inverted microscope.

(6) Remove protoplast suspension from petri dish with a Pasteur pipette and pass through a 60-80 μm mesh stainless steel sieve.

(7) Collect filtrate in a centrifuge tube with Parafilm^R or aluminum foil.

(8) Centrifuge at 50x g for 6 minutes.

(9) Remove the supernatant with a Pasteur pipette.

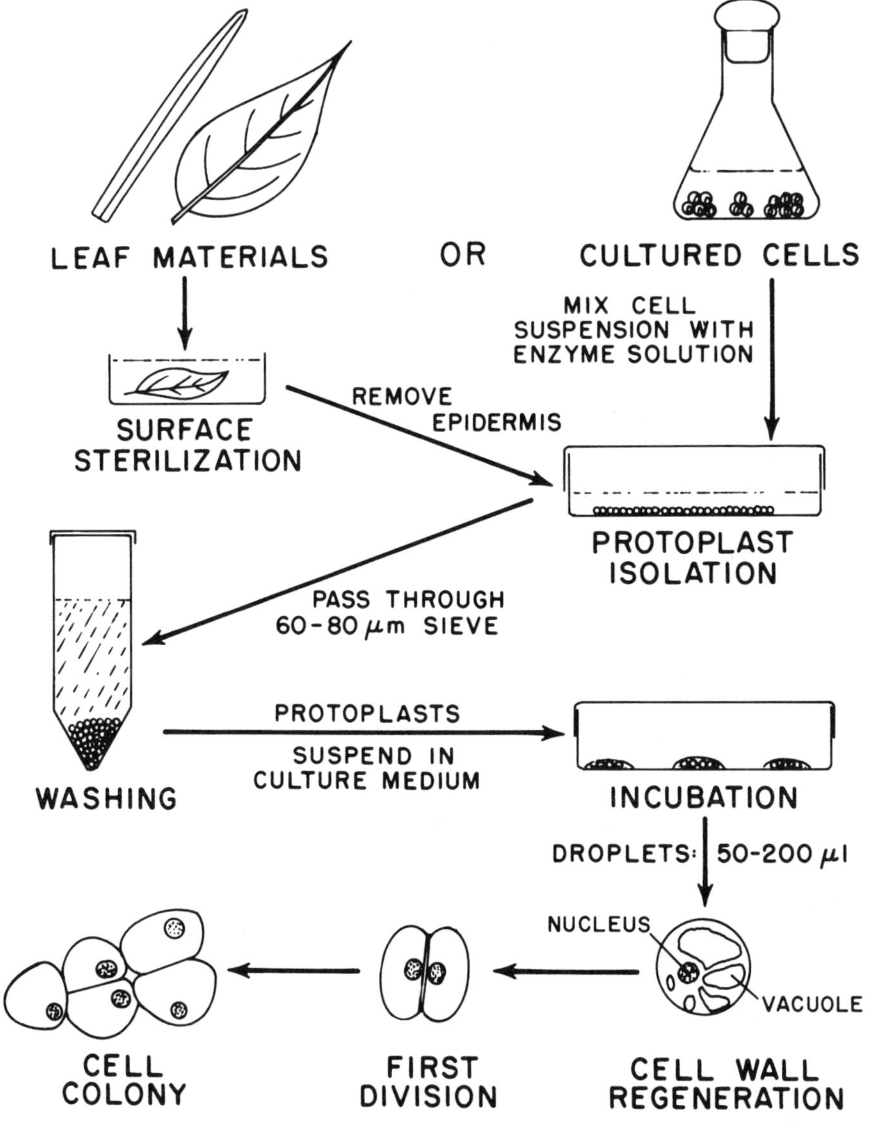

FIGURE 1. *Outline of the steps used in the isolation and culture of plant protoplasts.*

(10) Resuspend the protoplasts in 5-10 ml of PS-3 solution, centrifuge at 50x g for 5 minutes.

(11) Repeat *washing* once for fusion experiment and twice for culturing experiments.

(12) Resuspend the protoplast in 1 ml of PS-3 solution for fusion *or* in protoplast culture medium for culturing, density ca. 10^5 protoplasts/ml. (The protoplast concentration can be determined by using a Coulter Electronic Counter or a hemocytometer).

(13) For *culturing* distribute the protoplast suspension in ca. 150 µl droplets in 60 x-15 mm petri dishes, 6-8 droplets/dish.

(14) Seal petri dishes with Parafilm^R.

(15) Incubate in diffused light at 25°C in a covered plastic box humidified by a moist blotting paper in a beaker with 1% $CuSO_4$ solution.

2. From Cultured Suspension Cells.

(1) Mix equal volumes of cell suspension and enzyme solution (E-1 or E-2) in a 60 x 15 mm petri dish.

(2) Seal the dish with Parafilm^R, incubate at 25°C for 6-16 hours with slow shaking (50 rpm) on a gyrotory shaker.

(3) Follow the same procedure (Step 6 to Step 15) as for leaf protoplasts (filtering, washing, and culturing).

D. Procedure for Protoplast Fusion (Fig.2).

All operations should be performed aseptically in a laminar air flow cabinet.

(1) Place a 3 µl drop of autoclaved Silicone 200^R fluid in the center of a 60 x 15 mm petri dish using a 1 ml syringe.

(2) Place a 22 x 22 mm cover slip on the drop.

(3) Mix mesophyll protoplasts and cell culture protoplasts 1:1.

(4) Pipette 150 µl of the mixed protoplast suspension onto the cover slip with a Pasteur pipette.

(5) Allow protoplasts to settle onto the cover slip to form a thin layer (ca. 5 minutes).

(6) Slowly add, drop by drop, 450 µl (total ca. 6 drops) of PEG solution to the protoplast suspension. Observe the protoplast agglutination/adhesion on the stage of an inverted microscope.

(7) Incubate in the PEG solution for 5 to 25 minutes at room temperature.

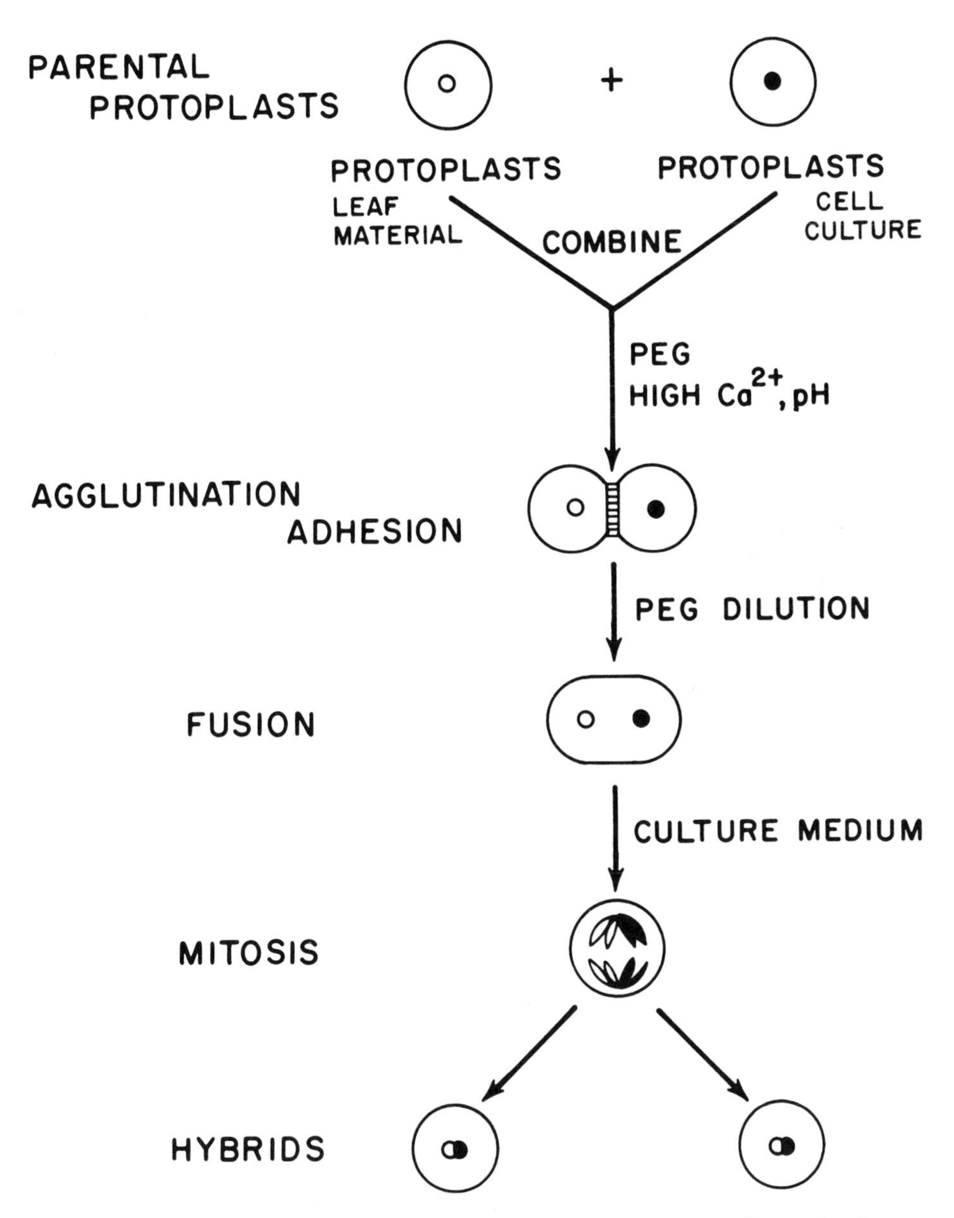

FIGURE 2. Scheme depicting stages in protoplast fusion and production of hybrids.

(8) Begin elution by gently adding 0.5 ml of Eluting solution (a+b, 1:1) to the mixture. After another 10 minutes add an additional 1 ml of Eluting solution.

(9) Wash the protoplasts 5 times with 1-2 ml (16-20 drops) aliquots of protoplast culture medium. Allow the protoplasts to remain in fresh medium for 5 minutes between washings.

(10) Leave ca. 500 μl of culture medium on the cover slip to maintain the fused and unfused protoplasts. Put 0.5-1.0 ml of medium in droplets in the petri dish to maintain humidity.

(11) Seal the petri dish with Parafilm^R.

(12) Incubate at 25°C in diffused light (60-100 lux) in a covered plastic box humidified by a moist blotting paper in a beaker with 1% $CuSO_4$ solution.

E. Cytological Observation (32)

(1) Cell wall digestion and regeneration (fused and unfused) protoplasts stained with Calcofluor^R in fluorescence microscope (Calcofluor white^R MZR, 0.1% in 0.4M sorbitol). Cellulose layers will fluoresce when irradiated with UV light at 366 nm.

(2) First division, usually occurs within one to two days. Colonies of cells should form within two weeks.

(3) Heterokaryocytes/products of a mesophyll + cell culture protoplasts, can be identified by the green plastids from the leaf protoplasts. Multinucleated protoplasts and heterokaryocytes can be determined by staining with carbol-fuchsin stain in 45% acetic acid. (See 5, below).

(4) Determine the frequency of surviving protoplasts with Evans blue.

(5) Determine the frequency of multinucleated protoplasts. Fix the protoplasts before staining them with carbol-fuchsin.

a. Fix samples of cell culture for 30 minutes. (Fix samples of protoplast preparation for 30 minutes).

b. Stain fixed samples of cell and protoplast preparations with carbol-fuchsin.

c. Determine the frequency of multinucleated protoplasts.

d. Determine the mitotic index of cell culture (protoplast source) and protoplasts during and after isolation.

(6) Determination of mitotic index.

a. Transfer a sample of fixed cells onto a slide and add an equivalent amount of stain.

b. Wait for a few minutes, cover preparation with a cover slip and heat the slide until small bubbles appear.

c. Cover slide with paper tissue and squash the preparation gently.

d. Count 1000 nuclei and determine the number of nuclei in mitosis.

$$MI = \frac{\textit{number of nuclei in mitosis}}{\textit{total number of nuclei observed}} \times 100$$

F. *Isolation, Proliferation and Plant Regeneration of Protoplasts of Potato and Cassava (Fig. 3)*

1. *Plant Materials.* Leaf materials from plants grown in growth chambers at 11 hour photoperiod of 4000 lux light intensity and 20°C constant temperature. However, the optimum temperature varies with different plant species. The following is a list of various temperatures used with different plant species:

Cassava	28°C	Soybean	28°C
Corn	20°C	Sweet Potato	28°C
Papaya	27°C	Tomatoes	20°C
Potato	20°C	Wheat	18°C

2. *Stock Solutions.* The required solutions are presented in Table VIII. Double distilled water should be used in preparation of all stock solutions. All stock solutions except the organics are stored at 4-5°C, the minor elements and iron solutions in brown bottles. Once the organic solution is made, 100 ml is poured into 10 small wash bottles and stored at sub-zero temperatures.

The salts (S) and organic combined solution is made up of 95 ml of salt solution mixed with 5 ml of 200X organics, shaken well and stored at 4-5°C.

Plant hormone stock solutions are prepared at a concentration of 0.1 mg/ml and stored at 4-5°C. The following chemicals are used:

	Dissolved in:
nathaleneacetic acid (NAA)	ethanol/H_2O
benzyladenine (BA)	dil. HCl/H_2O
indole-3-acetic acid (IAA)	KOH/H_2O
zeatin	dil. HCl/H_2O
gibberellic acid (GA)	ethanol/H_2O

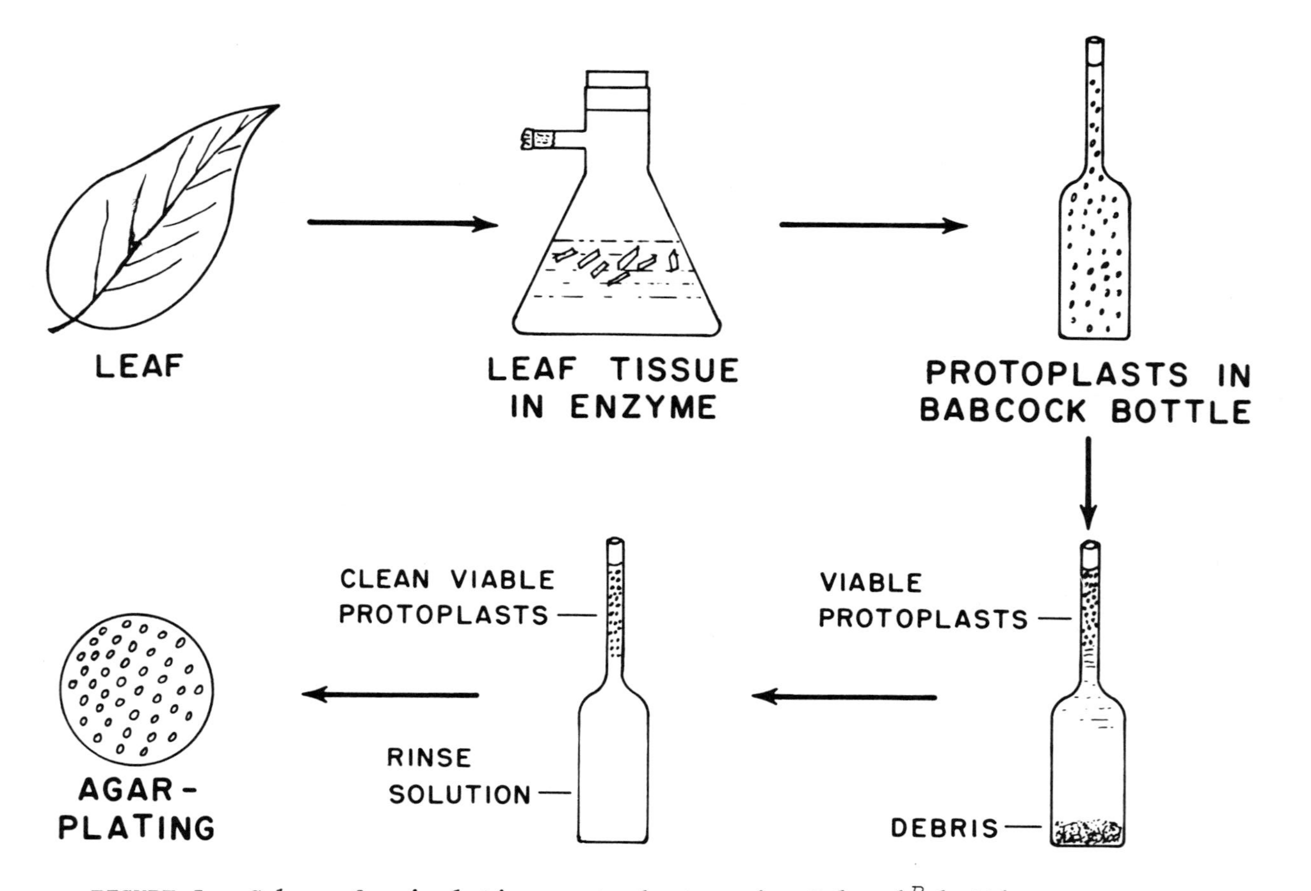

FIGURE 3. Scheme for isolating protoplasts using Babcock[R] *bottles.*

TABLE VIII. Stock Solutions for the Culture of Protoplasts of Potato and Cassava

Chemical	per 1000 ml
Major salts (S) (10X)	
KNO_3	19 g
$CaCl_2 \cdot 2H_2O$	4.4 g
$MgSO_4 \cdot 7H_2O$	3.7 g
KH_2PO_4	1.7 g
Minor elements I (100X)	
$ZnSO_4 \cdot 7H_2O$	920 mg
H_3BO_3	620 mg
$MnCl_2 \cdot 4H_2O$	1980 mg
Minor elements II (100X)	
KI	83 mg
$Na_2MoO_4 \cdot 2H_2O$	25 mg
$CuSO_4 \cdot 5H_2O$	2.5 mg
$CoSO_4 \cdot 7H_2O$	3.0 mg
Organics (200X)	
Folic acid	100 mg
Meso-inositol	20 g
Thiamine·HCl	100 mg
Glycine	400 mg
Nicotinic acid	1 g
Pyridoxine·HCl	1 g
Biotin	10 mg
Iron (10X)	
$Na_2EDTA \cdot 2H_2O$	373 mg
$FeSO_4 \cdot 7H_2O$	278 mg

3. *Procedures.*

a. *Preconditioning of plant materials.* This requires a preconditioning fertilizer (PF) of the following composition.

		Amount/1000 ml
Major salts:	1X S	100 ml of 10X stock solution
Minors:	Minor I	5 ml of 100X stock solution
	Minor II	5 ml of 100X stock solution
Iron		5 ml of 10X stock solution

pH = 5.6

Young plants (ca. 10 cm) obtained through cuttings or seeds should be placed in pots containing only vermiculite. Precondition the plants by placing them in growth chambers at low light intensity and short photoperiod. Fertilize with 400 ml/pot of PF once per two weeks, and keep the vermiculite moist by watering with distilled water. When the leaves become fully grown and well spread, start using them for protoplast isolation.

b. Floating solution.

	Amount/1000 ml
1 mM NH_4NO_3	10 ml 0.1 M stock solution
1 mM $CaCl_2 \cdot 2H_2O$	10 ml 0.1 M stock solution
1 ppm NAA	1.0 mg
5 ppm BA	5.0 mg

pH = 5.6

Remove leaf from parent plant, then place it on a float solution with its top side down floating on the solution. Keep in the dark at room temperature for 48 hours.

c. Sterilization and soaking (in a laminar flow hood).

(1) Immerse leaves in 10% commercial Chlorox[R] solution containing a few drops of Tween 20[R]. (Time: 3 minutes).
(2) Transfer the leaves to 70% ethanol. (Time: 30 seconds).
(3) Rinse four times in sterile distilled water.
(4) Place the leaves on a sterile paper towel and allow to dry for about 5 minutes.
(5) Using a nylon brush, gently stroke the lower leaflets surface until they appear light green, and then cut into squares (approximately 3 cm in diameter).
(6) Place the tissues (5 g) in a 500 ml evacuation flask containing 200 ml of the following soak solution:

Soak solution:	Amount/1000 ml
1/4X S	25.0 ml of 10X stock solution
0.5 mg/l BA	0.5 mg
1.0 mg/l NAA	1.0 mg
1/4X Organics	1.25 ml of 200X stock solution

pH = 5.6, autoclave for 15 minutes. Place the soak solution with the leaf tissue in refrigerator (4-5°C) for overnight.

d. Enzyme digestion.

Enzyme solution:	Amount/100 ml
0.1% macerozyme R-10	100 mg
0.5% cellulase R-10	500 mg
1.0% PVP-10 (MW 10,000)	1 g
1/4X S	2.5 ml of 10X stock solution
0.3 M sucrose	10.27 g
5 mM MES	1 ml of 0.5 M stock solution
10 mg/l casein hydrolysate	1 mg

pH = 5.6

Sterilize by filtration through Nalgene filter.

Replace the soak solution with that of the enzyme (approx. 100 ml), and then vacuum infuse enzyme solutions into leaf tissues. Afterwards, place the enzyme-treated tissues on a gyratory shaker at 28°C for 4-5 hours and 80 rpm.

Pour the digested leaf tissue into a funnel containing four layers of cheesecloth and collect protoplasts in Babcock bottles. Centrifuge the Babcock bottles for 10 minutes at 800x g in International Model HN-S. During centrifugation, the viable protoplasts will float to the top while the debris settles to the bottom or remains suspended.

e. Rinse solution.

Rinse solution	Amount/100 ml
1/2X S	5 ml of 10X stock solution
10 mg/l casein hydrolysate	1 mg
0.3 M sucrose	10.27 g

pH = 5.6, autoclave for 15 minutes.

Collect the protoplasts by using sterile Pasteur pipettes, and place them in a rinse solution in a Babcock bottle, then centrifuge as before, but for 5 minutes.

f. Holding solution.

Holding solution	Amount/100 ml
1/2X S	5 ml of 10X stock solution
10 mg/l casein hydrolysate	1 mg
0.2 M sucrose	6.84 g
0.025 M mannitol	456 mg
0.025 M xylitol	380 mg
0.025 M inositol	550 mg
0.025 M sorbitol	456 mg

pH = 5.6, filter sterile by Nalgene filter.

Collect the protoplasts and place them in a small volume (2 ml) of holding solution.

Confirm their numbers/ml by using a hemacytometer and dilute their numbers to 4 x 10^5/ml by using holding solution.

Incubate the protoplasts for 50-60 minutes in holding solution to allow their adjustment to the higher osmotic and mineral salt levels.

g. Plating.

Protoplast Plating Media		Amount/100 ml
Part I	1X organics	2.5 ml of 200X stock solution
	1/2X Minor I	2.5 ml of 100X stock solution
	1/2X Minor II	2.5 ml of 100X stock solution
	1/2X iron	25 ml of 10X stock solution
	50 mg/l casein hydrolysate	25 mg
	1 mg/l NAA	0.5 mg
	0.4 mg/l BA	0.25 mg
	Agarose #7	1 g
	pH = 5.6, autoclave for 15 minutes.	
Part II	1X S	12.5 ml of 10X stock solution
	0.25 M sucrose	10.7 g
	0.025 M mannitol	570 mg
	0.025 M xylitol	470 mg
	0.025 M inositol	560 mg
	0.025 M sorbitol	570 mg
	pH = 5.6, filter sterilize.	

Mix 5 ml of Part I with 20 ml of Part II. Allow the solution to cool down to around 28°C.

Add protoplasts to plating media at about 10000 protoplasts/ml.

Place the plating media and protoplast in a 60 x 15 mm plastic petri dish (3 ml/petri dish) and seal dishes with Parafilm[R]. Incubate under dim light (500 lux) at 22-24°C in a plastic box.

h. Proliferation medium (P.-medium).

	Amount/1000 ml = 50 petri dishes
1X S	100 ml of 10X stock solution
1/2X Minors I	5 ml of 100X stock solution
1/2X Minors II	5 ml of 100X stock solution
1X Organics	5 ml of 200X stock solution
1/2X Iron	50 ml of 10X stock solution
0.25% sucrose	2.50 g
0.3 M mannitol	54.175 g

40 mg/l adenine sulfate	40 mg
5 mM MES	10 ml of 0.5 M stock solution
0.5 mg/l BA	0.5 mg
400 mg/l casein hydrolysate	400 mg
2 mM NH_4Cl	20 ml of 0.1 M stock solution
1 mg/l NAA	1 mg
0.9% agar	9.0 g

pH = 5.6, autoclave for 15 minutes.

When medium cools down, pour (20 ml) into each petri dish. Three weeks after plating, transfer the plating media onto the P.-medium by using a small spatula or pipette. Spread the medium evenly over the surface of P.-medium. Incubate the plates at 22°C in continuous light of 4000 lux. Two weeks later, when the calluses become green in color and 0.1-1.00 mm in diameter, transfer them individually to fresh P.-medium. Incubate for 3 weeks at the same conditions as before.

i. Shoot-formation medium (S.-medium).

	Amount/400 ml = 20 plates
1X S	40 ml of 10X stock solution
1/2X iron	20 ml of 10X stock solution
1/2X Minor I	2 ml of 100X stock solution
1/2X Minor II	2 ml of 100X stock solution
1X Organics	2 ml of 200X stock solution
2 mM NH_4Cl	20 ml of 0.1 M stock solution
0.25% sucrose	1 g
0.2 M mannitol	14.6 g
80 mg/l adenine sulfate	32 mg
5 mM MES	4 ml of 0.5 M stock solution
0.1 mg/l IAA[1]	0.04 mg
100 mg/l casein hydrolysate	40 mg
1.5% Difco agar	6 g
0.1 mg/l zeatin[1]	0.04 mg

pH = 5.6, autoclave for 15 minutes.

Transfer calluses from P.-medium onto S.-medium by lifting them up with a scalpel. Incubate the plates at 400 lux light intensity, 24°C constant temperature, and 16 hour photoperiod. Shoot induction occurs within 4-5 weeks.

[1] *Add filter-sterilized IAA and zeatin to cool autoclaved medium.*

j. Rooting medium (R.-medium)

	Amount/1800 ml = 60 plates
1X S	180 ml of 10X stock solution
Organics	9 ml of 200X stock solution
1X KH_2PO_4	72 ml of 25X stock solution
0.5 mg/l thiamine·HCl	0.9 mg
1/2X Minor I	9 ml of 100X stock solution
1/2X Minor II	9 ml of 100X stock solution
1/2X Iron	90 ml of 10X stock solution
1% sucrose	18 g
40 mg/l adenine sulfate	72 mg
5 mM MES	18 ml of 0.5 M stock solution
0.2 mg/GA_3[1]	0.36 mg
0.05 mg/l NAA	0.09 mg
5 mM NH_4Cl	90 ml of 0.1 M stock solution
0.1 M mannitol	32.79 g
0.5% Sigma or Noble agar	3 g

pH = 5.6, autoclave for 15 minutes.

Transfer calluses with shoots from S.-medium, and push them to the bottom of the soft layer of R.-medium. Incubate at 24°C with 400 lux light intensity and 12 hour photoperiod. Shoot elongation and root development usually occurs within 4-5 weeks.

k. Plantlet formation. Transfer the small plantlets from R.-medium into small pots (10 cm) containing a mixture of vermiculite and sand (1:1 volume). Water with distilled water and fertilize once with Peter^R solution (1 g/1000 ml). Peter^R is a 20:20:20 (N,P,K) fertilizer. Place in moist chamber for 3-6 weeks, then plant them in greenhouse or field.

X. CONCLUDING THOUGHTS

In this chapter, the emphasis has been on the technical aspects of protoplast isolation, culture, development, and its role in hybrid production. The laboratory section gives detailed protocols, which should form an ideal starting point for those needing the technology to enter this field. Plant modification is an exciting area of research, in which protoplasts will continue to play a major role. However

[1]*Add filter-sterilized GA_3 solution to cool autoclaved medium. Use 100 X 25 mm petri dishes.*

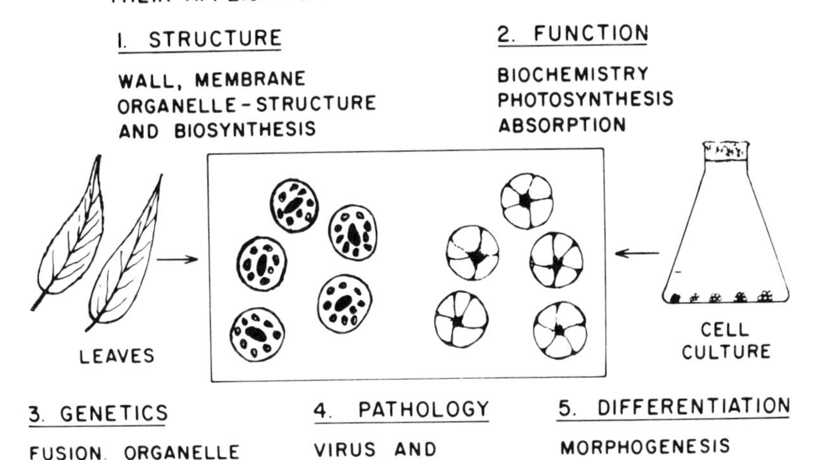

FIGURE 4. Areas of fundamental plant science in which the use of protoplasts is or will become a valuable research tool.

it is important that the researcher know his chosen material well before proceeding along this particular route of investigation. In addition to its role in studies on plant modification, protoplasts, regardless of whether they are derived from the intact plant or from cell cultures, have an important place in research on many areas of fundamental plant science. Such areas include studies in relation to the structure, function, genetics, pathology and differentiation of plant cells (Figure 4). Already many important advances are being made along these lines and we can expect more in the future.

REFERENCES

1. Arnold, S. von, and Eriksson, T., *Physiol. Plant. 36*, 193 (1976).
2. Belliard, G., Vedel, F., and Pelletier, G., *Nature 281*, 401 (1979).

3. Bhojwani, S.S., Power, J.B., and Cocking, E.C., *Plant Sci. Lett. 8*, 85 (1977).
4. Binding, H., and Nehls, R., *Z. Pflanzenphysiol. 88*, 327 (1978).
5. Binding, H., and Nehls, R., *Molec. gen. Genet. 164*, 137 (1978).
6. Binding, H., Nehls, R., Schieder, O., Sopory, S.K., and Wenzel, G., *Physiol. Plant. 43*, (1978).
7. Bogorad, L., *Genetic Engineering 1*, 181 (1979).
8. Bourgin, J., Chupeau, Y., and Missonier, C., *Physiol. Plant. 45*, 288 (1979).
9. Brar, D.S., Rambold, S., Gamborg, O.L., and Constabel, F., *Z. Pflanzenphysiol. 95*, 377 (1979).
10. Brar, D.S., Rambold, S., Constabel, F., and Gamborg, O.L., *Z. Pflanzenphysiol. 96*, 269 (1980).
11. Bui-Dang-Ha, D., Norreel, B., and Masset, A., *J. Exp. Bot. 26*, 263 (1975).
12. Cassells, A.C., and Barlass, M., *Physiol. Plant. 42*, 236 (1978).
13. Constabel, F., and Kao, K.N., *Can. J. Bot. 52*, 1603 (1974).
14. Constabel, F., Kirkpatrick, J.W., and Gamborg, O.L., *Can. J. Bot. 51*, 2105 (1973).
15. Constabel, F., Dudits, D., Gamborg, O.L., and Kao, K.N., *Can. J. Bot. 53*, 2092 (1975).
16. Constabel, F., Weber, G., Kirkpatrick, J.W., and Pahl, K., *Z. Pflanzenphysiol. 79*, 1 (1976).
17. Coutts, R.H.A., and Wood, K.R., *Plant. Sci. Lett. 9*, 45 (1977).
18. Davey, M.R., Bush, E., and Power, J.B., *Plant Sci. Lett. 3*, 127 (1974).
19. David, A., and David, H., *Z. Pflanzenphysiol. 94*, 173 (1979).
20. Dudits, D., *In* "Cell Genetics in Higher Plants" (D. Dudits *et al.*, eds.), p.153, Akademiai Kiado, Budapest, (1976).
21. Dudits, D., Kao, K.N., Constabel, F., and Gamborg, O.L., *Can. J. Genet. Cytol. 18*, 263 (1976).
22. Dudits, D., Hadlaczky, G., Levi, E., Fejer, O., Haydu, Z. and Lazar, G., *Theor. Appl. Genet. 51*, 127 (1977).
23. Dudits, D., Hadlaczky, G.Y., Basjszar, G.Y., Kones, C.S., Lazar, G., and Horvath, G., *Plant Sci. Lett. 15*, 101 (1979).
24. Durand, J., Potrykus, I., and Donn, G., *Z. Pflanzenphysiol. 69*, 26 (1973).
25. Eriksson, T., *In* "Plant Tissue Culture and its Bio-technological Application," (W. Barz, E. Reinhard and M.H. Zenk, eds.), p. 313. Springer-Verlag, Berlin (1977).

26. Evans, D.A., Wetter, L.R., and Gamborg, O.L., *Physiol. Plant. 48*, 225 (1980).
27. Fowke, L.C., and Gamborg, O., *Int. Rev. Cytol.* In press (1980).
28. Gamborg, O.L., *In* "CRC Handbook in Nutrition and Food", (M. Recheigl, Jr., ed.), p. 415. CRC Press, Inc., Boca Raton, Fla., (1977).
29. Gamborg, O.L., and Holl, F.B., *In* "Genetic Engineering for Nitrogen Fixation", (A. Hollaender *et al.*, eds.), p. 299, Plenum Publishing Corp., New York, (1977).
30. Gamborg, O.L., and Bottino, P.J., *In* "Adv. Biochem. Eng." (in press) (1980).
31. Gamborg, O.L., and Shyluk, J.P., *Bot. Gaz. 137*, 301 (1976).
32. Gamborg, O.L., and Wetter, L.R., (eds.), "Plant Tissue Culture Methods," National Research Council of Canada, Prairie Regional Laboratory, Saskatoon (1975).
33. Gamborg, O.L., Shyluk, J., and Kartha, K.K., *Plant Sci. Lett. 4*, 285 (1975).
34. Gamborg, O.L., Murashige, T., Thorpe, T.A., and Vasil, I. K., *In vitro 12*, 437 (1976).
35. Gamborg, O.L., Ohyama, K., Pelcher, L.E., Fowke, L.C., Kartha, K.K., Constabel, F., and Kao, K., *In* "Plant Cell and Tissue Culture Principles and Application," (W.R. Sharp *et al.*, eds.) p.371, Ohio State Univ. Press, Columbus (1979).
36. Gatenby, A.A., and Cocking, E.C., *Plant Sci. Lett. 8*, 275 (1977).
37. Gleba, Y.Y., and Hoffman, F., *Molec. gen. Genet. 164*, 137 (1978).
38. Gleba, Y.Y., and Hoffman, F., *Naturwiss. 66*, 547 (1979).
39. Glimelius, K., Eriksson, T., and Grafe, R., *Physiol. Plant. 44*, 273 (1978).
40. Grambow, H.J., Kao, K.N., Miller, R.A., and Gamborg, O.L. *Planta 103*, 348 (1972).
41. Iwai, S., Nagao, T., Nakota, K., Kawashima, N., and Matsuyama, S., *Planta 147*, 414 (1980).
42. Kao, K.N., *Molec. gen. Genet. 150*, 225 (1977).
43. Kao, K.N., and Michayluk, M.R., *Planta 115*, 355 (1974).
44. Kao, K.N., and Michayluk, M.R., *Z. Pflanzenphysiol. 96*, 135 (1980).
45. Kao, K.N., Gamborg, O.L., Miller, R.A., and Keller, W.A. *Nature (New Biol.) 232*, 124 (1971).
46. Kao, K.N., Constabel, F., Michayluk, M., and Gamborg, O. L. *Planta 120*, 215 (1974).
47. Kartha, K.K., Michayluk, M.R., Kao, K.N., and Gamborg, O. L., *Plant Sci. Lett. 3*, 265 (1974).
48. Kartha, K.K., Gamborg, O.L., Constabel, F., and Kao, K.N. *Can. J. Bot. 52*, 2435 (1974).

49. Keller, W.A., and Melchers, G., *Z. Naturforsch. 28c*, 737 (1973).
50. Kirby, E.G., and Cheng, T.Y., *Plant Sci. Lett. 14*, 145 (1979).
51. Koblitz, H., *Biochem. Physiol. Pflanzen. 170*, 287 (1976).
52. Krishnamurthi, M., *Euphytica 25*, 145 (1976).
53. Krumbiegel, G., and Schieder, O., *Planta 145*, 371 (1979).
54. Maliga, P., Kiss, Z.R., Nagy, A.H., and Lazar, G., *Molec. gen. Genet. 163*, 145 (1978).
55. Melchers, G., Sacristan, M.D., and Holder, A.A., *Carlsb. Res. Communications 43*, 203 (1978).
56. Nam, L.-S., Landova, B., and Landa, Z., *Biol. Plant. 18*, 389 (1976).
57. Pelcher, L.E., Gamborg, O.L., and Kao, K.N., *Plant Sci. Lett. 3*, 107 (1974).
58. Potrykus, I., Harms, C.T., Lorz, H., and Thomas, E., *Molec. gen. Genet. 156*, 347 (1977).
59. Rachie, K.O., *In* "Proc. Genetic Engineering in Plants," Rockefeller, New York, N.Y., (in press) (1980).
60. Rathnam, C.K.M., and Edwards, G.E., *Plant Cell Physiol. 17*, 177 (1976).
61. Raveh, D., and Galun, E., *Z. Pflanzenphysiol. 76*, 76 (1975).
62. Raveh, D., Huberman, E., and Galun, E., *In Vitro 9*, 216 (1973).
63. Savage, A.D., King, J., and Gamborg, O.L., *Plant Sci. Lett. 16*, 367 (1979).
64. Schell, J. *et al.*, *Proc. R. Soc. Lond. B. 204*, 251 (1979).
65. Schieder, O., *Molec. gen. Genet. 162*, 113 (1978).
66. Shahin, E.A., and Shepard, J.F., *Plant Sci. Lett. 17*, 459 (1980).
67. Sharp, W.R., Larsen, P.O., Paddock, E.F., and Raghavan, V., (eds.) "Plant Cell and Tissue Culture," Ohio State University Press, Columbus, (1979).
68. Shepard, J.J., and Totten, R.E., *Plant Physiol. 60*, 313 (1977).
69. Smith, H.H., Kao, K.N., Combatti, N.C., *J. Hered. 67*, 123 (1976).
70. Straus, A., and Potrykus, I., *Physiol. Plant. 48*, 15 (1980).
71. Strobel, G.A., *In* "Trends in Biochem. Sci." p.247 (1976).
72. Szabodos, L., and Dudits, D., *Expt. Cell Res.* (in press) (1980).
73. Thorpe, T.A. (ed.), "Frontiers of Plant Tissue Culture, 1978", University of Calgary Press, Calgary, Canada (1978).

74. Tsai, C., Chien, Y., Chou, Y., and Wu, S., *In* "Proc. Symp. Plant Tissue Culture", p. 317, Science Press, Peking, (1978).
75. Vardi, A., Spiegel-Roy, P., and Galun, E., *Plant Sci. Lett. 4*, 231 (1975).
76. Vasil, I.K. (ed.), "Perspectives in Plant Cell and Tissue Culture", *Int. Rev. Cytol. Suppl. 11B*, Academic Press, New York, (1980).
77. Vasil, V., and Vasil, I.K., *Z. Pflanzenphysiol. 92*, 379 (1979).
78. Vasil, I.K., Ahuja, R., and Vasil, V., *Adv. Genet. 20*, 127 (1979).
79. Zapata, F.J., Evans, P.K., Power, J.B., and Cocking, E.C., *Plant Sci. Lett. 8*, 119 (1977).
80. Zelcher, A.D., Aviv, D., and Galun, E., *Z. Pflanzenphysiol. 90*, 397 (1978).

MUTAGENESIS AND *IN VITRO* SELECTION

Walter Handro

Plant Tissue Culture Laboratory
Department of Botany
University of São Paulo
São Paulo
Brazil

I. INTRODUCTION

The improvement of the techniques for *in vitro* tissue and cell culture in the last 15 years, which allows for the recovery of whole plants from many cultures, has opened the possibility of using this experimental system in the production and selection of variant or mutant tissue and cell lines, and consequently, whole organisms. The greatest advantage of these *in vitro* systems is the ease of treating great numbers of cells with modifying agents, and/or submitting them to selective procedures in well controlled environmental conditions. This has provided more effective and rapid results than conventional techniques. On the other hand, haploid plants and their cells are now relatively easy to obtain in several species, making the screening of modifications, which are often recessive, less laborious.

It is important to remember that natural variability found among living organisms is a consequence of changes occurring in the genetic material. These changes, normally defined as mutations, can occur at the DNA chromosome level as well as in the extra-nuclear DNA found in the chloroplasts and mitochondria. Several levels of mutation, introducing new phenotypes, can be found in nature. These may include alterations of one or more nucleotides of the DNA structure up to changes in a whole set of chromosomes. Their frequencies in natural populations are very low and greatly

ISBN 0-12-690680-7

variable among different biological systems. However, it must be pointed out that these frequencies can be significantly increased by using chemical or physical mutagenic agents.

The application of mutagens to tissues or cell populations cultured *in vitro* to enhance the rates of spontaneous mutations and the use of direct selection for the screening of spontaneous mutants or variant lines, have been used in several laboratories in the last 10 years, with the aim of further recovery of mutated whole plants. Several authors have reviewed this subject (14, 26, 58, 59, 60, 102, 108), and they have pointed out the major problems involved and have listed the main new phenotypes selected by using *in vitro* techniques.

II. GENOTYPIC VARIABILITY IN TISSUE AND CELL CULTURES

One of the problems which may affect the use of cell or tissue cultures for the selection of variant phenotypes is the occurrence of high genotypic variability in such cultures (20, 21, 26, 89). In the meristematic tissues of whole plants, especially in the apical meristems, mechanisms of strict control in the sequence of DNA synthesis - mitosis together with continuous cell divisions, prevent respectively extra-duplication of the DNA (which produces somatic polyploidy) and the increase of the number of cells undergoing spontaneous chromosome structural changes. These facts allow for the maintenance of the genetic stability of meristematic tissues (20). On the other hand, in many plant species, polyploidy is present in non-meristematic tissues, due to the occurrence of endomitosis, as in the tobacco stem pith. In this case there is an increasing gradient of ploidy level from the apex to the base, and this has physiological and biochemical consequences (52, 72). Tissue cultures derived from different regions demonstrated distinct morphogenetic responses *in vitro*. Similar results have been observed in our laboratory, using haploid tobacco plants. Thus, the selection of the plant material for obtaining variants must be made carefully. Even in tissues of defined ploidy, when they are explanted and cultured *in vitro*, the cell environment is changed and the new conditions may lead to distinct cell behavior. Thus, the degree of ploidy shows a tendency to increase according to the culture age (68, 72). The former authors also demonstrated that the dynamics of cell proliferation may be different in haploid and diploid cell cultures (68). Aneuploidy is another very frequent condition observed in old cultures from plants with particular genetic constitutions such as *Saccharum* species, where a high degree of mosaicism may lead to the formation of variant clones (41, 57). Depending on the explanted material, one can obtain regeneration of plants of selected ploidy from callus cultures (77, 80, 86). Thus, tissue culture may provide chromosome variations allowing for the regeneration of plants with different phenotypes, which can be used to increase variability through sexual crosses or to obtain cell tissues with a high degree of variability, as in the case of sugar cane. On the other hand, variability or instability in tissue or cell cultures may also be a limitation in research, as well as in the stability of variant or mutated cell lines.

III. INDUCED MUTATION IN TISSUE CULTURES

Increase in genetic variability may be induced in large and homogeneous populations of plant cells or in callus tissues, by exposing cultures (or plants before tissue explanting) to physical or chemical mutagenic agents. These are capable of increasing the frequency of changes in the genetic material when the cultures are placed under conditions which allow for the rapid screening of the mutants. These conditions may also be used to select and recover spontaneous variants or mutants directly from untreated material. The main types of mutants are (1) *auxotrophic*, which require nutritional supplements for normal growth, (2) *resistant*, which resist specific drugs, antimetabolites, or abnormal environmental or nutritional conditions, and (3) *autotrophic* mutants, that grow in deficient media and are capable of synthesizing some substances normally required for wild lines. Other kinds of mutants exhibiting particular characteristics, hence differing from the parental lines, may also be selected. It must be made clear that, mutations are considered true only when, from variant tissue or cell lines, one can regenerate plants capable of producing similar mutated individuals. But this requirement has been achieved only in a few cases.

A. Physical Mutagens

Several kinds of radiation[1] are potentially mutagenic, such as electromagnetic radiation (UV, X- and γ-rays) and particles (electrons, neutrons, protons, β^- and α-particles). These radiations are mutagenic because they are capable of transferring enormous amounts of energy: while violet visible light produces 3.3 ev/photon, γ-rays from a ^{137}Cs source produces 600,000 ev/photon. As a result of the interaction between radiation and matter, atoms and molecules are excited or ionized. In the case of UV radiation, the bulk of the energy is transferred, to the living systems, in the form of excitation of nucleic acids. In the case of ionizing radiation, several kinds of events may occur such as disruptions to water molecules, DNA, enzymes, growth substances, etc., leading to disturbances in the mechanisms responsible for homeostasis in the cell. As a result, different responses can be recorded: for instance, some or all the disruptions may be repaired; the cell may survive carrying some particular kind of injury, which may be a physiological or a genetic disturbance, which in turn may be somatically or sexually propagated; or, the disturbance may be of such a kind that it results in mitotic inhibition leading to loss of differentiation capacity and cell death.

The degree to which higher plants are affected depends on: (1) the type and amount of radiation; (2) the environmental conditions prior to, during, or after irradiation (temperature, light, oxygen); and (3) the kind or stage of the material which is being irradiated (plant species, ontogenetic stage, ploidy level, phase of cell cycle, etc.). All these considerations may be applied to cell and tissue cultures. Effects of radiation in plant tissues have long been studied (45). However, regarding the use of radiation in the production of mutants, the few data in the literature show that this area has not been explored greatly.

By using radiation for mutant production, one must consider different situations. Firstly, *if the materials to be irradiated are embryoids or plantlets originated from tissue culture procedures, including anther culture.* In this case we are dealing with a sort of sophisticated shoot apex irradiation. The results of this treatment and its interpretation are in line with others obtained by

[1] *For details on radiation and biological processes, see Bacq and Alexander (3), Grosch (37), and Gunckel and Sparrow (38).*

traditional procedures such as those concerned with seed, meristem or whole plant irradiation. Using this approach, Nitsch (78) and Nitsch *et al.* (79) obtained different phenotypes of haploid tobacco plants, after irradiation of plantlets produced by anther culture. Anthers were also irradiated prior to their culture, but the results were less impressive. These authors however, did not verify if these altered characteristics were true mutations, able to be transmitted to their progenies. Similar results were obtained by Corduan working on callus and plantlets from anther culture of *Hyoscyamus niger* (18). Devreux and Nettancourt carried out some experiments by irradiating *Nicotiana* flower buds with X-rays and then culturing the anthers to get haploids (22). The aberrant phenotypes obtained were considered mutants. The tissue cultures of these mutants produced diploid plants (through chromosome doubling in culture), with the same aberrant phenotypes. The authors carried out additional experiments using anthers from isogenic diploids, and the results were confirmed: i.e., low doses of radiation indeed produced the mutations.

Secondly, *tissue cultures could be used to isolate and propagate (by cloning procedures) some somatic mutations of natural occurrence or induced by radiation.* This technique was referred to by Dulieu, in 1972, as a possibility for the future (26). Recently, Shigematsu and Hashimoto have described the induction of flower chimeras by γ-radiation on *Chrysanthemum*, and further propagation of mutated plants through the *in vitro* petal culture (90).

Finally, *tissue cultures could be irradiated (prior to explantation or as callus or cell suspensions) and then subjected to the in vitro selection procedures in order to isolate mutant cell lines or individuals.* The use of radiation in the search of genetic modifications in tissue or cell cultures was envisaged by Melchers and Bergmann when they were studying the effects of X-rays on cell suspensions of *Anthirrinum majus* (69). Eriksson used cell suspensions of *Haplopappus gracilis* to investigate the effects of X-rays and UV light, and could select strains with altered anthocyanin content (28). Additional work along this line has been carried out and several kinds of mutated or variant cell lines have been described recently. Schieder, by irradiating *Datura innoxia* protoplasts with X-rays isolated 10 mutant callus lines with altered pigmentation, but no mutants were isolated from non-irradiated protoplasts (88). Berlyn and Zelitch have treated cell suspensions from haploid *Nicotiana tabacum* with UV light and obtained several cell lines and further regenerated plants resistant to INH (isonicotinic acid hydrazide) (5). The calluses of these plants had a block in the conversion of glycine to serine. Resistance to $NaClO_3$ was obtained agter irradiation of rose cells by UV light (73). Valine resistant mutants of tobacco, in which the trait was transmitted as a Mendelian character, were obtained by Bourgin after UV irradiation on mesophyll protplasts (8). Finally, Tabata et al. reported that by irradiating tobacco cells with UV light it was possible to select several lines that were able to yield more nicotine than original lines (95). In spite of a decrease of the nicotine level after some cell generations, the high biosynthetic potential of these variant cell cultures was maintained in the regenerated plants.

In my opinion, radiation is potentially useful in producing plant mutants through tissue culture, even though only a few successful cases have been published to date. It is necessary to obtain a better knowledge of the various parameters that must be considered when this technique is used. These include radiosensitivity of the plant material, trials using a larger spectrum of doses, determination of ideal physiological and environmental conditions of the cultures to be irradiated, etc. Recently we have found that growth and morphogenetic responses of γ-irradiated tobacco callus tissues depend significantly on these conditions (42).

B. Chemical Mutagens

Compared with physical agents, chemical mutagens are perhaps more capable of leading to specific and predictable changes involved in a mutation, on account of the direct chemical action of such compounds on DNA molecules (for details, see 2). Some effects of chemical mutagens include: (1) the alteration in the sequence of nucleotides (substitutions), (2) direct changes in the nucleotide structure, such as the removal of the amino group from adenine and cytosine, (3) the addition of new chemical groups to the bases, leading to new possibilities in the pairing or replication, or (4) even inactivation of DNA. The most commonly used chemical mutagens, ethyl or methylmethanesulfonate (EMS or MMS), belong to the group of alkylating agents, that act by adding methyl or ethyl groups to guanine which can thus behave as a base analogue of adenine, with production of pairing errors. Base analogues such as 5-bromodeoxyuridine (BudR) and 5-bromouracil (BU) are also mutagens.

Chemical mutagens in tissue culture have been used by many investigators in search of variant lines. This approach is based on the acceptance that they would enhance the frequency of mutation. However, in most cases, controls without mutagens have not been included in the experiments to verify the frequencies of spontaneous mutations.

By using N-methyl-N'-nitro-N-nitrosoguanidine (NTG), Lescure and Péaud-Lenoel were able to get an auxin autotrophic sycamore cell line, which was stable (55), and the cells contained an altered auxin oxidase (53). Carlson selected auxotrophic mutants in tobacco haploid cells, by using a chemical mutagen (EMS), and microbiological techniques for screening (11).

In the last case and many others in the literature, no evidence was presented for an actual mutagen-inductive effect of the chemical agent used. However, many papers describe an enhanced production of mutants by the action of chemical agents. Sung treated soybean and carrot cells, with EMS and NTG, and increased the frequency of 5-fluorouracil and cycloheximide resistant lines 10-fold (92). Widholm showed a dramatic mutagenic effect of EMS, i.e., a 10-fold increase in the frequency of 5-methyltryptophan-resistant carrot cells, over a mutagenic effect of UV light (103). Müller and Grafe selected chlorate resistant tobacco cells lacking nitrate reductase by using N-ethyl-N-nitroso-urea (ENU) as a mutagenic agent (71). This mutation was transmitted to regenerated plants. This finding has allowed for research on fusion of protoplasts of deficient lines, producing hybrids capable of reducing nitrate (34). A list of chemical mutagens which have been used in tissue culture is presented in Table I.

C. Direct Selection

The most common way to obtain and select variant lines or mutants is to expose cells or callus to special conditions which allow for the survival of only a small fraction of the population presumptively consisting of spontaneous mutants adapted to these conditions. These conditions include: high concentration of metabolites, toxic drugs, metabolite analogues, absence of essential nutrients or hormones, environmental stress, etc. The main cases described in the literature are listed and summarized in recent reviews (58, 59, 60, 102, 108). Some of these findings will be described because of their application in fundamental and applied studies.

The selection of BUdR resistant lines of *N. tabacum* (65) allowed for recent studies on the mechanism of BUdR resistance (66). Work along the same line has been carried out by Ohyama using *Glycine max* cells (81, 83). Streptomycin

TABLE I. Use of Chemical Mutagens in Plant Tissue Culture

Mutagen	Concentration	Time of treatment	Culture[a]	Species	Treatment efficacy	Reference
EMS	1-100mg/l	continuous	a	*Nicotiana tabacum*	(-)	78
EMS	0.25%	2-3 h	cs	*Daucus carota*	(+)	102
EMS	0.075-1.5%	1 h	cs	*Nicotiana sylvestris*	(-)	24
EMS	2.5%	2 h	cs	*Glycine max* *Daucus carota*	(+)	92
EMS	0.25%	1 h	c,p	*Nicotiana tabacum*	?	12
EMS	0.25%	1 h	cs	*Nicotiana tabacum*	?	11
EMS	1%	1 h	cs	*Oryza sativa*	?	15
MMS	50 mg/l	?	cs	*Saccharum officinarum*	(+)	41
NTG	50-500 μg/ml	2 h (average)	cs	*Daucus carota*	(+)	92
NTG	5mg/l	4 h	cs	*Petunia hybrida*	(+)	76
NTG	100-400mg/l	5 min	cs	*Acer pseudoplatanus*	(+)	55
NTG	100-400mg/l	30 min	cs	*Acer pseudoplatanus*	(+)	9
NTG	0.1 μM	5 days	cs	*Nicotiana tabacum*	?	39
NTG	50 μg/l	16 h	p	*Glycine max*	?	81
NTG	10 μg/ml	?	p	*Datura innoxia*	?	51
ENU	0.25 mM	?	cs	*Nicotiana tabacum*	(+)	71

[a] a = anther, cs = cell suspension, c = cells, p = protoplasts.

resistant lines were obtained from tobacco plants by Maliga *et al.* (61), the character being transmissible to further generations of regenerated plants, but only maternally, indicating a cytoplasmic mutation (62).

Cell lines resistant to amino acid analogues have been described for several species. For instance, 5-methyltryptophan (5-MT) resistant *D. carota*, *N. tabacum* and *S. tuberosum* cell lines have been isolated, and most of the 5-MT resistant carrot and potato cell lines have been shown to be auxin autotrophic (104). These lines brought on new data concerned with the mechanism of resistance at the metabolic level, such as differences in enzymes in susceptible and resistant lines (13). Widholm (107) showed that 5-MT resistant lines were able to regenerate plants, and calluses from them retained the resistance and altered enzymatic activity. Sung *et al.* (94) verified that a 5-MT resistant carrot cell line exhibited an overproduction of IAA, being auxin autotrophic.

Overproduction of amino acids was observed in cell mutant lines of several species. Carrot, barley and rice cell lines resistant to aminoethylcysteine produced a high level of lysine (9, 87). Hibberd *et al.* working on maize callus cultures, selected lines that were resistant to high levels of L-threonine (43). These produced more methionine, lysine, threonine and isoleucine than did the parental lines. Widholm selected carrot cell lines resistant to amino acid analogues, which oversynthesised phenylalanine, methionine, lysine and tryptophan (106).

Selection of resistant cell lines for stressed environmental conditions have been described, such as resistance to chilling (24), high salt or ion concentrations, e.g., NaCl (19, 25), aluminum (70), etc. Temperature sensitive lines have also been described (64, 99). Another group of resistant mutants was that selected for resistance to herbicides. A well studied case is that of tobacco cells resistant to picloram (17). In this case resistant plants were regenerated, and the trait was demonstrated to be transmitted by a single dominant Mendelian allele. Disease and pathotoxin resistance have also been described for cell cultures. Heinz reported the high rates of resistant sugar cane plants to eye-spot disease (*Helminthosporium sacchari*) obtained from cell cultures treated with a toxin from the eye-spot fungus (40). *Pseudomonas tabaci* resistant tobacco plants were obtained from methionine sulfoximine resistant cells (12). Gengenbach *et al.* selected maize cell lines resistant to the pathotoxin from *Helminthosporium maydis* race T (33). From these cell lines, plants resistant to the disease were regenerated, and the resistance was shown to be inherited only by the female, indicating a mutation in the extranuclear genome.

It should be pointed out that many variant lines obtained in culture have not been shown to be true mutations, capable of being transmitted to regenerated plants or of being maintained out of the selective conditions. Many of these phenotypic changes are only epigenetic in nature. An account of some of the principal mutants or variant cell or tissue lines obtained by various methods of induction and selection, is given in Table II.

Despite the great number of mutants or variant lines selected through *in vitro* techniques that have been carried out to date, most of these are restricted to species of a few genera commonly used in plant tissue culture. These include *Daucus*, *Solanum*, *Nicotiana*, and *Datura*. The control of growth and differentiation in *in vitro* cultures of these species, which has been known for a long time, provided a necessary requirement for success in *in vitro* mutagenesis and selection of new phenotypes. With the widespread use of culture techniques by researchers interested in applied problems, especially crop plants, one may hope for success in the next few years in the production of several kinds of mutants of high economic value. For instance, the selection of salt resistant cell lines may be the first step towards the production of plants adapted for growth in saline soils, one of the commonest conditions in irrigated land in several countries. Another aspect to be emphasized is the obtaining of amino acid analogue resistant mutants with altered biosynthetic pathways, i.e., having an oversynthesis of some amino acids. A recent review by Widholm shows how cereal grains low in some amino acids could have improved nutritional value (105). The production of plants with increased net photosynthesis is another possibility to be considered, if cell lines with an altered glycolate pathway affecting photorespiration can be produced, as envisaged by Berlyn and Zelitch (5).

Resistance to herbicides and pesticides is also of great importance in crop plants. In spite of this, most cases of resistance obtained from cell lines have not been carried through a sexual cycle, and therefore the resistance cannot be verified. It must be emphasized that the *in vitro* search for these kinds of mutants require less effort and cost than work done in the field (35). Obtaining disease resistant plants is a field wide open to research.

Based on results from the last ten years, it can be concluded that *in vitro* systems have proven to be excellent for obtaining mutants or cell lines useful in fundamental

TABLE II. Main Variant Cell or Tissue Lines Selected through in vitro Techniques

Phenotype	*Species*	*Reference*
Resistance		
aminoethylcysteine	*Arabidopsis thaliana*	75
	Daucus carota	10, 106
	Oryza sativa	15, 87
	Nicotiana tabacum	101
	Zea mays	50
aluminum	*Lycopersicon esculentum*	70
aminopterin	*Datura innoxia*	67
8-azaguanine	*Acer pseudoplatanus*	9
	Haplopappus gracilis	48
	Nicotiana tabacum	54
azatidine-2-carboxylic acid	*Zea mays*	50
6-azauracil	*Haplopappus gracilis*	47
6-azauridine	*Haplopappus gracilis*	48
5-bromodeoxyuridine	*Acer pseudoplatanus*	9
	Glycine max	81
	Nicotiana tabacum	54, 65
carboxin	*Nicotiana tabacum*	85
chilling	*Capsicum annuum*	24
	Nicotiana sylvestris	24
chloramphenicol	*Nicotiana sylvestris*	(see 59)
chlorate	*Nicotiana tabacum*	71
	Rosa damascena	73
colchicine	*Acer pseudoplatanus*	(see 59)
cycloheximide	*Daucus carota*	36, 92
	Nicotiana tabacum	63
deltahydroxylysine	*Nicotiana tabacum*	101
ethionine	*Daucus carota*	101
	Nicotiana sylvestris	109
p-fluorophenylalanine	*Acer pseudoplatanus*	31
	Daucus carota	84
	Nicotiana tabacum	84
6-fluorotryptophan	*Petunia hybrida*	76
5-fluorouracil	*Daucus carota*	92
Helminthosporium toxins	*Saccharum officinarum*	41
	Zea mays	33

TABLE II (continued)

Phenotype	Species	Reference
Herbicides		
amitrole	*Nicotiana tabacum*	3
asulam	*Brassica napus*	29
atrazine	*Brassica campestris*	(see 35)
2,4-D	*Citrus sinensis*	(see 59)
	Daucus carota	102
	Nicotiana sylvestris	109
diaquat, paraquat	*Glycine max*	46
picloram	*Nicotiana tabacum*	17
hydroxyproline	*Daucus carota*	101
isonicotinic acid hydrazide	*Nicotiana tabacum*	5
kanamycin	*Nicotiana sylvestris*	25
mercury chloride	*Petunia hybrida*	76
methionine sulfoximine	*Nicotiana tabacum*	12
5-methyltryptophan	*Daucus carota*	93, 100
	Nicotiana tabacum	100
	Solanum tuberosum	104, 106
neomycin	*Nicotiana sylvestris*	(see 59)
norleucine	*Solanum tuberosum*	(see 59)
selenomethionine	*Daucus carota*	(see 102)
	Nicotiana tabacum	30
selenocystine	*Nicotiana tabacum*	30
sodium chloride	*Capsicum annuum*	23
	Nicotiana sylvestris	23, 109
	Nicotiana tabacum	74
streptomycin	*Nicotiana sylvestris*	(see 59)
	Nicotiana tabacum	61, 62, 98
	Petunia hybrida	7
threonine	*Nicotiana tabacum*	39
	Zea mays	43
valine	*Nicotiana tabacum*	8
Auxotrophy		
p-aminobenzoic acid	*Nicotiana tabacum*	11
arginine	*Ginkgo biloba*	97
	Nicotiana tabacum	11
biotin	*Nicotiana tabacum*	11
hypoxanthine	*Nicotiana tabacum*	11
lysine	*Nicotiana tabacum*	11

TABLE II (continued)

Phenotype	*Species*	*Reference*
proline	*Nicotiana tabacum*	11
	Zea mays	32
auxin (from autotrophic line)	*Acer pseudoplatanus*	111
Others		
auxin autotrophy	*Acer pseudoplatanus*	55
	Daucus carota	91, 94
glycerol utilization	*Nicotiana tabacum*	16
secondary products (altered production of carotenoids, anthocyanins, alkaloids)	*Several species*	95, 110
temperature sensitive	*Datura innoxia*	99
	Glycine max	99
	Nicotiana tabacum	64

research on biochemistry, genetics and cell physiology. The utilization of all these results and of a well developed methodology to solve many practical problems of crop improvement is now becoming routine. These advances offer great encouragement for the future.

IV. GENERAL LABORATORY AND EXPERIMENTAL PROCEDURES

In this section some general information related to the laboratory work on induction and selection of mutants or variant lines will be given. It must be pointed out that for the development of work along these lines it is necessary to have a background of the *in vitro* culture conditions of the tissues or cell suspensions being used and their requirements for continuous growth and differentiation (see Chapters 1, 2 and 3). These conditions are variable depending on the species considered, but for many species there are some available data in the literature. For untried species, the establishment of this background is essential prior to the beginning of mutagenic or selective treatments.

A. *The Use of Physical Mutagens*

1. Ionizing Radiation. γ-rays are usually supplied by ^{137}Cs or ^{60}Co sources, the last being much more energetic (1.33 Mev, against 0.6 Mev of ^{137}Cs source). This provides the same dose of radiation with a smaller time of exposure to the biological materials. X-rays are produced by X-ray machines where highly accelerated electrons (cathodic rays) are beamed onto a metallic target (anode). Depending on the voltage produced, one obtains "soft" or "hard" rays. Hard rays may be as energetic as γ-rays. Voltages of around 100 kV are sufficient to irradiate biological material.

Ionizing radiation can be used to irradiate cells or tissues before isolation or explanation, or material already in culture tubes, petri dishes, Erlenmeyer flasks, etc. Cells or tissues irradiated by ionizing radiation when in culture, must be immediately transferred to fresh medium after irradiation, because of the harmful effects of radiation on medium components (1).

Radiation doses must be chosen after previous experiments to determine the radiosensitivity of the plant material. This may be highly variable. Tobacco callus tissue can grow after γ-irradiation, but doses of 10 kR inhibit about 50% of the growth, and doses higher than 15 kR inhibit growth completely (49). Tobacco cells irradiated with γ-rays at doses of 0.5-2 kR are able to produce callus and regenerate buds, and can survive under 8 kR doses (27). In work on mutagenesis, Devreux and Nettancourt used X-rays to irradiate tobacco microspores, with doses in the range of 5-20 rads and 500-2000 rads (22). Nitsch *et al.* used doses from 300-3000 rads, to irradiate haploid plantlets, with effective results being obtained in the range of 1500-2500 rads (79). The radiation source was not mentioned but the dose rate (900 rads/min) indicates that it was probably ^{60}Co. Shigematsu and Hashimoto used 4 kR and obtained flower chimeras in *Chrysanthemum* (90). Krumbiegel treated *Datura innoxia* and *Petunia hybrida* protoplasts with X-rays (500-3000 rads) and verified that haploids were more sensitive than diploids (51). A dose of 1000 rads allowed for a survival of about 50% of the haploid protoplasts and this dose was employed as a mutagenic treatment. Doses of 3000 rads allowed for a survival of only 2-3% of the protoplasts.

2. UV Light. This is generally provided by germicidal lamps of far UV light with a wavelength of 254 nm. UV light must be used directly on the biological material, without barriers which can absorb the radiation. Even the thickness of the cell layer and medium to be irradiated must be considered in improving the efficiency of the treatment. Erikkson treated thin layers of cells sedimented on a quartz plate by irradiating them below, the UV light being completely transmitted through the quartz (28). With respect to dosage, Eapen reports the capacity of tobacco cells to regenerate buds after irradiation with 1000-8000 $ergs/mm^2$ of UV light (27). Widholm obtained 5-MT resistant carrot cells by treatment with UV light. However, he did not indicate the dosage used (103). Murphy *et al.* showed that 42 J/m^2 were effective in enhancing the frequency of $NaClO_3$ resistant rose cells (73).

In the case of UV light, substitution of the medium after irradiation is not essential, because the low doses used probably produce little or no effect on medium components. Irradiation with shorter waves may be ionizing and in these cases the medium must be replaced.

B. *The Use of Chemical Mutagens*

The chemical mutagens which have been used in plant tissue and cell culture belong to two main classes of compounds:

(1) alkylalkane sulfonates: ethyl-methane-sulfonate (EMS), and methyl-methane-sulfonate (MMS);
(2) nitrosocompounds: N-methyl,N',nitro, N-nitrosoguanidine (NG, NTG or MNNG), and N-ethyl-N-nitrosourea (ENU).

These compounds are used in several concentrations and are added to the culture medium or suspension culture after sterilization through filtration. After the treatment, the medium must be replaced, or the cell suspension washed. Table I shows the concentration and treatment time intervals used by various researchers.

The most used chemical mutagenic agents are highly genotoxic (carcinogenic) and must be handled carefully. Some of the most current safety precautions are listed below. These are adapted from Trieff (96).

(1) All operations with genotoxic agents, such as the opening of containers, preparation of solutions, treatment of contaminated glassware, etc., should be conducted under a fume-hood with good ventilation.
(2) Protective eye glasses and a fully fastened laboratory coat should always be used.

(3) Contact of mutagens with skin must be avoided. Hands must be immediately washed after completion of any procedures in which genotoxic agents have been used. It must be pointed out that skin should not be rinsed with organic solvents especially not DMSO, which greatly facilitates absorption of the genotoxic compound into the skin.

(4) Glassware cleaning must be done in a fresh "cleaning solution" (sodium or potassium dichromate in strong sulfuric acid); after 1-2 days genotoxic compounds are destroyed. Specific genotoxic compounds can also be destroyed by methods which take advantage of their particular type of chemical reactivity. Alkylating agents (EMS, MMS) may be destroyed by suitable nucleophiles such as $Na_2S_2O_3 \cdot 5H_2O$. The treatment for MMS destruction is at least one hour, for EMS, 20 hours. Material contaminated with EMS can also be decontaminated by rinsing with a solution containing 0.5% mercaptoacetic acid in 1 M NaOH or KOH (56). For destruction of NTG, one can use $Na_2S_2O_3$, 2% in phosphate buffer pH 8-9, at least for one hour.

C. *Outlines for Mutagenesis and/or Selection Procedures*

1. *Mutagenic Treatment*

a. Physical mutagens. The first step is the irradiation of the plant material. Tissues prior to explantation may be treated by direct irradiation of larger organ pieces, followed by normal procedures of explantation and *in vitro* culture. Tissues or cells in culture to be irradiated with γ- or X-rays are treated directly by exposing the culture flasks to the source of radiation. After irradiation the tissues or cells must be transferred to fresh medium. In the case of irradiation with UV light, tissues or cells must be exposed directly to UV light, irradiation occurring under aseptic conditions. Cells or protoplasts may be irradiated in opened petri dishes, with the cells constituting a thin layer. In the case of γ- and X-rays the effective dose applied is defined precisely, and is based on the type of the source or radiation, the distance between the biological material and the source, and the time of exposure. UV light doses cannot be expressed precisely because the amount of effective radiation depends upon the thickness of the tissues, cell layers, surrounding medium, etc. The type of UV lamp, distance and energy emitted (measured by a dosimeter) should always be indicated.

b. Chemical mutagens. Chemical mutagens are employed to treat cells or tissues in culture. Normally mutagens are added to the autoclaved medium through filtration, in defined amounts to give a final desired concentration. All mutagens mentioned may be diluted in water or in a fraction of the culture medium before addition to the autoclaved medium. Cell or protoplast suspensions are normally treated during a defined time interval (see Table I), then the material is rinsed and cultured in mutagen-free medium. Callus or explanted tissues are cultured in solid medium where the mutagen has been incorporated, the time of treatment being much more variable (days or months).

c. The choice of doses. The dosage to be used in search of mutations must be selected after preliminary experiments to characterize the responses produced by a given kind of mutagen, especially in the case of radiation. Cultures or plant material to be cultured must be subjected to different doses or concentrations of mutagens which will define the lethal and sublethal conditions and those which have no effects on growth rates. As a first approach, doses given in this chapter and their multiples may be tested. The dosage used must allow for a survival of 40-60% of the population, in comparison with the untreated cultures. Stronger doses which allow for little survival rates (e.g., 1%), may be more effective on mutagenesis, but in some cases, depending upon the type of culture, the final yield in the recovery of mutants may be extremely low.

After the mutagenic treatment, cultures may be placed on normal medium for growth and subsequent subculturing, or placed directly under selective conditions which allow for the recovery of the mutants.

2. Selection of Mutants or Variant Lines. Selection procedures used for mutagen-treated cultures are the same used in direct selection of spontaneous mutants from untreated cultures or tissues. The desired phenotypic variants are screened according to the specific conditions of the medium, and prepared for the selection of resistant, autotrophic, auxotrophic or other variant phenotypes. The main general steps that can be followed in the selection of different kinds of mutants are indicated below.

a. Resistant mutants.

(1) Incubation of cell or protoplast suspensions in a drug-added medium (herbicides, base analogues, toxins, salts,

etc.), or under stressed environmental conditions (chilling, high temperatures, etc.);

(2) Recovery of the surviving fraction; subculturing, until a stable population adapted to the selective conditions is obtained.

(3) Test for resistance in non-selective medium. Test also for the occurrence of other characteristics associated with resistance (oversynthesis of metabolites, altered metabolic pathway, etc.);

(4) Plating on agar medium to obtain calluses (cells may be diluted in gelled medium heated at $\pm 38^{o}C$ and then poured in petri dishes, ca. 10 ml/plate, or directly poured on the gelled medium);

(5) Test for resistance of the calluses;

(6) Regeneration of plants from calluses;

(7) Test for resistance of tissues or cells from regenerated plants;

(8) Sexual crosses to define how the new trait is transmitted.

Note: Selection may be started directly from Step (4), if the drug is added onto agar plates, or if they are incubated under selective environmental conditions. Freshly explanted and callus tissues also may be cultured directly from Step (4).

In the case of cells or protoplasts, procedures may be highly variable, depending on the cell cycle, growth rates of culture, responses of cells to the selective condition, etc.

The completed sequence presented above is infrequently achieved. Resistance sometimes is expressed only under the stressed environmental conditions, and more rarely in regenerated plants.

b. Auxotrophic mutants.

(1) Incubation of cell suspension in a minimal medium containing BUdR (10^{-5}M), in the dark, during the time of one cell generation (previously determined). Wild cells which can grow in minimal medium divide and incorporate BUdR (a thymidine analogue) into the DNA. The presumptive mutants that exist in the cell population, do not divide in the minimal medium, and consequently do not incorporate BUdR.

(2) Washing of cell suspension in fresh BUdR-free medium and plating on minimal medium (agar plates), under light. Cells which incorporate BUdR are killed. The surviving plated cells may form calluses, and those which do not possess normal growth rates (as compared to wild lines) are presumptively auxotrophic mutants.

(3) Culturing of "mutant" calluses in different nutritionally supplemented media to define the nature of auxotrophy. Readers are referred to the screening table presented by Holiday (44).

(4) Test for the maintenance of the trait through several cultures, regeneration of plants, etc.

c. Others. Several kinds of mutants or variant lines may be selected by using the steps mentioned above. Some variant phenotypes may be detected through their altered pigmentation, especially those related to anthocyanins, carotenoids and chlorophyll. Autotrophic lines are selected in media lacking some components, especially growth substances. All these kinds of variant lines may appear spontaneously or may be induced by mutagens.

V. PROTOCOL FOR ISOLATION OF BUdR-RESISTANT CELLS FROM SOYBEAN PROTOPLASTS

The following step-wise production and selection procedure for the isolation of BUdR-resistant cell lines from protoplasts of soybean is adapted from Ohyama (82) with permission.

A. Equipment

(1) Small centrifuge, swinging bucket type.
(2) Water bath, 35°C.
(3) Stainless steel mesh filter, 44 μm.

B. Materials and Reagents

(1) Protoplasts prepared from suspension cultures of soybean (*Glycine max*). For preparation see Chapter 4 (this volume).
(2) N-Methyl-N-nitro-N-nitrosoguanidine (NTG, 1 mg/ml in water).
(3) O-B5-Sb medium (see Chapter 2 for O-B5 medium). B5 (2,4-D- and sucrose-free) containing 0.28 M sorbitol, pH 5.5.
(4) 1-B5-CA medium. (see Chapter 2 for 1-B5 medium). 1-B5 containing 0.1% casamino acids, pH 5.5. (Bacto Vitamin-free Casamino Acids (Difco)).

(5) 1-B5-CA-Sb medium.
1-B5 containing 0.1% casamino acids and 0.28 M sorbitol, pH 5.5.

(6) 1-B5-CA agar (0.8%) medium.
1-B5-CA containing BUdR (20 μg/ml) and uridine (20 μg/ml); 20 ml layered into sterile disposable petri dishes (15 x 100 mm). These plates are prepared 24 h prior to use to allow evaporation of excess moisture.

(7) 1-B5-CA agar (0.5%) medium.
1-B5-CA containing BUdR (20 μg/ml) and uridine (20 μg/ml). Maintain at 35°C.

C. *Procedures*

1. *NTG-Treatment of Protoplasts.*

(1) Suspend protoplasts in 9.5 ml 0-B5-Sb medium (approximately 6 x 10^6 protoplasts/ml).

Note: See Chapter 4 (this volume) for methods for determining protoplast concentration.

(2) Add 0.5 ml NTG solution (final concentration, 50 μg/ml) and incubate at 28°C for 16 h in the dark.

(3) Wash protoplasts three times with 10 ml 0-B5-Sb medium. Centrifuge at 100 x g for 3 min.

2. *Culturing of NTG-Treated Protoplasts.*

(1) Suspend protoplasts in 1-B5-CA-Sb medium (10 ml) and culture at 28°C for 3-4 weeks as droplets (0.1 ml) in petri dishes (15 x 100 mm) sealed with Parafilm®. Protoplast survival <1%.

(2) Collect cells on 44 μm stainless steel mesh filter and wash with 1-B5-CA medium.

3. *Plating Cells on BUdR Medium.*

(1) Suspend cells in 1-B5-CA agar (0.5%) medium at 35°C (1,500 cell clusters/ml).

(2) Pour 5 ml of the cell suspension onto solidified 1-B5-CA agar (0.8%) medium and spread evenly.

(3) After the agar solidifies, seal plates with Parafilm® and incubate in the dark at 28°C for 4-5 weeks.

(4) Transfer individual colonies to 1-B5-CA agar (0.8%) medium.

Note: Individual cell colonies can be transferred using the broad end of sterile toothpicks.

(5) After 1-2 weeks, transfer individual cell colonies to 5 ml 1-B5-CA medium containing BUdR (20 μg/ml) and uridine (100 μg/ml). The cells should be subcultured as often as needed in the above liquid medium.

ACKNOWLEDGMENTS

I thank Drs. Kurt G. Hell and Hamilton J. Targa for their helpful suggestions and revision of the manuscript, especially in the topics on ionizing radiation. I thank also Dr. Walkyria R. Monteiro for the English revision.

REFERENCES

1. Amirato, P.V., and Steward, F.C., *Dev. Biol. 19*, 87 (1969).
2. Auerbach, C., "Mutation Research. Problems, Results and Perspectives". Chapman & Hall, London, (1976).
3. Barg, R., and Umiel, N., *Z. Pflanzenphysiol. 83*, 437 (1977).
4. Bacq, Z.M., and Alexander, P., "Fundamentals of Radiobiology". Pergamon Press, Oxford, (1961).
5. Berlyn, M.B., and Zelitch, I., *Abstr. Int. Congress Plant Tissue and Cell Culture., 4th, Calgary*, p. 134 (1978).
6. Binding, H., *Z. Pflanzenzuchtg. 67*, 33 (1972).
7. Binding, H., Binding, K., and Straub, J., *Naturwissenschaften 57*, 138 (1970).
8. Bourgin, J.P., *Mol. gen. Genet. 161*, 225 (1978).
9. Bright, S.W.J., and Northcote, D.H., *Planta 123*, 79 (1975).
10. Bright, S.W.J., and Miflin, B.J., *Abstr. Int. Congress Plant Tissue and Cell Culture, 4th, Calgary*, p. 140 (1978).
11. Carlson, P.S., *Science 168*, 487 (1970).
12. Carlson, P.S., *Science 180*, 1366 (1973).
13. Carlson, J.E., and Widholm, J.M., *Physiol. Plant. 44*, 251 (1978).
14. Chaleff, R.S., and Carlson, P.S., *Ann. Rev. Gen. 8*, 267 (1974).

15. Chaleff, R.S., and Carlson, P.S., *In* "Genetic Manipulations with Plant Material" (L. Ledoux, ed.), p. 351. Plenum Press, N.Y., (1975).
16. Chaleff, R.S., and Parsons, M.F., *Genetics 89*, 723 (1978).
17. Chaleff, R.S., and Parsons, M.F., *Proc. Natl. Acad. Sci. (USA) 75*, 5104 (1978).
18. Corduan, G., *In* "Genetic Manipulations with Plant Material" (L. Ledoux, ed.), p. 578. Plenum Press, N.Y., (1975).
19. Croughan, T.P., Stavarek, S.J., and Rains, D.W., *Crop Sci. 18*, 959 (1978).
20. D'Amato, F., *In* "Applied and Fundamental Aspects of Plant Cell, Tissue and Organ Culture" (J. Reinert and Y.P.S. Bajaj, eds.), p. 343. Springer-Verlag, Berlin, (1977).
21. D'Amato, F., *In* "Frontiers of Plant Tissue Culture 1978" (T.A. Thorpe, ed.), p. 287. University of Calgary Press, Calgary, (1978).
22. Devreux, M., and de Nettancourt, D., *In* "Haploids in Higher Plants. Advances and Potential" (K.J. Kasha, ed.), p. 309. University of Guelph, Guelph, (1974).
23. Dix, P.J., and Street, H.E., *Plant Sci. Lett. 5*, 231 (1975).
24. Dix, P.J., and Street, H.E., *Ann. Bot. 40*, 903 (1976).
25. Dix, P.J., Joo, F., and Maliga, P., *Molec. Gen. Genet. 157*, 285 (1977).
26. Dulieu, H., *Phytomorphology 22*, 283 (1972).
27. Eapen, S., *Protoplasma 89*, 149 (1976).
28. Eriksson, T., *Physiol. Plant. 20*, 507 (1967).
29. Flack, J., and Collin, H.A., *Abstr. Int. Congress Plant Tissue and Cell Culture, 4th, Calgary*, p. 171 (1978).
30. Flashman, S.M., and Filner, P., *Plant Sci. Lett. 13*, 219 (1978).
31. Gathercole, R.W.E., and Street, H.E., *New Phytol. 77*, 29 (1976).
32. Gavazzi, G., Nava-Racchi, M., and Tonelli, C., *Theor. Appl. Genet. 46*, 339 (1975).
33. Gengenbach, B.G., Green, C.E., and Donovan, C.M., *Proc. Natl. Acad. Sci. USA 74*, 5113 (1977).
34. Glimelius, K., Eriksson, T., Grafe, R., and Muller, A.J., *Physiol. Plant. 44*, 273 (1978).
35. Gressel, J., Zilkah, S., and Ezra, G., *In* "Frontiers of Plant Tissue Culture 1978" (T.A. Thorpe, ed.), p. 427. University of Calgary Press, Calgary (1978).
36. Gresshof, P.M., *Theor. Appl. Genet. 54*, 14 (1979).
37. Grosch, D.S., "Biological Effects of Radiations". Blaisdell Publ. Co., New York (1968).

38. Gunckel, J.E., and Sparrow, A.H., *In* "Encyclopaedia of Plant Physiology" (W. Ruhland, ed.), Vol. XVI, p. 555. Springer-Verlag, Berlin, (1961).
39. Heiner, Y.R., and Filner, P., *Biochim. Biophys. Acta 215,* 152 (1970).
40. Heinz, D.J., *In* "Induced Mutation in Vegetatively Propagated Plants", p. 53. Intern. Atomic Energy Agency, Vienna, (1973).
41. Heinz, D.J., Krishnamurthi, M., Nickell, L.G., and Maretzki, A., *In* "Applied and Fundamental Aspects of Plant Cell, Tissue and Organ Culture" (J. Reinert and Y.P.S. Bajaj, eds.), p. 3. Springer-Verlag, Berlin, (1977).
42. Hell, K.G., Handro, W., and Kerbauy, G.B., *Environ. Expt. Bot. 18,* 225 (1978).
43. Hibberd, K.A., Green, C.E., Walter, T.J., and Gengenbach, B.G., *Abstr. Int. Congress Plant Tissue and Cell Culture, 4th, Calgary,* p. 141 (1978).
44. Holiday, R., *Nature 178,* 987 (1956).
45. Howland, G.P., and Hart, R.W., *In* "Applied and Fundamental Aspects of Plant Cell, Tissue and Organ Culture" (J. Reinert and Y.P.S. Bajaj, eds.), p. 731. Springer-Verlag, Berlin, (1977).
46. Hughes, K.W., *Abstr. Int. Congress Plant Tissue and Cell Culture, 4th, Calgary,* p. 170 (1978).
47. Jones, G.E., and Hann, J., *Theor. Appl. Genet. 54,* 81 (1979).
48. Jones, G.E., and Horsch, R.B., *Abstr. Int. Congress Plant Tissue and Cell Culture, 4th, Calgary,* p. 143 (1978).
49. Kerbauy, G.B., and Hell, K.G., *Int. J. Rad. Biol. 35,* 273 (1979).
50. King, P.J., Potrykus, I., and Thomas, E., *Physiol. Vég. 16,* 389 (1978).
51. Krumbiegel, G., *Environ. Expt. Bot. 19,* 99 (1979).
52. Lavee, S., and Galston, A.W., *Plant Physiol. 43,* 1760 (1968).
53. Lescure, A.M., *Bull. Soc. Chim. Biol. 52,* 953 (1970).
54. Lescure, A.M., *Plant Sci. Lett. 1,* 375 (1973).
55. Lescure, A.M., and Péaud-Lenoel, C., *C.R. Acad. Sci. Paris 265,* 1803 (1967).
56. Lewis, E.B., and Bacher, F., *Drosophila Information Service 43,* 193 (1968).
57. Liu, M., and Chen, W., *Euphytica 25,* 393 (1976).
58. Maliga, P., *In* "Cell Genetics of Higher Plants" (D. Dudits, G.L. Farkas and P. Maliga, eds.), p. 59. Akademiai Kiado, Budapest, (1976).

59. Maliga, P., *In* "Frontiers of Plant Tissue Culture 1978" (T.A. Thorpe, ed.), p. 381. University of Calgary Press, Calgary, (1978).
60. Maliga, P., *In* "*Int. Rev. Cytol.*" *Suppl. 11A* (I.K. Vasil, ed.), p. 225. Academic Press, New York (1980).
61. Maliga, P., Sz.-Breznovits, A., and Marton, L., *Nature New Biol. 244,* 29 (1973).
62. Maliga, P., Sz.-Breznovits, A., and Marton, L., *Nature 255,* 401 (1975).
63. Maliga, P., Lazar, G., Svab, G., and Nagy, F., *Molec. Gen. Genet. 149,* 267 (1976).
64. Malmberg, R.L., *Abstr. Int. Congress Plant Tissue and Cell Culture, 4th, Calgary,* p. 137 (1978).
65. Marton, L., and Maliga, P., *Plant Sci. Lett. 5,* 77 (1975).
66. Marton, L., Nagy, F., Gupta, K.C., and Maliga, P., *Plant Sci. Lett. 12,* 333 (1978).
67. Mastrangelo, I.A., and Smith, H.H., *Plant Sci. Lett. 10,* 171 (1977).
68. Mathews, P.S., and Vasil, I.K., *Z. Pflanzenphysiol. 77,* 222 (1975).
69. Melchers, G., and Bergmann, L., *Ber. dt. Bot. Ges. 71,* 459 (1958).
70. Meredith, C.P., *Plant Sci. Lett. 12,* 25 (1978).
71. Müller, A.J., and Grafe, R., *Molec. Gen. Genet. 161,* 67 (1978).
72. Murashige, T., and Nakano, R., *Amer. J. Bot. 54,* 963 (1967).
73. Murphy, T.M., Wright, L.A., and Street, H.E., *Abstr. Int. Congress Plant Tissue and Cell Culture, 4th, Calgary,* p. 149 (1978).
74. Nabors, M.W., Daniels, A., Nadolny, L., and Brown, C., *Plant Sci. Lett. 4,* 155 (1975).
75. Negrutiu, I., Cattoir-Reynaerts, A., and Jacobs, M., *Abstr. Int. Congress Plant Tissue and Cell Culture, 4th, Calgary,* p. 138 (1978).
76. Nijkamp, H.J.J., Colijn, C.M., and Kool, A.J., *Abstr. Int. Congress Plant Tissue and Cell Culture, 4th, Calgary,* p. 145 (1978).
77. Nishi, T., and Mitsuoka, S., *Jap. J. Gen. 44,* 341 (1969).
78. Nitsch, J.P., *Phytomorphology 19,* 389 (1969).
79. Nitsch, J.P., Nitsch, C., and Péreau-Leroy, P., *C.R. Acad. Sci. Paris 269,* 1650 (1969).
80. Novak, F.J., and Vyskot, B., *Z. Pflanzenzuchtg. 75,* 62 (1975).
81. Ohyama, K., *Expt. Cell Res. 89,* 31 (1974).

82. Ohyama, K., *In* "Plant Tissue Culture Methods" (O.L. Gamborg and L.R. Wetter, eds.), p. 36. N.R.C. Canada, Saskatoon. Publ. 14383 (1975).
83. Ohyama, K., *Environ. Expt. Bot. 16*, 209 (1976).
84. Palmer, J.E., and Widholm, J., *Plant Physiol. 56*, 233 (1975).
85. Polocco, J.C., and Polocco, M.L., *Ann. N.Y. Sci. 287*, 385 (1977).
86. Sacristan, M.D., and Melchers, G., *Molec. Gen. Genet. 105*, 317 (1969).
87. Schaeffer, G.W., *Abstr. Int. Congress Plant Tissue and Cell Culture, 4th, Calgary*, p. 140 (1978).
88. Schieder, O., *Molec. Gen. Genet. 149*, 251 (1976).
89. Sheridan, W.F., *In* "Genetic Manipulations with Plant Material" (L. Ledoux, ed.), p. 263. Plenum Press, N.Y., (1975).
90. Shigematsu, K., and Hashimoto, S., *Abstr. Int. Congress Plant Tissue and Cell Culture, 4th, Calgary*, p. 135 (1978).
91. Sung, Z.R., *Plant Physiol. 56 (suppl.)*, 37 (1975).
92. Sung, Z.R., *Genetics 84*, 51 (1976).
93. Sung, Z.R., Smith, J., and Signer, E.R., *In* "Genetic Manipulations with Plant Material" (L. Ledoux, ed.), p. 574. Plenum Press, N.Y., (1975).
94. Sung, Z.R., Liu, S.-T., and Sell, S., *Abstr. Int. Congress Plant Tissue and Cell Culture, 4th, Calgary*, p. 138 (1978).
95. Tabata, M., Ogino, T., Yoshioka, K., Yoshimura, N., and Hiraoka, N., *In* "Frontiers of Plant Tissue Culture 1978" (T.A. Thorpe, ed.), p. 381. University of Calgary Press, Calgary (1978).
96. Trieff, N.M., *In* "1st Annual Course in the Principles and Practices of Genetic Toxicology", University of Texas, Galveston, Texas, (1976).
97. Tulecke, W., *Plant Physiol. 35*, 19 (1960).
98. Umiel, N., and Goldner, R., *Protoplasma 89*, 83 (1976).
99. Weber, G., and Lark, K.G., *Abstr. Int. Congress Plant Tissue and Cell Culture, 4th, Calgary*, p. 134 (1978).
100. Widholm, J.M., *Biochim. Biophys. Acta 279*, 48 (1972).
101. Widholm, J.M., *Can. J. Bot. 54*, 1523 (1976).
102. Widholm, J.M., *In* "Plant Tissue Culture and Its Bio-technological Applications" (W. Barz, E. Reinhard, and M.H. Zenk, eds.), p. 112. Springer-Verlag, Berlin, (1977).
103. Widholm, J.M., *In* "Molecular Genetic Modification of Eucaryotes" (I. Rubenstein, R. Phillips, and C.E. Green, eds.), p. 57. Academic Press, New York, (1977).
104. Widholm, J.M., *Planta 134*, 103 (1977).

105. Widholm, J.M., *Crop Sci. 17*, 597 (1977).
106. Widholm, J.M., *J. Exp. Bot. 29*, 111 (1978).
107. Widholm, J.M., *Abstr. Int. Congress Plant Tissue and Cell Culture, 4th, Calgary*, p. 138 (1978).
108. Widholm, J.M., *In* "Propagation of Higher Plants through Tissue Culture" (K.W. Hughes, R. Henke, and M. Constantin, eds.), p. 189. U.S. Dept. of Energy. Conf. 7804111 (1979).
109. Zenk, M.H., *In* "Haploids in Higher Plants. Advances and Potential" (K.J. Kasha, ed.), p. 339. University of Guelph, Guelph, (1974).
110. Zenk, M.H., El-Shagi, H., Arens, H., Stöckigt, J., Weller, E.W., and Deus, B., *In* "Plant Tissue Culture and Its Bio-technological Application" (W. Barz, E. Reinhard and M.H. Zenk, eds.), p. 27. Springer-Verlag, Berlin, (1977).
111. Zyrd, J.P., *Plant Sci. Lett. 6*, 157 (1976).

MERISTEM CULTURE AND CRYOPRESERVATION -- METHODS AND APPLICATIONS[1]

K.K. Kartha

Prairie Regional Laboratory
National Research Council
Saskatoon, Saskatchewan
Canada

I. MERISTEM CULTURE

Shoots of all angiosperms and gymnosperms grow by virtue of their apical meristems. The apical meristem is usually a dome of tissue located at the extreme tip of a shoot and measures ca. 0.1 mm in diameter and ca. 0.25 to 0.3 mm in length. These tissues comprise cells which are anatomically indistinguishable, but are commonly grouped into arbitrary zones such as tunica, corpus, central mother cells, flank meristem and rib meristems. The apical meristems are first formed during embryo development and they remain, except for period of dormancy, in an active state of division throughout the vegetative phase of the plant. The totipotency of apical meristem cells form the basis of meristem culture technique.

Meristem culture has been extensively used, during the last two decades, in the clonal propagation of plants, where in reality the meristem and subjacent tissue is used. The rapidity at which plants are propagated through the technique has found its application in the horticultural industry. Moreover, since the constituent cells of meristems are genetically stable, the regenerated plants would be genetically identical to the donor plants. Genetic stability is very desirable and very much needed

[1]*NRCC No. 18200*

ISBN 0-12-690680-7

particularly in experimental material. The most important application of meristem culture is the production of pathogen-free plants, especially viruses, and also the longterm storage of such virus-free germplasm through cryopreservation techniques.

Limasset and Cornuet (30) demonstrated that in systemically infected plants, the virus was unevenly distributed in the host and that the virus titer decreased as the vegetative point is reached. This significant observation paved the way for the pioneering work of Morel and Martin (36) who envisaged that plants regenerated by the *in vitro* culture of meristems would be virus-free. They demonstrated their concept in 1952 by culturing the meristem tips of virus-infected dahlias and regenerating virus-free shoots which subsequently were grafted to healthy dahlia stocks to produce virus-free plants. Ever since this pioneering work, meristem culture has been widely used in vegetative propagation, clonal multiplication and production of virus-free plants from a wide range of plant species. Some of the results of the elimination of viral pathogens from food crops by meristem culture are listed in Table I.

Plant regeneration from meristems could successfully be accomplished for vegetatively propagated as well as seed-propagated crops. In vegetatively propagated crops, the use of infected planting material causes the main spread of virus to the progeny in addition to insect vectors and other means of transmission. It is estimated that about 1 in 10 of the known plant viruses is transmitted also through the seeds of infected host plants. Seed transmission of viruses is very prevalent in legume crops such as pea *(Pisum sativum)*, cowpea *(Vigna unguiculata)*, soybean *(Glycine max)*, and bean *(Phaseolus vulgaris)*. Since virus multiplication and host metabolism are very closely associated, attempts to selectively interfere viral replication without adversely affecting normal metabolic processes of host cells have mainly been unsuccessful. Therefore, the most efficient way of eliminating viral pathogens from crop plants is through meristem culture technique.

A. *The Technique*

The following section describes the procedures employed in the *in vitro* culture of meristems from seed-propagated and from vegetatively propagated crops. Specific examples are chosen to demonstrate the methodology.

1. Seed-Propagated Crops. (e.g., pea, soybean, cowpea, bean, chickpea, and coffee).

a. Preparation of material. Many of the seeds are externally contaminated with seed-borne organisms such as fungi and bacteria and require sterilization. The seeds are immersed in cold water for a few seconds and the floating ones discarded. The seeds are then thoroughly rinsed in 70% ethanol for 1 min followed by disinfection in a 20% commercial bleach solution (1.2% sodium hypochlorite) for 15 to 20 min on a shaker at 200 to 250 rpm. They are rinsed 3 to 4 times with sterile distilled water to remove all traces of chlorine. The seeds are aseptically germinated in glass jars on wet absorbant cotton. Immediately upon germination (4 to 5 days depending upon the species) the shoot apical meristems measuring ca. 0.3 to 0.5 mm in length are aseptically isolated.

b. Isolation of meristems. Isolation of meristems is to be performed aseptically under a stereo microscope preferably in a laminar air flow hood. It can also be carried out under semi-sterile conditions since the meristematic domes are enclosed in many whorls of leaves and usually are free of contaminants. All the dissecting instruments viz. razor-knives, needles and forceps are sterilized by immersion in 70% ethanol. They are rinsed in sterile distilled water and dried under several folds of sterile filter/blotting paper. Disinfection of the instruments should be done as frequently as possible. The dissecting microscope and the stage should also be disinfected with 70% ethanol taking care that the ethanol does not enter the optical system of the microscope. The shoot tip is held under the microscope with a pair of forceps under a workable magnification (10X, 20X) and the outer whorls of leaves are removed, one by one, with either the needle or the knife until the meristem dome and a pair of leaf primordia are reached. With the knife held steady, a V-shaped cut is applied 0.3 to 0.5 mm below the tip of the dome and the excised tissue (explant) removed along with portions of procambial tissue. It is immediately planted on 2.5 ml solidified culture medium contained in small pyrex tubes (10 x 1.2 cm). The tubes are plugged with cotton and incubated under appropriate environmental conditions (temperature and photoperiod).

Meristems from a wide range of species have successfully been cultured on MS (40) and B5 medium (11) solidified with 0.6 to 0.8% Difco Bacto agar (21). Differentiation of shoots, roots or both is regulated by the interaction of

TABLE I. Viruses Eliminated from Food Crops through Meristem Culture

Food crops	Viruses eliminated	Reference
Asparagus (*Asparagus officinalis*)	viruses designated A-,B-,C-	83
Banana (*Musa* spp.)	cucumber mosaic virus	6
	unidentified virus	
Cassava (*Manihot esculenta*)	Indian and African mosaic virus	19
	African mosaic and brown streak	18
	frog-skin	Roca (unpublished)
Cauliflower (*Brassica oleracea*)	cauliflower mosaic virus	76
	turnip mosaic virus	
Garlic (*Allium sativum*)	mosaic virus	38, 53
Gooseberry (*Ribes grossularia*)	vein banding	17
Horse radish (*Armoracia lapathifolia*)	turnip mosaic virus	38
Pea (*Pisum sativum*)	seed-borne mosaic virus	20
Potato (*Solanum tuberosum*)	paracrinkle virus, virus X	26
	S, Y	50, 84
	X, Y, S	70
	A, X, S	27
	X, S	64
	spindle tuber	65
	X, Y, S, leaf roll virus	38
	A, S	63
	S, M	47
	X	46
Raspberry (*Rubus ideaus*)	mosaic virus	49

TABLE I (continued)

Food crops	Viruses eliminated	Reference
Rhubarb (*Rheum rhaponticum*)	turnip mosaic virus, cherry leaf roll virus, cucumber mosaic virus, strawberry latent ring spot virus	74
Strawberry (*Fragaria* spp.)	crinkle, vein banding	35
	yellows virus complex	51
	latent C virus	33
	yellow edge, vein chlorosis crinkle, latent A virus	73
	latent A virus	5
	virus complex	38
	mild yellow edge, pallidosis mottle	39
Sugar cane (*Saccharum officinarum*)	mosaic virus	15, 38, 29
Sweet potato (*Ipomoea batatas*)	internal cork virus, rugose mosaic, feathery mottle	38, 44
	unidentified virus	1, 45
Yams		
Taro (*Colocasia esculenta*)	dasheen mosaic virus	14
Cocoyam (*Xanthosoma brasiliense*)	unspecified virus	14, 66

nutrients, growth hormones and environmental conditions. A summary of the conditions for plant regeneration from meristems of some plant species is given in Table II.

2. *Vegetatively Propagated Crops*

a. Cassava (Manihot esculenta). Cassava is an annual crop which is vegetatively propagated by stem cuttings. Meristems taken from sprouting cuttings are better suited for plant regeneration than from mature plants. Dormant stakes are cut into 10 cm sections with at least 2 nodes each, the upper ends sealed with paraffin, and potted in a mixture of vermiculite, peat moss and sand (3:2:1). The material is grown at 26°C, in 16 h photoperiods at a light intensity of ca. 8000 lux from combined fluorescent and incandescent lamps. The cuttings sprout within 5 to 7 days. Shoot tips developing from these stakes are used as source of meristems. Shoots developing from the axial buds could also be used thereby providing a continuous supply of meristems for culturing. The stem apices (1-2 cm long) are sterilized by immersion in 70% ethanol for 60 sec followed by washing them 3 times in sterile distilled water. Disinfection with commercial bleach solution is not necessary since the meristem domes are located within several whorls of leaves. The procedure for the isolation of meristems is similar to the one described for seed-propagated crops. The isolated meristems are cultured on agar-solidified MS medium supplemented with benzyladenine (BA), naphthaleneacetic acid (NAA) and gibberellic acid (GA_3) at concentrations of 0.5, 1.0 and 0.1 µM, respectively. The cultures are incubated at a constant temperature of 26°C, 16 h photoperiods and 4000 lux intensity and 70% relative humidity.

b. Strawberry (Fragaria X ananassa). Dormant strawberry plants removed from cold storage were potted in a mixture of vermiculite, peatmoss and sand (3:2:1) and grown in a greenhouse. The flowers were regularly pinched off to promote production of runners. Runner-tips (5 cm long) were excised from the runners and disinfected in a 20% solution (by volume) of commercial bleach (1.2% sodium hypochlorite) for 20 min and washed several times with sterile distilled water. Runner-tip meristems (0.4 to 0.5 mm) were aseptically isolated and cultured on agar solidified MS medium supplemented with 1 µM BA, 1 µM indolebutyric acid (IBA) and 0.1 µM GA_3 and the cultures incubated at 26°C with 16 h photoperiods at a light intensity of 4000 lux and 70% relative humidity. Under these conditions the meristems differentiate into shoots in 3 to 4 weeks. These shoots are aseptically recultured on agar-solidified MS medium supplemented with 10 μM BA and incubated under similar environmental conditions to induce mass proliferation of shoots. Reculturing the shoots on the same nutrient medium but containing lower levels of BA (1.0 μM) and NAA (1.0 μM) or on medium devoid of growth hormones promote root differentiation. By this technique, several thousands of plantlets could be grown within a short period of time.

B. *Morphogenetic Responses of Meristems in Culture*

Meristems exhibit a wide range of morphogenetic responses in culture (Table II). The differentiation process of meristems into either shoots, roots, callus, multiple buds or complete plants is governed by a number of factors, viz. (1) the size of meristem-tips, (2) type of culture medium, (3) the kind and concentration of growth hormones, (4) the environmental conditions (light, temperature and photoperiod) and (5) other factors such as the seasonal fluctuations of the donor plants and the type of containers used

1. *Size of Meristem Tips.* The size of meristem tips is one of the crucial decisive factors which governs the regenerative capacity of meristems and also in increasing the probability of recovering virus-free plants. In the case of cassava, only meristems exceeding 0.2 mm in length regenerated into plantlets and those which are less than that size developed either callus or roots. Similarly, soybean meristems differentiated into plantlets only if their size exceeded 0.4 mm. Generally, the larger the size of the meristem, the greater will be the number of plants regenerated. But if the experiments are aimed at eliminating viral pathogens, the reverse is the rule; i.e., the smaller the size of the meristem, the greater will be the chance of eliminating the virus.

2. *Culture Medium.* Murashige and Skoog medium has been very satisfactory for culturing meristems from seed-propagated and vegetatively-propagated crops. The medium devised by Gamborg *et al.* (11). B5 has been successfully used for culturing pea meristems (21). Meristems grow well on nutrient medium solidified with 0.6 to 0.8% agar. But meristems from plants like orchids (e.g., *Cattleya*) release large quantities of phenolic compounds into the medium which eventually become cytotoxic. In cases like that or in instances where pigments are released into the medium from

TABLE II. *Morphogenetic Responses of Meristems in Culture*

Species	Medium	Hormone Conc. (μM)	Environmental conditions Temp./photoperiod	Morphogenetic responses	Reference
Pea (*Pisum sativum*)	B5	BA 0.5	$26^{o}C/16$ h	40-50% shoot regeneration	
	"	"	$20^{o}C$ day, $15^{o}C$ night/16 h	100% shoot regeneration	
	½B5	NAA 1.0	"	Root differentation on shoots	
	B5	NAA 1.0	"	whole plants	21, 24
Chickpea (*Cicer arietinum*)	MS	BA 1 to 5	$26^{o}C/16$ h	Multiple shoots	20
	"	IBA 1.0	"	Root differentiation on shoots	
Soybean (*Glycine max*)	MS	BA 0.1 + NAA 1.0	"	Whole plants	20
Cowpea (*Vigna unguiculata*)	MS	BA 0.001 + NAA 1.0	"	Whole plants	20
Bean (*Phaseolus vulgaris*)	MS	BA 5.0 + NAA 10	"	Multiple shoots and roots	Kartha (unpublished)
	"	NAA 1.0	"	Whole plants	Kartha (unpublished)
Peanut (*Arachis hypogaea*)	MS	BA 0.1 + NAA 1 to 10	"	Whole plants	

TABLE II (continued)

Species	Medium	Hormone Conc. (µM)	Environmental conditions Temp./photoperiod	Morphogenetic responses	Reference
Sweet potato (*Ipomoea batatas*)	MS	BA or zeatin or kinetin (1 to 5) + similar levels of NAA/IAA	"	Multiple shoots	Kartha (unpublished)
Tomato (*Lycopersicon esculentum*)	MS	BA, NAA 0.1 to 1.0 or 0.1 to 10 zeatin + same levels of IAA/NAA	"	Whole plants	23
Cassava (*Manihot esculenta*)	MS	BA 0.5, + NAA 1.0 + GA_3 0.1	"	Whole plants	22

the explants, use of a liquid medium is preferable since it facilitates change of culture medium at periodic intervals. Concentrations of growth hormones utilized with solidified media may, at times, be deleterious when a liquid medium is used, as has been reported for potato meristems (34). In such cases it may be advisable to use the hormones at a lower concentration since the initial uptake by the tissues may be much faster than from agar.

The pH of the culture medium has also been reported to affect meristems in culture. Generally an acidic medium with a pH of 5.6 to 5.8 supports growth of most of the meristems in culture. A shift in pH from 5.7 to 5.4 within a week in unused medium has been observed (34). Cassava meristems after an initial sign of growth completely cease further development when cultured on medium with a pH of 4.8 (Kartha, unpublished). Carnation meristems grow better on medium with a pH of 5.5 than at pH 6.0 (68).

3. Growth Hormones. A proper balance of cytokinin and auxin is expected to result in the regeneration of whole plants. Pea meristems cultured on B5 medium supplemented with various levels of BA and NAA regenerated only shoots. The amount of endogenous hormones present in the explant could alter the morphogenetic responses as is evident in the case of pea and bean meristems. In pea meristems, only shoot regeneration is achieved in the presence of exogenously applied BA or BA in conjunction with NAA, and the differentiated shoots fail to develop roots even after prolonged incubation on the same medium. A totally different response, differentiation of shoots and roots (whole plants) is achieved using 1.0 μM NAA as the only exogenously applied hormone. This indicates that the endogenous level of cytokinins synthesized or present in the meristem approaches the required cytokinin:auxin balance (in association with the exogenously applied NAA) to bring about whole plant regeneration. Any additional supply of exogenous BA to the culture medium disrupts this balance thereby resulting in the differentiation of shoots only. The induction of roots on differentiated shoots could not be accomplished by means of the same nutrient medium supplemented with any concentrations of NAA or other auxins. However, if the nutrient medium is employed at half its strength and NAA supplied at levels varying from 0.1 to 1.0 μM, the shoots readily differentiate roots. An explanation for this phenomenon is difficult to make except to hypothesize the role of mineral salt inhibition of the rooting process (21).

Among the cultured meristems of leguminous crops, chick-pea (*Cicer arietinum*), bean (*Phaseolus vulgaris*), cowpea

(*Vigna unguiculata*), soybean (*Glycine max*), and peanut (*Arachis hypogaea*), multiple shoot differentiation was predominant in the case of chickpea and bean (Kartha, unpublished). Shoots differentiated from chickpea meristems failed to form roots when auxins like IAA or NAA were employed alone or in conjunction with various concentrations of BA. However, rooting could be induced by reculturing the shoots on MS medium supplemented with 1.0 μM indolebutyric acid (IBA). The bean meristems under the influence of BA alone (5 μM) or in combination with 10.0 μM NAA proliferated and produced multiple buds (10 to 15 buds/explant). But the growth of these buds into shoots was retarded until gibberellic acid (GA_3) was applied (0.1 μM). On the other hand, when NAA (1.0 μM) was employed as the only exogenously supplied hormone, vigorously growing plantlets were regenerated. Here again, the effect of NAA alone on whole plant regeneration is identical to the one observed in the case of pea meristems. Soybean meristems have a tendency to form callus from the basal cut surface, the extent of callus formation depending upon the concentration of cytokinin in the medium. Callus production occurs at the expense of shoot differentiation. Soybean meristems differentiated into plantlets with a frequency of 30 to 50% when BA and NAA were applied at concentrations of 0.1 and 1.0 μM, respectively. On the other hand 100% plant regeneration from cowpea meristems was obtained at a remarkably low level of BA (0.001 μM) plus 0.1 μM of NAA. Similarly, although peanut meristems differentiate into whole plants in a wide range of concentrations of BA and NAA, a low level of BA (0.1 μM) and a high level of NAA (10.0 μM) has been found to be best suited for maximum plant regeneration (100%). Under a wide range of cytokinins and auxins, tomato and sweet potato meristems were induced to form complete plants. Although multiple bud formation was not observed in the case of tomato, sweet potato meristems formed multiple buds under high levels (5 to 10 μM) of either BA, zeatin or kinetin.

Coffee meristems develop only shoots under all concentrations and combinations of cytokinins alone or in combination with auxins such as NAA, IAA or IBA; the shoot regeneration frequency approaching 100%. Sporadic rooting after prolonged incubation exceeding 2 months in MS medium supplemented with 1.0 μM IBA has been recently observed (Kartha, unpublished). Incorporation of activated charcoal in the culture medium did not produce any positive effect with regard to rooting.

The stimulatory effect of GA_3 in association with cytokinins and auxins in shoot differentiation is very well exemplified in the *in vitro* culture of cassava (*Manihot esculenta*) meristems (22, 43). A comparative study was carried out to determine the effect of growth regulators (BA, zeatin, kinetin, 2iP, NAA in the presence and absence of GA_3) on plant regeneration from cassava meristems (43). Out of 4 cytokinins, BA was best suited for plant regeneration followed by zeatin, kinetin and 2iP. In conjunction with 1.0 or 0.1 μM NAA and 0.1 μM GA_3, all levels of BA induced plant regeneration in 100% of the meristems. Shoot development was poor in most combinations involving higher concentrations of NAA. Addition of GA_3 improved plant regeneration and its effect was more pronounced at higher concentrations of NAA. The ideal concentration for achieving 100% plant regeneration from meristems of several genotypes of cassava is: BA 0.5 μM, NAA 1.0 μM, GA_3 0.1 μM (19, 22).

4. *Environmental Conditions (Temperature, Light and Photoperiod).* Incubation of cultures at a constant temperature of 26°C, 16 h photoperiods at 3000 to 4000 lux intensity was quite satisfactory for culturing meristems of chickpea, soybean, cowpea, bean, peanut, sweet potato, coffee (Kartha, unpublished), cassava (19, 22, 43) and tomato (23). But, at an optimal concentration of 0.5 μM BA, only 40 to 50% of pea meristems regenerated shoots when incubated at a constant temperature of 26°C with 16 h photoperiods at 4000 lux light intensity (21). On the other hand, when cultured at an alternating day and night temperature of 20 and 15°C and 16 h photoperiods, 100% of the meristems regenerated shoots (24). The maximum plant regeneration (100%) from cowpea meristems was achieved at a constant temperature of 26°C with 16 h photoperiods as opposed to low frequency plant regeneration (25 to 30%) at an alternating temperature of 20°C day and 15°C night. The critical portion of light spectrum for shoot induction is the blue region while root initiation is stimulated by red light. Therefore, grolux lamps or a combination of white fluorescent and incandescent lamps should be ideal for regeneration of whole plants.

5. *Seasonal Fluctuations.* Seasonal fluctuations of donor plants affect the morphogenetic potential of meristems in culture. Carnation meristems survive better when taken from plants in early spring and autumn than in winter or summer (68). Similarly for most potato varieties meristems isolated in spring and early summer rooted more readily than those taken later in the year (34, Quak, unpublished). Meristems from bulbs and corms are best taken at the end of their dormancy period. However, seasonal variations can be eliminated if the donor plants are grown under controlled environmental conditions.

6. Culture Containers. A number of glass containers have been used for the culture of meristems. The most commonly used ones are pyrex tubes of 10 x 1.2 cm size. We observed that under identical cultural conditions, cassava meristems differentiate directly into plantlets with minimal callusing when they are cultured in small tubes contrary to profuse callusing and delayed shoot regeneration when cultured on 30 ml of nutrient medium contained in 120 ml glass jars. Similarly gladiolus meristems grew better in small tubes than in larger ones (71).

C. *Elimination of Viruses by Meristem Culture*

The most important application of meristem culture lies in the prospect of producing virus-free plants which are genetically identical. Many plant species have been freed of viral pathogens either by meristem culture alone or meristem culture in combination with thermotherapy. There are few review articles in the literature dealing with various aspects of virus elimination by meristem culture techniques (16, 52, 74, 75).

Some of the work done on eliminating viruses from food crops by meristem culture alone or coupled with thermotherapy are listed in Table I.

1. Elimination of Cassava Mosaic Disease by Meristem Culture. Mosaic disease of cassava poses serious problems to the crop and is prevalent in most of the cassava growing areas of the world. The spread of the disease is through the use of infected cuttings. Although the disease is also transmitted by the whitefly (*Bemesia tabaci*), under field conditions the rate of transmission is very low (Nair, unpublished). The exact etiological agent has not been conclusively proved to be a virus although mechanical transmission of the disease has recently been reported (7). The Indian and Nigerian cassava mosaic disease are identical with regard to the type of symptoms produced on susceptible cultivars, host range and transmission either by grafting or through the insect vector.

Meristems were isolated from mosaic-diseased cuttings of the Indian and Nigerian cultivars, Kalikalan and Ogunjobi respectively. They were cultured on MS medium supplemented with B5 vitamins (11) and solidified with 0.8% Difco Bacto agar and containing growth hormones at indicated concentrations: BA 0.5 μM, NAA 1.0 μM and GA_3 0.1 μM. Plantlets derived from the meristems were potted in 3:2:1 mixture of vermiculite, peatmoss and sand and kept under observation for the presence or absence of mosaic symptoms. More than 60% of the plantlets regenerated from meristems of up to 0.4 mm in length did not exhibit any mosaic symptoms, whereas those from meristems exceeding 0.4 mm did. The freedom from the mosaic disease was confirmed by graft inoculation tests using scions from meristem-derived plants and grafting them onto healthy cassava stocks.

In order to increase the recovery of mosaic disease-free plants, the infected cuttings were grown at an elevated temperature (35°C) for 30 days. Upon sprouting all the cuttings exhibited mosaic symptoms, but as the incubation period progressed the symptoms were masked accompanied by profuse vegetative growth. Within 30 days the plants apparently looked healthy, although such plants redeveloped mosaic symptoms when transferred to lower temperature (22°C). Meristems isolated from plants which were held at 35°C for 30 days were cultured *in vitro* as described above. All the regenerated plants (100%) from both cultivars were mosaic disease-free as was evident by visual observation and confirmed by transmission tests. Moreover meristem tips of up to 0.8 mm in length could be cultured to produce 100% mosaic disease free plants. Plantlets regenerated from meristem tips exceeding 0.8 mm showed mosaic symptoms.

The experiments clearly indicated that cassava mosaic disease prevalent in India and Nigeria could be eliminated by meristem culture alone or coupled with heat therapy. The success and frequency of producing mosaic disease-free plants appear to be governed by the size of the meristem tips (maximum 0.4 mm). Apparently, the cassava mosaic agent is prevalent in shoot apical regions of a diseased plant below 0.4 mm. The higher temperature apparently favored plant growth and may have retarded the multiplication or infection by the causative agent. Plant hormonal balance may be a determining factor for stimulating plant growth and providing adverse conditions for the pathogen. The disappearance of symptoms on plants grown at 35°C does not appear to be the result of inactivation of the pathogen since symptoms reappeared on the foliage after the plants were exposed to reduced temperature.

2. *Elimination of Pea Seed-borne Mosaic Virus by Meristem Culture.* Pea seed-borne mosaic virus is transmitted mechanically and also through the seeds. Infection of seeds by the virus in the breeding lines (germplasm) has been observed to be as high as 80%. The use of virus-infected seeds for planting facilitates early infection of the crop and subsequent mechanical spread of the disease. In nature the disease is also transmitted by aphid vectors. Early

infection results in stunting of the plants and concomitant reduction in yield.

Using pea meristem culture technique, seed-borne mosaic virus has been eliminated from over 100 lines of pea. Virus-free plants were regenerated by culturing meristems of 0.3 to 0.4 mm in length without heat treatment. On artificial inoculation with the sap expressed from diseased plants, the virus induces systemic symptoms on healthy pea and necrotic local lesions on *Chenopodium amaranticolor*. Plants regenerated from meristems were bioassayed for the presence or absence of the virus using *C. amaranticolor* as the indicator plant. More than 90% of the plants regenerated from meristems were indexed to be free of seed-borne mosaic virus.

Meristem culture techniques may also be employed for eliminating seed-borne viruses from cowpea, bean and soybean.

II. CRYOPRESERVATION

The continuing search for high yielding varieties of crop plants with resistance to pathogens and pests warrants the availability and maintenance of large collections of germplasm. For crop species which are vegetatively propagated, preservation of germplasm heavily taxes the manpower and land resources. Moreover germplasm collections are subject to attacks by pathogens and pests. For species which are propagated through seeds, although the most efficient way of germplasm storage is in the form of seeds, it is not always feasible because (a) some plants do not produce fertile seeds, (b) some seeds remain viable only for a limited duration, (c) some seeds are very heterozygous and therefore, not suitable for maintaining true-to-type genotypes and (d) seeds of certain species deteriorate rapidly due to seed-borne pathogens.

As plant cells are inherently totipotent, tissue culture techniques could, in conjunction with careful manipulation of cryobiological methods, profitably be used for the storage and preservation of important crop species. Since meristem culture techniques, as discussed earlier, have been widely used for the clonal propagation and production of genotypically identical progenies in a virus-free state, plant meristems offer an excellent system for the long-term preservation of germplasm. Long-term storage of plant tissues should be done at extremely low temperature, preferably at that of liquid nitrogen (-196°C) so as to immobilize all or most of the metabolic activities.

Discovery of chemicals with cryoprotective properties has contributed significantly toward the development of cryopreservation procedures for biological material. The protective action of glycerol on spermatozoa against the adverse effects of freezing was first demonstrated by Polge *et al.* (48). Since this pioneering work a number of low molecular weight neutral solutes have been found to have potential properties of cryoprotection. Among them, the most commonly used one is dimethyl sulfoxide (DMSO) which was originally used to prevent freezing damages to human and bovine red blood cells and to bull spermatozoa (31).

Earlier work of cryopreservation concentrated on material such as cultured mammalian cells (59), rat ovarian tissue (8), chick embryo (54) and mammalian embryos (77). But during the last decade an upsurge of interest in the freeze-preservation of plant cells, tissues and organs is discernible as evidenced by the increasing number of reports on the successful cryopreservation of plant cells and meristems (Table III and IV).

Plant meristems are ideal candidates for cryopreservation aimed at long-term storage of germplasm in a virus-free state. Cultured plant cells are less than ideal systems for this purpose because of the possible genetic changes which may occur during the growth phase before and after freezing.

A. *Cryopreservation Techniques*

The following materials can be considered for cryopreservation:

(1) Shoot apical and lateral meristems.
(2) Cultured plant cells and somatic embryos.
(3) Protoplasts.
(4) Other plant organs such as embryos, endosperms, ovules, nucleus, anthers/pollens and seeds.

The basic techniques of cryopreservation for all materials are more or less similar, however, modifications may have to be made to suit each experimental system. For convenience the steps involved in freeze-preservation are described in 3 protocols, i.e., for plant meristems, cells cultured as suspension and freshly isolated protoplasts.

1. *Plant Meristems.* The technique of a successful cryopreservation of plant meristems involves the following steps:

(1) Aseptic isolation of shoot apical and lateral meristems.

(2) *In vitro* cold conditioning of meristems, if required.

(3) Pre-culturing of meristems on appropriate culture medium supplemented with cryoprotectants for specified periods of time.

(4) Addition of cryoprotectants at specimen preparation stage prior to freezing.

(5) Controlled cooling and freezing of specimens.

(6) Storage in liquid nitrogen.

(7) Rapid thawing of frozen meristems.

(8) Removal of cryoprotectants.

(9) Reculture of retrieved meristems and induction of plant regeneration.

Essentially 2 types of freezing, either by rapid or slow cooling, have been employed for plant meristems (Table IV).

a. Freezing by Rapid Cooling. The first report on the successful freeze-preservation of shoot-apices/meristems was

TABLE III. List of Plant Cells, Callus Cultures and Protoplasts Frozen in Liquid Nitrogen

Species	*Reference*
Cell cultures	
Acer pseudoplatanus	42, 69, 79, 82
Atropa belladona	42
Capsicum annum	82
Catharanthus roseus	Kartha (unpublished)
Datura stramonium	2
Daucus carota	2, 9, 28, 41, 82
Glycine max	2
Nicotiana tabacum	2
Oryza sativa	58
Zea mays	81
Callus cultures	
Populus glerica	56
Saccharum sp.	72
Protoplasts	
Bromus inermis	32
Daucus carota	32, 82
Winter black locust	62

TABLE IV. Summary of the Methodology for the Cryopreservation of Plant Organs

Species	Specimen conditioning and cryoprotectant treatments	Freezing method	Survival (%)	Storage duration	Reference
Apple (Winter buds) (*Malus pumila* - McIntosh red)	hardening at -3° for 20 days	slow freezing -30 to -40°C and immersion in liquid nitrogen (LN)	100% regrowth, 77% budding success	23 months	55
Carnation (shoot-tips) (*Dianthus caryophyllus*)	DMSO 5%	rapid freezing in LN	>30% callus growth, plant	2 months	60
Carnation (shoot-tips) (*Dianthus caryophyllus*)	cold treatment of donor plants at 4°C for 3 days, DMSO 5%	rapid freezing in LN	60% differentiation	1 week to 6 months	61
Carrot (somatic embryos and clonal plantlets) (*Daucus carota*)	dehydration of DMSO-treated (2.5 to 20%) embryos and plantlets	slow freezing 1 to 5°C/min to -100°C frozen in aluminum foil, immersion in LN	>60% plant regeneration	1 hour	80
Cassava (meristems) (*Manihot esculenta*)	preculture on nutrient medium 4-6 days followed by 5% DMSO treatment	rapid dry freezing in LN	30-40% callus growth	1 hour	Kartha (unpublished)

TABLE IV. (continued)

Species	Specimen conditioning and cryoprotectant treatments	Freezing method	Survival (%)	Storage duration	Reference
Cassava (meristems) (*Manihot esculenta*)	Glycerol 10% plus sucrose 5%	rapid freezing in LN	13% plantlets 8% callus + roots	-	3
Chick pea (meristems) (*Cicer arientinum*)	1 day preculture on medium with 4% DMSO freezing in 4% DMSO	slow freezing 0.6°C/min to -40°C followed by immersion in LN	20-30% plant regeneration, multiple shoots	1 hour	20
Pea (meristems) (*Pisum sativum*)	preculture on nutrient medium with 5% DMSO	slow freezing 0.6°C/min to -40°C, immersion in LN	73% plant regeneration after 1 h in LN; >60% plant regeneration after 26 weeks in LN	1 year	24
Potato (shoot-tips) (*Solanum goniocalyx*)	preculture on nutrient medium for 72 h DMSO 10%	rapid freezing in LN	20% survival (callus and shoots)	4 weeks	12

TABLE IV. (continued)

Species	Specimen conditioning and cryoprotectant treatments	Freezing method	Survival (%)	Storage duration	Reference
Strawberry (runner-apex) (*Fragaria X ananassa*)	DMSO 16%	rapid freezing in LN on coverglass preceded by freezing to -20 to -30°C	60-80% shoot regeneration	-	57
Strawberry (meristems from in vitro propagules)	2 day preculture on nutrient medium with 5% DMSO, freezing solution 5% DMSO	slow freezing 0.84°C/min to -40°C, immersion in LN	95% plant regeneration followed by mass proliferation of shoots	8 weeks	25
Tomato (seedlings) (*Lycopersicon esculentum*)	DMSO 10 to 15%	LN vapor phase freezing	>30% callus and plants	1 day	13

made by Seibert (60) using carnation (*Dianthus caryophyllus*) shoot-tips. Freshly isolated apices were incubated for 4 days in the dark at 26°C on agar-solidified MS nutrient medium supplemented with IAA and kinetin at 0.1 and 0.5 mg/l respectively. At the end of the incubation period, the shoot-apices were placed in a 4 ml vial containing 0.5 ml of a freezing solution consisting of 5% DMSO in liquid nutrient medium. Freezing was accomplished by pouring liquid nitrogen directly into the freezing vial and then dipping the vial into an open Dewar flask filled with liquid nitrogen. The frozen meristems were thawed by plunging the vials into a 37°C water bath. The survival was determined by reculturing the meristems on nutrient medium and was scored on the basis of callus formation, growth or chlorophyll formation and shoot regeneration. Based on the above criteria, a survival rate of 33% including one plant regeneration was obtained.

The rapid freezing technique, with modifications, has also successfully been used with potato and strawbery apices (12, 57). When employing the rapid freezing technique, the viability of meristematic cells appeared to be maintained by preventing the growth of intracellular ice crystals formed during rapid cooling by rapidly passing the tissue through the temperature zone in which lethal crystal growth occurs (Luyet's mechanism) (61). However, the rapid freezing technique cannot be applied to all plant species.

b. *Freezing by slow cooling.* Slow cooling permits the flow of water from the cell to the outside thereby promoting extracellular ice formation instead of the lethal intra-cellular freezing. We compared slow and rapid freezing with pea and strawberry meristems and concluded that slow freezing is preferable to rapid freezing in order to retain maximum viability (24, 25). The following examples explain the methodology developed for the cryopreservation of pea and strawberry meristems.

i. Pea meristems. Meristems isolated from aseptically germinated pea seeds were precultured for 48 hrs on nutrient medium (B5 + 0.5 μM BA) supplemented with 5% DMSO. The cultures were incubated at a diurnal temperature of 20°C day and 15°C night at 16 h photoperiods and 70% relative humidity. At the end of the preculture, the jars containing the meristems were kept in an ice bath for 2 hrs after which time the meristems were removed aseptically and transferred to a centrifuge tube containing 10 ml of chilled cryoprotectant-free liquid B5 medium in an ice bath. Equal volumes of chilled culture medium containing 10% DMSO were added step-wise over a 30 min period until the desired final concentration of 5% was reached, and allowed to equilibrate for another 10 min. Batches of 20 meristems were then removed with a Pasteur pipette and transferred to 1.2 ml sterile prescored glass ampoules along with exactly 1 ml of the culture medium containing 5% DMSO (freezing solution). All the ampoules were flame-sealed except one to which the temperature probe was to be inserted. Carefully controlled cooling and freezing was achieved in a Linde BF-4-1-6 freezing system (Union Carbide). The freezing chamber was precooled to 0°C prior to the introduction of the ampoules. In the freezing chamber the ampoules were arranged at equal distances around the one containing the temperature probe. The samples were frozen at cooling velocities ranging from 0.5 to 1.0°C/min down to a terminal freezing temperature of -40°C. The ampoules were then removed and stored in liquid nitrogen immersion phase (-196°C) in a Linde LR-40 refrigerator (Union Carbide). After 1 h storage in liquid nitrogen one batch of ampoules was taken out and thawed rapidly for 90 sec in a water bath at 37°C without agitation and immediately returned to an ice bath while other sets of ampoules were left in storage for periodic assays of viability. The contents of the thawed ampoules (meristems+freezing solution) were transferred to a centrifuge tube in an ice bath and diluted with 6 equal volumes of chilled B5 medium step-wise over a period of 30 min with periodic shaking. Finally the meristems were washed 4 times with B5 medium and cultured in 2.5 ml of the culture medium solidified with 0.8% agar in 10 x 1.2 cm pyrex tubes and incubated in a growth cabinet as described earlier. The same procedure was followed for culturing the meristems stored in liquid nitrogen over extended periods of time.

One hundred per cent plant regeneration was obtained from meristems which were not subjected to freezing. Preculturing the meristems in the presence of 5% DMSO and frozen at a rate of 0.6°C/min resulted in highest survival rate (73%). Even after 26 weeks of storage in liquid nitrogen more than 60% of the meristems regenerated plants which subsequently were grown to maturity. Cooling rates which were above or below 0.6°C/min resulted in lesser recovery rate. Rapid freezing was lethal to pea meristems. Cryoprotectants such as glycerol or ethylene glycol were ineffective.

ii. Strawberry meristems. As discussed under 'Meristem culture' a population of strawberry plantlets was mass propagated *in vitro* from a few runner-tip meristems. The procedure utilized for the slow freezing was identical to the one described for pea meristems. The strawberry meristems isolated from *in vitro* shoots were precultured on medium supplemented with 5% DMSO or glycerol as cryoprotectants, frozen either slowly at cooling velocities of 0.5 to 1.0^{o}C/min to -40°C or rapidly and stored in liquid nitrogen. Maximum viability and plant regeneration (95%) was obtained when the meristems were precultured on medium supplemented with 5% DMSO, and then frozen at a cooling rate of 0.84^{o}C/min. Viability of 35% was observed when the meristems precultured on 5% glycerol-supplemented medium for 3 days were frozen at a cooling rate of 0.94^{o}C/min. Rapid freezing and rapid dry freezing using either DMSO or glycerol resulted in reduced viability. Viability and plant regeneration frequency of 95% was obtained after the meristems were in storage for one week in liquid nitrogen. From samples assayed after 8 weeks of storage in liquid nitrogen, more than 55% of the meristems regenerated into plantlets.

2. *Cryopreservation of Cell Cultures.* A number of cell cultures have successfully been frozen and stored in liquid nitrogen (Table III). Various steps involved in the cryopreservation of cell cultures are basically the same as those explained for meristems. Initially a fast dividing cell culture line is established and the cytotoxicity to various cryoprotectants determined. The cells are concentrated to a dense mass in an ice bath by allowing the suspension to settle and removing the excess culture medium or by centrifugation for 3 min at 100 x g. One ml of the sample is distributed into each cryogenic ampoule and the supernatant medium pipetted out. Appropriate cryoprotectants in 1 ml quantities at chosen concentrations are gradually added to each ampoule in 4 installments over ca. 30 min and allowed to equilibrate for another 15 to 30 min. The ampoules are flame-sealed and frozen according to the procedure explained earlier at cooling rates of 1 to 2°C/min down to either -80 or -100°C and stored in liquid nitrogen. Thawing and washing are done as explained earlier. The cells are recultured on appropriate nutrient medium on a shaker at 120 rpm maintaining a cell density of 10^5 to 10^6 cells/ml. The technique employed for assessing the viability is discussed later.

3. *Freeze-Preservation of Protoplasts.* Very little work has been done on the freeze-preservation of plant protoplasts. Since protoplasts are bound by only plasma membrane, they are very sensitive to freeze-preservation

procedures and also to osmotic shock and mechanical injury. The use of plant protoplasts in studies on the freeze-preservation of plant cells affords a system which permits elucidation of the role the cell wall plays in cell survival at low temperatures.

Mazur and Hartmann (32) studied the freezing of plant protoplast and the effect of cell wall removal using 2 species of plants, *Bromus inermis* and *Daucus carota*. Protoplasts isolated from 2 or 3 day old subcultures were resuspended by drop-wise addition of the osmoticum containing the chosen concentrations of glycerol. One half of the resulting protoplast suspension was placed into 16 x 150 mm glass screw cap test tubes and allowed to equilibrate with the cryoprotectant for 15 min. Glycerol and DMSO were utilized as cryoprotectants.

After equilibration the tubes were placed into a refrigerator at $4^{o}C$ for 3 hrs, and were frozen in a 3-step sequence (-20°, -60° and $-196^{o}C$). The results indicated that 68% of the carrot protoplasts survived when 0.7 M glycerol was utilized. A survival of 44% was obtained when DMSO was employed as cryoprotectant.

B. Viability Tests

1. Meristems. The most convincing test for viability of frozen-thawed meristems would be their ability to regenerate into plants upon return to culture. Viability based on the ability of meristem cells to turn green or form callus cannot be considered as absolute proof from an applied point of view. If only a few constituent cells of meristems survive, they might regrow into callus and after a long period of incubation differentiate shoots. This reaction would introduce the possibility of genetic instability. Recovering callus from frozen-thawed meristems may not be of much value in systems which are recalcitrant to the induction of plant regeneration *in vitro*.

2. Cell cultures. A number of tests are available to determine the viability of frozen-thawed cells.

a. Fluorescein diacetate (FDA) staining. The FDA staining technique was originally devised by Widholm (78) and is based on the observation that only living cells would be stained with FDA and emit fluorescence under UV light. A 0.5% stock solution of FDA is prepared in acetone and stored at $-20^{o}C$. 0.5 ml of this solution is added to 25 ml of chilled culture medium in order to obtain a final concentration of 0.01%. A drop of this solution is mixed with a drop of frozen-thawed cell suspension on a slide, the slide is covered with a cover slip and examined after 15 to 20 min with a research microscope with UV attachment. Initially the cells and cell aggregates are scored under tungsten light and then switched to UV. Since only the viable cells emit fluorescence, the viability is expressed as per cent survival.

b. Triphenyl tetrazolium chloride (TTC) method. The TTC method was adapted by Steponkus and Lanphear (67). The survival is based on the reduction of TTC by viable cells by mitochondrial activity, to form formazan, a water insoluble red compound. This reduction is considered to be quantitative and the formazan is made soluble by addition of ethanol and the supernatant can thus be examined spectrophotometrically. The procedure involves the following steps (4):

(1) Prepare a 0.6% TTC solution in 0.5 M phosphate buffer (8.9 g/l Na_2HPO_4 + 6.8 g/l KH_2PO_4).
(2) Incubate ca. 100 mg cells in 3 ml TTC solution for 15 hrs at 30°C.
(3) Remove the TTC solution with a Pasteur pipette and wash the cells with double distilled water.
(4) Centrifuge the cells and extract with 7 ml of 95% ethanol in a 80°C water bath for 5 min.
(5) Cool the extract and make up the volume to 10 ml with 95% ethanol.
(6) Record the absorbance of the pink solution with a spectrophotometer at 530 nm.
(7) Express the absorbance values of the frozen-thawed cells as per cent survival over the control.

Other parameters such as mitotic indices, cell number, cell culture volume, dry and fresh weight, plasmolysis and deplasmolysis, plating efficiency, leakage of electrolytes, could also be used and compared with the unfrozen control in order to assess the viability. A combination of various assays is advisable for a true representation of viability results.

3. Protoplasts. The FDA staining could be used to assess viability in addition to the plasmolysis and deplasmolysis tests (62). Mazur and Hartmann used tryptan blue dye exclusion method for the viability assay of frozen-thawed protoplasts (32). Live protoplasts exclude dye while dead protoplasts do not.

4. *Factors affecting the Viability of Frozen Tissue/ Cells*. The most important factors which influence the success of cryopreservation of plant meristems or cultured cells are: (1) nature, type and physiological status of cells prior to freezing, (2) type of cryoprotectants, (3) cooling rates and freezing methods, (4) terminal freezing temperature prior to storage in liquid nitrogen, (5) storage temperature and (6) thawing methods.

The physiological state of either the meristems or the cultured cells would determine, to a great extent, the success or failure of a cryopreservation program. In general, actively dividing meristematic cells or cells in an exponential phase with dense cytoplasm are more suitable and such cells can withstand the freezing procedures, better than cells in the late lag phase or cells with large vacuoles. Meristems isolated from plants which tolerate cold conditions in nature can survive freezing better than those from tropical plants. The amount of cellular water contained in the cells would also play a role in the success of cryopreservation; partly dehydrated cells being better than succulent or highly turgid cells. Although many factors contribute to the freeze-killing of cells the most important one which plays a key role is the intracellular ice formation during progressive cooling.

Preculturing the meristems on nutrient medium supplemented with 5% DMSO for a period of 48 hrs followed by slow freezing at a rate of less than 1.0^{o}C/min down to -40°C has been found to be optimal for the maximum survival of pea as well as strawberry meristems (24, 25). The cooling rate optima should be worked out for each species in conjunction with the concentration and exposure duration of the cryoprotectant in the freezing solution. Although rapid freezing by direct immersion of cryoprotectant-treated meristems in liquid nitrogen has resulted in the survival of carnation shoot tips (60), shoot apices of potato (12), and runner apices of strawberry (57), this method was not suitable for pea and chickpea meristems. However, some survival (5 to 6%) was obtained with meristems from *in vitro* shoots of strawberry (25). The majority of available information on the cryopreservation of cell cultures and findings from this laboratory strongly indicate that slow freezing is preferable to rapid freezing. During slow freezing, depending upon the permeability of the plasma membrane to water, the cells equilibrate by transfer of water to external ice whose vapor pressure is lower than the supercooled cell interior. On the other hand, during rapid freezing, if the cells are not sufficiently permeable to water, they will not remain in equilibrium with external ice but will continue to supercool until they eventually equilibrate by internal freezing. Rapid freezing, in addition to producing intracellular ice crystals, also produces small crystals which are likely to enlarge due to recrystallization or grain growth during warming because of their high surface energies.

The extent of cytotoxicity varies with the type and concentration of the cryoprotectant and the nature of plant species used. Cultured cells exhibit delayed toxicity to cryoprotectants as opposed to freshly isolated meristems. Since DMSO permeates very rapidly into the cells in comparison to glycerol, a concentration of 5 to 10% (by volume) and an exposure duration of 30 min in the freezing solution is advisable for most systems. In the case of glycerol the concentration and exposure duration may be increased anywhere from 30 to 60 min or as necessary since it permeates into the cells only slowly. A combination of cryoprotectants has been demonstrated to protect the cell cultures more effectively than single cryoprotectants alone (10).

The terminal freezing temperature prior to storage in liquid nitrogen could also affect the viability. The terminal freezing temperature would vary according to the rate at which the specimen is cooled (Kartha, unpublished). Nevertheless, the terminal freezing temperature should be the one at which all the freezable water from the cell is excluded and allowed to freeze extracellularly. For pea and strawberry meristems, this temperature has been determined to be $-40^{\circ}C$ for a cooling rate of less than $1.0^{\circ}C/min$. For cell cultures the terminal freezing temperature could vary from -40 to -80 or $-100^{\circ}C$ for cooling rates of 1 to $5^{\circ}C/min$.

Storage temperature is very critical for assuring viability for extended storage of germplasm. At the temperature of liquid nitrogen ($-196^{\circ}C$) almost all the metabolic activities of cells are arrested and are affected little by passage of time. Our experiments with pea and strawberry meristems indicated that meristems remain viable in liquid nitrogen storage for extended periods of time. On the other hand, storing the frozen samples at $-86^{\circ}C$ resulted in reduced viability (Kartha, unpublished).

Rapid thawing in a water bath at 37 to $40^{\circ}C$ gives maximum viability for samples which were frozen by slow cooling. On the other hand, slow warming has been beneficial for some of the animal cells. Most of the work on cryopreservation of cultured cells and meristems support the beneficial effects of rapid thawing.

III. SUMMARY

Meristem culture is a proven technique for the production of disease-free plants and their continued vegetative propagation in a genetically stable state. Many important food crops have been freed of systemic viral pathogens and a wide range of plants of horticultural importance has been mass propagated by this method. Plantlets regenerated by the *in vitro* culture of meristems may also be used as test-screening and selection systems for pathogens, chemicals and minerals causing toxicity problems in nature. Short-term storage of germplasm of vegetatively propagated crops can successfully be accomplished by meristem culture technique. But only cryopreservation of meristems can offer an ideal and realistic method for the long-term storage of germplasm in a genetically stable and at the same time in a virus-free condition. This, in addition to being able to establish disease-free gene banks, also permits the international exchange of virus-free germplasm.

REFERENCES

1. Alconero, R., Santiago, A.G., Morales, F., and Rodriguez, F., *Phytopathology 65*, 769 (1975).
2. Bajaj, Y.P.S., *Physiol. Plant. 37*, 263 (1976).
3. Bajaj, Y.P.S., *Crop Imp. 4*, 48 (1977).
4. Bajaj, Y.P.S., and Reinert, J., *In* "Applied and Fundamental Aspects of Plant Cell, Tissue, and Organ Culture" (J. Reinert and Y.P.S. Bajaj, eds.), p. 757. Springer-Verlag, Berlin, Heidelberg, New York, (1977).
5. Belkengren, R.O., and Miller, P.W., *Plant Dis. Reptr. 46*, 119 (1962).
6. Berg, L.A., and Bustamante, M., *Phytopathology 64*, 320 (1974).
7. Bock, K.R., and Guthrie, E.J., *Plant Dis. Reptr. 62*, 580 (1978).
8. Deansley, R., *Proc. Roy. Soc. London. Ser. B 147*, 412 (1957).
9. Dougall, D.K., and Wetherell, D.F., *Cryobiology 11*, 410 (1974).
10. Finkle, B.J., and Ulrich, J.M., *Plant Physiol. 63*, 598 (1979).
11. Gamborg, O.L., Miller, R.A., and Ojima, K., *Exp. Cell Res. 50*, 151 (1968).

12. Grout, B.W.W., and Henshaw, G.G., *Ann. Bot. 42*, 1227 (1978).
13. Grout, B.W.W., Westcott, R.J., and Henshaw, G.G., *Cryobiology 15*, 478 (1978).
14. Hartmann, R.D., *Phytopathology 64*, 237 (1974).
15. Hendre, R.R., Mascarenhas, A.F., Nadgir, A.L., Pathak, M., and Jagannathan, V., *Indian Phytopath. 28*, 175 (1975).
16. Hollings, M., *Ann. Rev. Phytopath. 3*, 367 (1965).
17. Jones, O.P., and Vine, S.J., *J. Hort. Sci. 43*, 289 (1968).
18. Kaiser, W.J., and Teemba, L.R., *Plant Dis. Reptr. 63*, 780 (1979).
19. Kartha, K.K., and Gamborg, O.L., *Phytopathology 65*, 826 (1975).
20. Kartha, K.K., and Gamborg, O.L., *In* "Diseases of Tropical Food Crops" (H. Maraite and J.A. Meyer, eds.), P. 267. Proc. Int. Symp. Universite Catholique, Louvaine-La-Neuve, Belgium (1978).
21. Kartha, K.K., Gamborg, O.L., and Constabel, F., *Z. Pflanzenphysiol. 72*, 172 (1974).
22. Kartha, K.K., Gamborg, O.L., Constabel, F., and Shyluk, K., *Plant Sci. Lett. 2*, 107 (1974).
23. Kartha, K.K., Champoux, S., and Pahl, K., *J. Amer. Soc. Hort. Sci. 102*, 346 (1977).
24. Kartha, K.K., Leung, N., and Gamborg, O.L., *Plant Sci. Lett. 15*, 7 (1979).
25. Kartha, K.K., Leung, N., and Pahl, K., *J. Amer. Soc. Hort. Sci. 105*, 481 (1980).
26. Kassanis, B., *Ann. Appl. Biol. 45*, 422 (1957).
27. Kassanis, B., and Varma, A., *Ann. Appl. Biol. 59*, 447 (1967).
28. Latta, R., *Can. J. Bot. 49*, 1253 (1971).
29. Leu, L.S., *Rep. Taiwan Sugar Exp. Sta. 57*, 57 (1972).
30. Limasset, P., and Cornuet, P., *Compt. Rend. 228*, 1971 (1949).
31. Lovelock, J.E., and Bishop, M.H.W., *Nature 183*, 1394 (1959).
32. Mazur, R.A., and Hartmann, J.X., *In* "Plant Cell and Tissue Culture, Principles and Applications" (W.R. Sharp, P.O. Larsen, E.F. Paddock and V. Raghavan, eds.), p. 876, Ohio State Univ. Press, Columbus (1978).
33. McGrew, J.R., *Phytopathology 55*, 480 (1965).
34. Mellor, F.C., and Stace-Smith, R., *Can. J. Bot. 47*, 1617 (1969).
35. Miller, P.W., and Belkengren, R.O., *Plant Dis. Reptr. 47*, 298 (1963).

36. Morel, G., and Martin, C., *Compt. Rend. 235*, 1324 (1952).
37. Morel, G., Martin, C., and Muller, J.F., *Ann. Physiol. Végétale 10*, 113 (1968).
38. Mori, K., *Japan Agric. Res. Quart. 6*, 1 (1971).
39. Mullin, R.H., Smith, S.H., Frazier, N.W., Schlegel, D.E. and McCall, S.R., *Phytopathology 64*, 1425 (1974).
40. Murashige, T., and Skoog, F., *Physiol. Plant. 15*, 473 (1962).
41. Nag, K.K., and Street, H.E., *Nature 245*, 70 (1973).
42. Nag, K.K., and Street, H.E., *Physiol. Plant. 34*, 254 (1975).
43. Nair, N.G., Kartha, K.K., and Gamborg, O.L., *Z. Pflanzenphysiol. 95*, 51 (1979).
44. Nielsen, L.W., *Phytopathology 50*, 841 (1960).
45. Over de Linden, A.J., and Elliott, R.F., *New Zealand J. Agr. Res. 14*, 720 (1971).
46. Pennazio, S., and Redolfi, P., *Potato Res. 17*, 333 (1974).
47. Pett, B., *Z. Pflanzenschutz 10*, 81 (1974).
48. Polge, C., Smith, A.U., and Parkes, A.S., *Nature 164*, 666 (1949).
49. Putz, C., *Ann. Phytopath. 3*, 493 (1971).
50. Quak, F., *Advan. Hort. Sci. Appl. 1*, 144 (1961).
51. Quak, F., *Meded. Dir. Tuinbouw. 27*, 545 (1964).
52. Quak, F., *In* "Applied and Fundamental Aspects of Plant Cell, Tissue, and Organ Culture" (J. Reinert and Y.P.S. Bajaj, eds.), p. 598. Springer-Verlag, Berlin, Heidelberg, New York (1977).
53. Quiot, J., and Messiaen, C., *Actas Congr. Uniao Fitopatol Mediterr. 1972*, 429 (1972).
54. Rey, L.R., *Actualité Sci. Industr. Paris.* p. 1279 (1959).
55. Sakai, A., and Nishiyama, Y., *HortScience 13*, 225 (1978).
56. Sakai, A., and Sugawara, Y., *Plant Cell Physiol. 13*, 1129 (1972).
57. Sakai, A., Yamakawa, M., Sakata, D., Harada, T., and Yakuwa, T., *Low Temp. Sci. Ser. B36*, 31 (1978).
58. Sala, F., Cella, R., and Rollo, F., *Physiol. Plant. 45*, 170 (1979).
59. Scherer, W.F., and Hoogasian, A.C., *Proc. Soc. Exp. Biol. Med. 87*, 480 (1954).
60. Seibert, M., *Science 191*, 1178 (1976).
61. Seibert, M., and Wetherbee, P.J., *Plant Physiol. 59*, 1043 (1977).
62. Siminovitch, D., *Plant Physiol. 63*, 722 (1979).
63. Sip, V., *Potato Res. 15*, 270 (1972).

64. Stace-Smith, R., and Mellor, F.C., *Phytopathology 58*, 199 (1968).
65. Stace-Smith, R., and Mellor, F.C., *Phytopathology 60*, 1857 (1970).
66. Staritsky, G., *Trop. Root Tuber Crops Newsletter 7*, 38 (1974).
67. Steponkus, P.L., and Lanphear, F.O., *Plant Physiol. 42*, 1423 (1967).
68. Stone, O.M., *Ann. Appl. Biol. 52*, 199 (1963).
69. Sugawara, Y., and Sakai, A., *Plant Physiol. 54*, 722 (1974).
70. Svobodova, J., *Proc. 10th Intern. Botan. Congr. Edinburgh* 485 (1964).
71. Tramier, R., *Acad. Agr. Fr. 6 Oct.* 918 (1965).
72. Ulrich, J.M., Finkle, B.J., Moore, P.H., and Ginoza, H., *Cryobiology 16*, 550 (1979).
73. Vine, J., *J. Hort. Sci. 43*, 293 (1968).
74. Walkey, D.G.A., *In* "Frontiers of Plant Tissue Culture 1978" (T.A. Thorpe, ed.), p. 245. University of Calgary Press, Calgary (1978).
75. Walkey, D.G.A., *In* "Towards Plant Improvement by In Vitro Methods" (P.K. Evans, ed.), Academic Press, New York (in Press).
76. Walkey, D.G.A., Cooper, V.C., and Crisp, P., *J. Hort. Sci. 49*, 273 (1974).
77. Whittingham, D.G., Leibo, S.P., and Mazur, P., *Science 178*, 411 (1972).
78. Widholm, J.M., *Stain Tech. 47*, 189 (1972).
79. Withers, L.A., *Cryobiology 15*, 87 (1978).
80. Withers, L.A., *Plant Physiol. 63*, 460 (1979).
81. Withers, L.A., and King, P.J., *Plant Physiol. 64*, 675 (1979).
82. Withers, L.A., and Street, H.E., *Physiol. Plant. 39*, 171 (1977).
83. Yang, H.J., and Clore, W.J., *HortScience 11*, 474 (1976).
84. Yora, K., and Tsuchizaki, T., *Ann. Phytopathol. Soc. Jap. 27*, 219 (1962).

CYTOGENETIC TECHNIQUES

David A. Evans
Sandra M. Reed

Campbell Institute for Research and Technology
Cinnaminson, New Jersey

I. INTRODUCTION

Basic cytogenetic techniques for use with higher plants have been summarized in recent texts (18, 82). These texts are particularly useful for preparation of solutions used in staining and fixation, and include schedules for most cytogenetic methods that can be easily modified for use with higher plants. Unfortunately, neither of the above texts deals exclusively with plants or with the unique aspects of plant cells *in vitro*. Despite the appearance of interesting chromosomal abnormalities in cell cultures and in plants regenerated from cell cultures, cytogenetic analysis has not proceeded with this unique material. It is hoped that as interesting hybrid material is recovered from protoplast fusion experiments, and as large numbers of plants are regenerated from cultured somatic tissues or anthers that cytogenetic analysis will be used more frequently. In this chapter we will review cytogenetic techniques for use with cultured plant cells and with plants regenerated from cultured cells, as well as discuss the cytogenetic phenomena most often encountered in cultured cells.

ISBN 0-12-690680-7

II. LITERATURE REVIEW

A. *Cytogenetic Analysis of Cultured Plant Cells*

Chromosome instability of plant callus and cell suspension cultures has often been reported (17). These instabilities include numerical or structural changes in chromosomes. Numerical changes have been examined most often (Fig. 1A). A detailed assessment of chromosome rearrangements or other structural changes requires sophisticated karyotype analysis or banding techniques. Callus cells have particularly variable chromosome numbers, some of which are listed in Table I. Chromosome variability *in vitro* may appear as early as the first subculture (34), but is more often associated with long-term cultures (83). Numerical variation may be euploid or aneuploid. Endomitosis and endoreduplication are principal causes of polyploidy (17), while aneuploidy may result from translocations, mitotic non-disjunction, or deletions. Nearly all of the species listed in Table I have both euploid and aneuploid changes in chromosome number. Bayliss has discussed growth rates of diploid versus tetraploid cells in mixoploid cultures (4). These experiments suggest that under conditions of phosphate-limited growth and extended subculture interval the frequency of tetraploid mitoses increases. In some long-term cell cultures, aneuploid cells, with a selective advantage over the diploid cells, may completely overrun the culture resulting in a new stabilized chromosome number (24). Kao *et al.* have reported a culture of soybean (2n = 40) in which all cells had 37 chromosomes (42).

Structural changes and concomitant loss of genetic material may be an even more frequent source of variation in cultured cells than changes in chromosome number (91). Unfortunately, detailed analysis of chromosomal rearrangements have been completed only in a few cases with species with a small number of large chromosomes. Nonetheless, abnormal chromosomes, with structural alterations, have been reported for many plant species. Dicentric chromosomes have been reported in *Triticum monococcum* and *Triticum aestivum* (42), *Haplopappus gracilis* (91), *Daucus carota* (4), and *Hordeum* spp. (67). Anaphase bridges and acentric fragments in mitosis, indicative of translocations, have been reported for numerous species, e.g., *Allium sativum* (63). Microchromosomes and ring chromosomes have been reported in *Haplopappus gracilis* (84). Additional unique chromosomes have been reported in suspension cultures (megachromosomes [42]; isochromosomes

TABLE I. Chromosome Instability in Callus and Suspension Cultures

Species	2N of plant	2N of callus	Reference
Acer pseudoplatanus	52	50-250	5
Allium cepa	16	16-64	11
Allium sativum	16	8-32	63
Allium tuberosum	32	21-31	11
Cryptomeria japonica	18	15-18	51
Daucus carota	18	9-72	55
D. carota	18	14-50	88
D. carota	18	18-72	3
Haplopappus gracilis	4	3-32	84
Haworthia spp.	14	14-28	65
Hordeum vulgare x H. jubatum	21	19-150	67
Lilium longiflorum	24	24-25	83
Nicotiana tabacum	48	48-192	15
N. tabacum	48	27-126	61
N. tabacum	48	35-193	76
Paeonia suffruticosa	10	10-40	19
Pisum sativum	14	14-106	93
Prunus amygdalus	16	9-32	52
Saccharum spp.	108-128	71-300	36
Triticum aestivum	42	21-84	64
Vicia faba	12	11-60	74
Vicia hajastana	10	10-36	85
Zea mays	20	20-80	2
Haploid Species	*N of plant*		
Crepis capillaris	3	3-10	75
Pelargonium zonale	9	9-36	6
Nicotiana suaveolens x N. langsdorffii	25	10-53	34

[24]). Extensive karyotypic analysis has been completed only for *Crepis capillaris*, 2n = 6 (75), and to some extent with *Haplopappus gracilis*, 2n = 4 (84). In both species, each chromosome pair has unique morphology. Abnormal karyotypes were common in cultures of *H. gracilis*. In *H. gracilis*, deletions occurred most often in the long arm of chromosome

II, and among aneuploid cells an extra chromosome I appears more often than a chromosome II. These two observations support the hypothesis that some genetic material on chromosome I is advantageous for survival *in vitro* (84). In *C. capillaris*, chromosome identification is aided by presence of a satellite on one pair of chromosomes. Seven abnormal karyotypes, which arose via translocations, were identified in callus cultures (75). From data presented, it was obvious that the distribution of rearrangements was not random. One particular region of one chromosome (SAT) was preferentially involved in chromosome breaks in *Crepis* (75). Chromosome rearrangements have also been detected in callus cultures of *Haworthia* spp. (65).

B. *Cytogenetic Analysis of Plants Regenerated from Cell Cultures*

In many cases, plants regenerated from somatic cell cultures contain a normal chromosome complement (Fig. 1B) (29). In some species, only diploid plants have been recovered from mixoploid callus cultures (e.g., 52, 55), suggesting that diploid cells are selectively favored during plant regeneration. Nonetheless, aneuploid and polyploid plants have been recovered from a number of species cultured *in vitro* (Table II).

TABLE II. Chromosome Numbers of Plants Regenerated from Callus or Suspension Cultures

Species	*2N of plant*	*2N of regenerated plant*	*Reference*
Asparagus officinalis	20	20, 40	92
Brassica oleracea	18	18, 36	37
Lilium longiflorum	24	24, 25, 48	83
Nicotiana alata	18	18, 19	61
N. tabacum	48	57-90	76
Lolium multiflorum x L. perenne	14	15-30	1
Oryza sativa	24	24, 48	60
Saccharum spp.	108-128	17-120	36
Triticum aestivum	28	6-56	7
Hordeum vulgare x H. jubatum	21	15-30	67

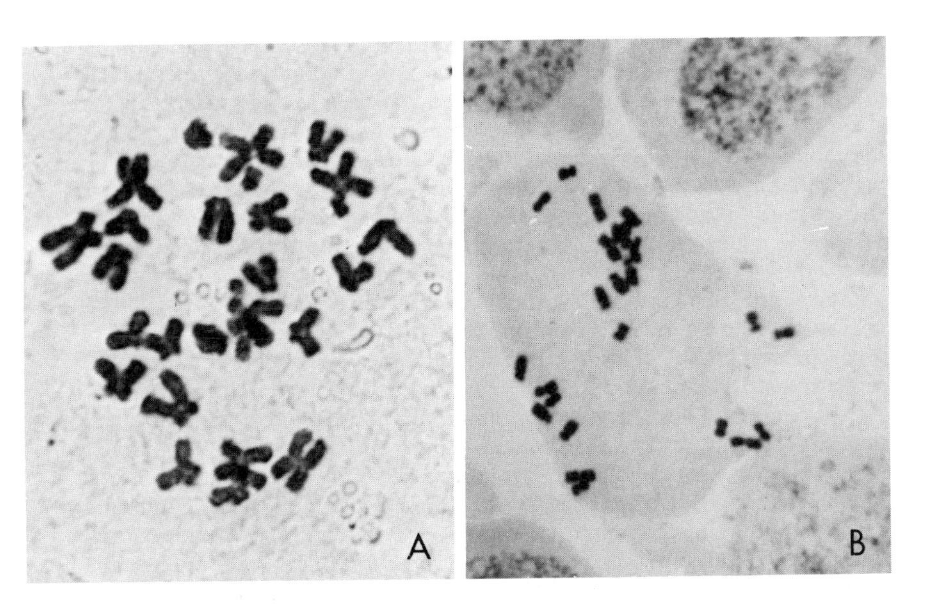

FIGURE 1. Mitotic analysis of cultured cells. (A) Sample metaphase preparation from aneuploid suspension culture of Vicia hajastana (2n = 10) with 26 chromosomes. (B) Root tips of plants regenerated from haploid suspension cultures of Nicotiana tabacum (2n = 24).

Tetraploid plants have been recovered from cultures initiated from diploid tissues of *Asparagus officinalis* (92), *Brassica oleracea* (37), *Lilium longiflorum* (83), *Nicotiana tabacum* (20), *Oryza sativa* (60), and *Pelargonium zonale* (6). Aneuploid plants have been reported less frequently. Aneuploids have been recovered from *L. longiflorum* (83), *N. tabacum* (76), *Saccharum* spp. (36), and *Lolium multiflorum* (1). Increased ploidy has been more common from cultured anthers (16). Diploid and tetraploid plants are common among plants regenerated from cultured anthers. Polyploid plants may arise from non-haploid cells that resulted from fusion of pollen grain generative and vegetative nuclei or growth of unreduced microspores (16). Once non-haploid cells arise, they may have selective advantage over haploid cells *in vitro*. Aneuploid plants have been recovered from anthers of *Solanum nigrum* (35).

Cytogenetic analysis has been reported for a small number of plants regenerated from cell cultures. Sheridan has reported the presence of an extra isochromosome, presumably representing a secondary trisomic, in a regenerated line of *Lilium longiflorum* (83). The trisomic condition ($2n + 1 = 25$) was observed in the callus culture used for plant regeneration. Sacristan and Melchers reported that aneuploid plants were recovered in high frequency from long-term cultures of both tumorous and non-tumorous lines of *N. tabacum* (76). The range of chromosome variability was greater for callus cells than for regenerated plants. The callus cultures used were up to 8 years old. Fragmented chromosomes were common in both callus cultures and regenerated plants, and all regenerated plants were sterile.

Orton has also reported that aneuploidy is greater in suspension and callus cultures than in regenerated plants of *Hordeum vulgare* x *H. jubatum* hybrids (67). By examining the ratio of long to short arms of chromosomes, Orton concluded that chromosome rearrangements had occurred in regenerated plants. In addition, metaphase I of meiosis of regenerated plants contained a higher frequency of chromosome associations, i.e., bivalents and multivalents, than the original hybrid tissue. The multivalents probably reflect presence of reconstructed chromosomes.

Meiotic analysis has also been completed for regenerates of ryegrass (1). A higher frequency of multivalents was observed for regenerated plants than for sexually produced hybrids of *Lolium multiflorum* x *L. perenne*. The presence of hexavalents was unique in sexual hybrids. One trisomic plant ($2N + 1 = 15$) was isolated. This plant was male sterile and upon chromosome doubling contained hexavalents. This parasexual trisomic was completely different from primary trisomics isolated by conventional means, suggesting a combination defect: trisomy and undefined chromosomal rearrangement. Chromosome rearrangements were evident from meiotic analysis of a number of regenerated ryegrass plants (1).

Despite the paucity of mitotic and meiotic analysis of plants regenerated from cell cultures, the published reports emphasize the usefulness of cytogenetic studies with regenerated plants.

Cultures of plant protoplasts are subject to the same sources of chromosome variation as are callus or cell suspension cultures. Nonetheless, plants with diploid chromosome number have been regenerated from mesophyll protoplast cultures (23), as well as protoplasts derived from cell cultures (29).

During protoplast fusion experiments cells may remain in culture for up to six months before plants are recovered (32). In this hybrid material, chromosome variability may be greater than the variability observed in plants regenerated directly from cell cultures. Examination of a summary of plants recovered from protoplast fusion experiments (Table III) suggests that chromosome variability, particularly with wide species combinations, is great. Aneuploid plants have been documented in plants recovered from at least 14 interspecific protoplast fusion experiments. It has been suggested that some of these plants may be the result of multiple fusion events (87), but if the chromosome counts published to date are accurate, aneuploidy is quite common following protoplast fusion.

Cytogenetic methods have been used to verify and analyse somatic cell hybrids. Differential nuclear staining of early fusion products has been used to estimate fusion frequencies. For example, using dilute carbol fuchsin, nuclei of *Nicotiana glauca* stain much darker than nuclei of *Glycine max* within the same cell (41).

Presence of heterochromatin has been observed in somatic hybrids between the *Nicotiana* species, *N. knightiana* + *N. tabacum* (49) and *N. otophora* + *N. tabacum* (Evans, unpublished),suggesting that hybridization has occurred. Chromosome number has been assayed in somatic hybrid plants as a method to verify hybridity. Unfortunately, the expected summation chromosome numbers were not observed, and the aneuploid chromosome numbers obtained are not as conclusive. Nonetheless, in somatic hybrids, morphological differences may be evident between chromosomes of different species. Chromosomes with distinct, different morphology have been simultaneously visible in cells of many somatic hybrid plants (26, 32, 47). Kao has used cytogenetic techniques to analyse cell fusion products of *Nicotiana glauca* + *Glycine max* (41). Although no plants were regenerated in this fusion experiment, separate clonal lines were isolated. Chromosome morphology is quite distinct between these species as chromosomes of *N. glauca* are at least twice the size of *G. max* chromosomes. In somatic cell hybrids, *N. glauca* chromosomes were preferentially eliminated.

TABLE III. Chromosome Numbers of Interspecific Somatic Hybrid Plants

Parent species	Number of plants	Expected chromosome number	Observed chromosome number	Reference
Nicotiana glauca + *N. langsdorffii*	23	42	56-64	87
N. glauca + *N. langsdorffii*	16	42	28-143	13
N. tabacum + *N. knightiana*	8	72	44-126	49
N. tabacum + *N. rustica*	10	60	60-91	57
N. tabacum + *N. glutinosa*	33	72	50-88	58
N. tabacum + *N. alata*	2	60	66-71	58
N. tabacum + *N. glauca*	20	72	72	26
N. tabacum + *N. nesophila*	15	96	96	25
Daucus carota + *D. capillifolius*	12	36	34-54	22
Datura innoxia + *D. stramonium*	8	48	46-72	77
Datura innoxia + *D. discolor*	3	48	46-48	77
Petunia parodii + *P. hybrida*	8	28	24-28	70
P. parodii + *P. parviflora*	3	32	31-40	71
Lycopersicon esculentum + *Solanum tuberosum*	10	48	50-72	53
Arabidopsis thaliana + *Brassica campestris*	5	60	30-60	32
Datura innoxia + *Atropa belladonna*	7	120	84-175	47

III. METHODOLOGY

A. *Cultured Cells*

1. *Mitotic Index.* Callus cultures contain few mitotically active cells. Mitotic index (MI), is often below 1% (e.g., 63). A low MI may require examination of a large cell population before suitable mitotic figures are obtained. Suspension cultures, on the other hand, are subcultured more often and have higher MI shortly after subculture. The MI of suspension cultures may be as high as 10% (24), or 12% (12).

Suspension cultures are usually initiated from recently subcultured callus cultures, but may be initiated directly from explants (30). When suspension cultures become established or when cultures are initiated from rapidly growing callus, the length of subculture can be reduced. Using a four day subculture regime, adequate mitoses are available at any time. Maximum MI usually occurs at 24 hours after subculture. For suspension cultures with longer subculture regime, the maximum MI will be reduced. In suspension cultures with extended culture periods, MI may be as low as in callus cultures.

MI can be determined by the method of Partanen (69). Cells are fixed without pretreatment. Fixed cells are rinsed and hydrolyzed in pectinase or HCl. Hydrolyzed cells are rinsed and placed in stain. MI is the ratio of mitotic cells, including prophase, to total cells. Care should be taken that if cells are partially synchronized, the MI will vary with the time of day. In cultures with low MI, cultures may require preliminary screening to insure sufficient chromosomes for karyotypic analysis.

2. *Chromosome Number of Cell Cultures.* Rapidly dividing cell suspension cultures have been used for numerous cellular genetic studies including gene transformation (46), protoplast fusion (26), and secondary product synthesis (96). Chromosomes have been examined from numerous species using the generalized protocol presented in Table IV (24).

For rapidly dividing suspension cultures colchicine is not required. Colchicine should not be used if anaphase abnormalities are to be examined (91). As colchicine contracts chromatin, such a pretreatment is ideal for chromosome counts. Analysis of chromosome structure may require no pretreatment or treatment with a mitotic arresting agent that does not condense chromatin (18).

TABLE IV. Protocol for Mitotic Analysis of Plant Suspension Cultures

1. *Prepare 10 mg/ml solution of colchicine dissolved in hormoneless culture medium. Colchicine should be made fresh or alternately stored for 1-2 weeks at -10°C.*
2. *Add 0.5 ml of colchicine solution to 4.5 ml of rapidly dividing suspension culture in sterile centrifuge tube (e.g., Falcon #2095). Final concentration of colchicine is 1 mg/ml.*
3. *Cells should be placed on a shaker for 2-6 hours. The 6 hour limit should not be extended as colchicine mitotic block is released, resulting in polyploid cells.*
4. *Cells are centrifuged at 100 g for 3 minutes and the colchicine is decanted.*
5. *Rinse cells in distilled water using centrifugation (Step 4).*
6. *Resuspend cells in Farmer's fixative, 3 parts 95% ethanol to 1 part acetic acid. Cells may remain in fixative for 2-12 hours.*
7. *Cells may be stored at 0°C after fixation by resuspending centrifuged cells in 80% ethanol. Alternately, cells can be hydrolyzed directly.*
8. *Following removal of fixative by centrifugation, cells are rinsed in distilled water.*
9. *Following rinse, cells are resuspended in 1 N HCl for hydrolysis at room temperature for 2-6 hours. Length of hydrolysis is dependent on the tissue or species.*
10. *Acid is removed and cells are rinsed twice in 0.1 M Na-acetate by centrifugation prior to staining.*
11. *Cells are resuspended in 0.5-2.0 ml stain for 2-12 hours prior to viewing. Cells placed on a microscope slide with Pasteur pipette are squashed beneath a coverslip until chromosomes are well spread. Seal coverslip with melted wax.*
12. *If hydrolysis is insufficient, remove stain by washing with distilled water and repeat steps 8-10.*

The fixation and hydrolysis treatments are critical for successful staining. Feulgen method is particularly sensitive to these steps. Acetic alcohol fixative (Farmer's fixative) yields maximum staining intensity with feulgen stain and carbol fuchsin stain. All fixatives should be freshly prepared. Acetic alcohol fixative may be stored in a freezer for up to one week. Chilled fixative is often preferred (94). Tissue cannot be stored in fixative but should be transferred to 70% ethanol for storage. Even in 70% ethanol, stainability is reduced with storage. Newcomer has suggested use of a mordant that permits long-term storage of tissue in the refrigerator (59). Root tips stored for 5-11 months in Newcomer's Fluid (6 parts isopropyl alcohol, 3 parts propionic acid, 1 part petroleum ether, 1 part acetone, and 1 part dioxane) retain maximum stainability. Stringham has suggested addition of ferric acetate to Newcomer's fixative fluid for examination of small chromosomes (90).

Numerous stains have been used for chromosomal material. The three stains most often used for chromosome analysis are feulgen, aceto-carmine, and carbol fuchsin. Methods for preparation and use of each stain have been published and are summarized (Table V).

An alternate method to prepare aceto-carmine stain has been proposed (89). In this procedure, a more concentrated solution of aceto-carmine is prepared initially and diluted to 2% with 85% ethanol after boiling. This solution is then filtered. This preparation results in more intense chromosome staining in some species.

Aceto-carmine stain is usually added directly to the microscope slide. The slide, with stain and tissue, is often heated prior to viewing. Addition of iron to the stain may increase the intensity of color. A few drops of a ferric chloride or ferric acetate solution may be added to the stain. Care should be taken to avoid excess iron.

The Feulgen method requires multiple rinses after fixation to remove excess fixative (18). Hydrolysis should be carried out at 60°C for 6 minutes. Root tips should be stained for 2-3 hours before viewing.

Carbol fuchsin stain, as described (Table V) can also be used for staining of plant protoplasts. This stain has long shelf life and appears useful for a number of plant species.

A technique for examination of plant protoplasts has been developed (39). As protoplasts burst readily, all solutions used should contain 0.3 M sorbitol. Farmer's solution, mixed with 0.3 M sorbitol, is mixed with protoplasts (9:1 dilution). Protoplasts are fixed overnight. Excess fixative may be decanted following fixation, or alternately, modified carbol fuchsin stain solution is added directly to the protoplast fixative mixture on a microscope slide (39). The protoplast mixture need not be squashed.

During protoplast fusion with polyethylene glycol (PEG), a single layer of protoplasts is attached to a coverslip (40). Attachment facilitates examination of parental and

TABLE V. Preparation of Commonly used Chromosome Stains

I. *Aceto-carmine (after Sharma and Sharma, 82)*
1. *Combine 45 ml of glacial acetic acid with 55 ml of distilled water to make 45% acetic acid solution.*
2. *Heat the acetic acid solution to boiling.*
3. *Add 0.5 g carmine slowly to the boiling solution. Stir with a glass rod.*
4. *Continue boiling until the dye dissolves.*
5. *Cool to room temperature.*
6. *Filter and store in a bottle with a glass stopper in the refrigerator. Stain stored in refrigerator is good for at least 1 yr.*

II. *Feulgen (after Van't Hof, 94)*
1. *Add 1 gm of basic fuchsin and 1.9 gm of potassium metabisulfite to 100 ml of 0.15 N HCl.*
2. *Allow stain to sit for 24 hrs.*
3. *Add 0.5 gm activated charcoal to the mixture.*
4. *Filter through Whatman #1 filter paper. Filter paper should be premoistened with 1 N HCl.*
5. *Stain should be colorless. If color remains, add more charcoal and refilter.*
6. *Store in the refrigerator at 2-5°C in stoppered flask. Stain will remain good for 2-4 months.*

III. *Carbol Fuchsin (after Kao, 39)*
1. *3 gm of basic fuchsin (Gurr) are added to 100 ml of 70% ethanol (Solution A).*
2. *10 ml of Solution A are added to 90 ml of 5% phenol dissolved in distilled water (Solution B). This solution must be used within 2 wks.*
3. *45 ml of Solution B are added to a mixture of 6 ml glacial acetic acid and 6 ml of 37% formaldehyde (Solution C).*
4. *5 ml of Solution C are added to a 95 ml mixture of 1.8 g sorbitol in 45% acetic acid. The sorbitol should be dissolved in water before adding acetic acid.*
5. *Stain is aged 2 wks prior to use and may then be stored at room temperature for at least 2 yrs.*

fused protoplasts. Following fusion, the coverslip with protoplasts can be removed from the culture dish and dipped overnight into fixative with sorbitol. Following fixation, the coverslip is dipped into stain, then inverted onto a microscope slide for examination.

B. *Mitotic Analysis of Regenerated Plants*

Cytogenetic analysis of regenerated plants requires use of classical techniques of mitotic and meiotic analysis. Regenerated plants arise via organogenesis or somatic embryogenesis. In each case, root formation is induced *in vitro*. For both plantlet survival and for mitotic analysis, extensive root development is necessary. Once roots have developed, plantlets can be transferred from culture vessels to the greenhouse. The transition to *in vivo* conditions can be facilitated by placing young plantlets into Jiffy-7^{R} peat pellets under high humidity. Peat pellets are also convenient for mitotic analysis as root tips can be collected from pellets prior to transfer to larger pots in a greenhouse. Alternately, plants at a later stage of development can be uprooted ca. 1 week after transplantation to collect young roots. A sample protocol for root tip mitotic analysis is in Table VI.

Numerous variations to the above protocol have been used for root tip chromosome analysis. For example, 8-hydroxyquinoline has been used extensively as an alternative to colchicine pretreatment. Collins has recommended treatment with 3 mM 8-hydroxyquinoline for 4-5 hours at 16-18°C, with *Nicotiana* root tips (14). Lamm has suggested using a 0.1% aqueous solution of Bromisovalum for 6-8 hours at 10°C to arrest cells of *Pisum sativum* in mitosis (48). Farmer's fixative, though used in most reports, is not always used if cells are stained with aceto-carmine. Carnoy's fixative, 6 parts ethanol: 3 parts chloroform: 1 part acetic acid, may be preferred with preparations to be stained with aceto-carmine. Although HCl is used most often for hydrolysis, the concentration, temperature, and duration of treatment have all been varied. In general, high temperature (60°C) is used in combination with shorter time intervals and with lower concentrations of HCl. Hydrolysis may be accomplished enzymatically with pectinase and cellulase. 0.5% pectinase and cellulase for 1-2 hours is sufficient to hydrolyze plant cells (39).

Mitotic chromosomes have been prepared primarily from root tips of regenerated plants. Removal of root tips is usually non-destructive to the plant. Mitotic figures have been obtained from *Nicotiana* flowers (10), as well as from shoot tips of protoplast derived plants of *Datura innoxia* (28). Burns' technique has not been applied to regenerated plants. As regenerated plants may include chromosomal chimeras (66), it may be desirable to obtain mitotic chromosome counts from cells not restricted to the root tip.

TABLE VI. Protocol for Mitotic Analysis of Root Tips Derived from Regenerated Plants

---

1. *Use well developed young root tips that have just grown through the netting of Jiffy-7[R] peat pellets. Roots are generally cut 2 mm from the tip to include mitotic figures.*
2. *Treat root tips with 1 mg/ml of colchicine at room temperature for 2-6 hours (4 hours optimum). Colchicine may be dissolved in water or in culture medium.*
3. *Place root tips in fixative (3 parts 95% ethanol: 1 part acetic acid) and store in the cold for 2-12 hours.*
4. *Remove fixative with pipette and replace with 0.1 M sodium acetate, pH 4.5.*
5. *Remove sodium acetate and replace with 1 N HCl for 2-6 hours at room temperature for hydrolysis.*
6. *Wash HCl with sodium acetate solution (0.1 M).*
7. *Place root tips into stain for 2-12 hours prior to viewing. Root tips are squashed beneath a coverslip. Slides may be temporarily sealed with melted wax or may be made more permanent using the method described below.*

Protocol for Making Permanent Slides from Squash Preparations of Root Tips

1. *This procedure can be used with stained material for preparation of permanent slides or may be used in combination with more complicated staining procedures, such as Giemsa banding (79).*
2. *Stained slide is placed on a block of dry ice, with the coverslip turned up.*
3. *When the slide is frozen, 5-10 minutes, a white frost will appear on the coverslip. Remove the coverslip from the frozen slide by lifting with a scalpel blade or razor blade. The coverslip should "pop" off.*
4. *Dip slide in 95% ethanol 2-3 times.*
5. *Transfer slide to absolute ethanol for 20 minutes. This procedure should partially destain the cytoplasm.*
6. *Mount in a drop of Permount. Place a clean coverslip on the Permount and carefully remove the air bubbles.*

Corolla cells of *Nicotiana* have reduced MI and chromosomes are difficult to spread. In an effort to alleviate these difficulties, pretreatment of corollas is accomplished in a 8-hydroxyquinoline-maltose solution for 5-6 hours at 18-20°C. Hydrolysis is accomplished in a mixture of 3 parts Carnoy's fixative and 2 parts of 10% HCl (10).

Cytogenetic analysis of regenerated plants should include chromosome banding methods. Greatest success to date has been reported with Giemsa C-bands reflecting constitutive heterochromatin. The value of chromosome banding in identification of break-points involved in translocations has been demonstrated for *Vicia faba* (21). Degree of difficulty in application of chromosome banding methods appears to be species dependent. In at least one species, *Nicotiana otophora*, bands are obtained quite easily (54). This method, which is ineffective with other *Nicotiana* species, appears restricted to *N. otophora*, which has large blocks of heterochromatin. Chromosomes of cereal plants examined in root tips have been most amenable to C-banding (e.g., 62). Unfortunately, plant regeneration *in vitro* from cereals has lagged behind other plant families (45). Nonetheless, procedures have been published for Giemsa C-banding of plant chromosomes from root tips of numerous species (e.g., 79). A generalized protocol for Giemsa C-banding of root tip chromosomes is presented in Table VII.

Giemsa C-banding has been used to distinguish the karyotype of many plant species, including *Vicia faba* (21), *Tulipa* (27), *Secale cereale* (31), *Trillium grandiflorum* (79), *Fritillaria* (79), *Hordeum vulgare* (62), and various wheat species (38). Giemsa C-banding techniques have been successfully applied also to callus cultures of *Vicia faba* (68).

While C-bands have been quite useful, conventional Giemsa G-bands have not been observed in plants. It has been suggested the G-bands are not visible as plant chromosomes are too contracted (33).

The hydrolysis treatment is often varied in the published C-banding techniques. HCl treatment at 60°C interferes with the activity of Giemsa stain (18). Gill and Kimber have suggested using 5% pectinase and cellulase in 1 N HCl at room temperature for hydrolysis (31). Barium hydroxide treatment is used for chromosome denaturation, while SSC is used with most plant species for renaturation. Staining is achieved with diluted Giemsa stain adjusted to pH 6.8-6.9. Stain has been applied for as short as 1-2 minutes (31) or as long as 24 hours (79). Achieving sufficient chromosome spreading with adequate staining requires variation of both hydrolysis and staining procedures.

Other differential staining methods have been used to examine higher plant species. (1) Hoechst 33258 dye in combination with 5-bromodeoxyuridine for one round of replication and thymidine for a second round of replication has been used to examine sister chromatid exchanges in root tips of *Vicia faba* (44). (2) The chromosomal DNA

TABLE VII. Composite Protocol for Giemsa C-Banding of Root Tip Chromosome Preparations

1. *Roots are excised and pretreated in 0.05-0.1% colchicine for 1-5 hours.*
2. *Fixation is usually achieved in Farmer's solution and may be stored up to 2 days in the refrigerator.*
3. *Hydrolysis is not always used. Most often dilute HCl (0.2-1.0 N) is used for 3-8 minutes.*
4. *Following rinsing, root tips are transferred into a drop of 45% acetic acid for squashing.*
5. *Gently heat the microscope slide, then cover slide with coverslip and squash.*
6. *Microscope coverslip is removed by dry ice method described in Table VI.*
7. *The slide is passed through 2 changes of 90% ethanol, then absolute ethanol and air dried.*
8. *The slide may be stored dry at room temperature following ethanol treatment.*
9. *The preparation is transferred to a freshly prepared, prewarmed, saturated aqueous solution of* $Ba(OH)_2$. *Treat from 5-60 minutes at room temperature.*
10. *Rinse in 2-3 changes of distilled water to remove film of* $BaCO_3$.
11. *Preparation is incubated in 2X-SSC (0.3 M NaCl plus 0.03 M trisodium citrate, pH 7.0) at* 60^oC *for 30-120 minutes.*
12. *Slide is washed with distilled water and stained in freshly prepared Giemsa stain. Giemsa stain is prepared by using Gurr R66 stock solution diluted 50X with Sorensen phosphate buffer pH 6.9.*
13. *Microscope slides should be examined to ascertain optimum duration of staining. Slides are usually stained ca. 30 minutes, but longer treatments have been reported.*

associated with formation of the nucleolus, termed the nucleolus organizer region (NOR), has also been examined using differential staining. Silver staining has been used to preferentially view NORs in *Vicia faba* (78) and *Phaseolus coccineus* (81). Cells to be examined are incubated in $AgNO_3$ at 65^oC for 12.5 hours. (3) Schweizer has demonstrated the value of using sequential treatment with fluorescent and Giemsa staining methods to examine plant chromosomes (80).

C. *Meiotic Analysis*

Analysis of meiotic chromosomes in higher plants is most frequently performed with squash preparation of pollen mother cells (PMCs). A general outline of this procedure is presented in Table VIII. Anthers are collected and fixed while still enclosed within immature floral buds. Buds should be collected from unstressed plants to avoid problems with flattening and spreading of the chromosomes. For species with large flowers, individual floral buds are collected. It is necessary to collect a range of bud sizes in order to insure that anthers containing PMCs at the appropriate stage of meiosis for analysis will be obtained. When the inflorescence is represented by a panicle, umbel or head, the entire inflorescence is collected and fixed. In this case, a range of developmental stages will be represented within one inflorescence by florets at different stages of maturity.

Before placing the floral buds in fixative, any excess calyx and corolla tissue should be removed. This will allow immediate penetration of the fixative into the anthers. The most commonly used fixatives for meiotic material are Carnoy's and Farmer's solutions. In both fixatives, propionic acid can be substituted for the acetic acid. A few drops of iron

TABLE VIII. Protocol for Meiotic Analysis of Regenerated Plants

---

1. *Collect immature floral buds from vigorous plants.*
2. *Place buds in fixative and store at room temperature for 12-24 hours.*
3. *Buds can be stored in 70% ethanol after fixation at 0-4°C for up to two months. Alternately, slides can be prepared immediately after fixation.*
4. *To prepare slides, place a single anther on a slide with a drop of aceto-carmine stain. Using a blunt instrument, squash the anther in the stain. Remove debris and place a coverslip over stain.*
5. *Heat preparation by passing slide over the flame of an alcohol burner several times. Do not allow the stain to boil.*
6. *Cover slide with blotting paper and apply pressure.*
7. *If chromosomes are not sufficiently spread and flattened, apply more stain along the edge of the coverslip and repeat steps 4-6.*

---

acetate or ferric chloride can be added to the fixative as a mordant. The ferric chloride serves a dual purpose in that it also softens the cytoplasm and allows for easier flattening of the cells.

Fixation is carried out at room temperature for 12-24 hours (18). A change to fresh fixative after a few hours is recommended as this removes autolytic enzymes that have been released into the fixative (8).

After fixation, preparations can be made immediately or the buds can be stored in 70% ethanol at 0-4°C. In *Nicotiana*, storage in ethanol for at least one week facilitates the spreading of the chromosomes. Storage should not last longer than 2 months. Fixation can also be followed directly by staining in alcoholic hydrochloric acid-carmine (89).

To prepare the slide, dissect an anther from a bud and place in a small drop of stain on a slide. Aceto-carmine is the most frequently used stain, but aceto-orcein may be preferable with some materials. Using a blunt instrument, such as a glass rod or the end of a metal dissecting needle, squash the anther in the stain. For large anthers, cut the anther in half before squashing. Remove all debris with forceps. If a mordant is required, unplated iron instruments can be used in the squashing procedure. Otherwise, use only stainless steel, nickle-plated, plastic, glass or teflon-coated instruments.

Place coverslip over stain. If correct amount of stain was used, no excess stain will be present outside the coverslip. Gently heat slide over flame of an alcohol burner. This heating flattens the cells, sticks them to the slide and spreads the chromosomes (50). Heat slide by making successive passes over the flame. Test heat of slide by passing it against back of hand between passes through flame. Do not allow stain to boil.

After placing several sheets of filter or bibulous paper over slide, flatten preparation by pressing down on coverslip with thumbs. The amount of flattening achieved is dependent on the amount of stain present, the amount of material on the slide, and on the amount of pressure applied (18). Be careful not to allow lateral movement of the coverslip while pressing. The preparation is now ready to be viewed under low magnification of the light microscope.

If the slide contains cells at an earlier or later stage than was desired for analysis, repeat procedure using anthers from a larger or smaller bud or floret. If cells are at the correct meiotic stage, but chromosomes are not sufficiently spread and flattened, add more stain along the edge of the coverslip. After the stain has moved under the coverslip, heat slide slightly and repeat flattening procedure.

Evaluations of meiotic chromosomes are used to analyze chromosome pairing relationships. Depending on the material being studied, meiotic analysis can be used to verify aneuploidy, detect chromosome pairing irregularities caused by aneuploidy, distinguish between different types of aneuploidy with the same somatic chromosome numbers, identify the particular aneuploid chromosome(s), detect and identify structural changes in chromosomes and determine the amount of homology that exists between genomes of different species.

The most commonly analysed meiotic stages are pachytene and metaphase I. Pachytene is usually the best stage for studying chromosome associations; however, in many species the pachytene chromosomes are too indistinct to allow a complete analysis to be made. In corn and tomato, individual chromosomes and chromosome arms can be recognized, thereby making these two species ideal subjects for pachytene analysis. For many other species, a study of chromosome pairing relations must be made with late prophase or metaphase I PMCs. While chromosome configurations are distinct at this stage, many of the associations that were present at pachytene will have fallen apart by metaphase. Useful information can also often be obtained from observations of anaphase I and II bridges and fragments and telophase I and II micronuclei.

Aneuploidy is occasionally encountered in plants regenerated from tissue culture. The expected meiotic chromosome configurations of some of the more commonly encountered aneuploids are given in Table IX. For a more detailed description of the cytology and breeding behavior of aneuploids, the reader is referred to Khush (43).

A plant with one less than the normal complement of chromosomes is referred to as a monosomic. The monosomic condition can usually be detected at metaphase I by the presence of a univalent. This univalent can most easily be observed in equatorial views as a chromosome lying off the metaphase plate. In cases where monosomy causes disturbances in meiotic pairing, a phenomenon known as "univalent shift", several univalent chromosomes may be observed. In rare cases, a trivalent may be formed when the monosome pairs with two homologous chromosomes. Micronuclei are frequently observed in telophase II configurations in monosomic PMCs, particularly when a univalent shift has occurred.

Identification of the monosomic chromosome is facilitated in species where a complete set of monosomics is available. In these cases, morphology of each of the monosomics can be compared to that of the unidentified monosomic. Pollen fertility can also often be helpful in identifying monosomics.

TABLE IX. Expected Meiotic Configurations of Frequently Occurring Aneuploids

Aneuploid condition	*Chromosome number*	*Meiotic Configuration*[a]
Monosomy	*2N - 1*	*Metaphase I: n-1 II + 1 I or n-2 II + 1 III*
Double monosomy	*2N - 2*	*Metaphase I: n-2 II + 2 I*
Nullisomy	*2N - 2*	*Metaphase I: n-1 II*
Primary trisomy	*2N + 1*	*Pachytene: n-1 II + 1 III* *Metaphase I: n-1 II + 1 III, n II + 1 I or n-1 II + 3 I*
Secondary trisomy	*2N + 1*	*Metaphase I: n II + 1 ring I; n-1 II + 1 III or n-1 II + 3 I*
Tertiary trisomy	*2N + 1*	*Pachytene: n-1 II + 1 III, n-2 II + 1 V or n II + 1 I* *Metaphase I: n-2 II + 1 V, n-2 II + 1 IV + 1 I, n-1 II + 1 III, n-2 II + 1 III + 2 I, n II + 1 I or n-1 II + 3 I*

[a]*Abbreviations: I=univalent; II=bivalent; III=trivalent; IV=quadrivalent; V=pentavalent.*

In *Nicotiana*, each of the monosomics exhibits a characteristic pollen fertility (86). The commonly used technique for estimating pollen fertility is based on percent aceto-carmine stainable pollen (Table X; Fig. 2A). In general, pollen fertility is a good indicator of possible abnormalities in chromosome number or structure and should be routinely measured with all unique plant materials.

In species of a polyploid origin, plants missing two complete chromosomes are occasionally observed. If the two missing chromosomes are non-homologous, the plant is termed a double monosomic. In a nullisomic, the two missing chromosomes are homologous. Although the somatic chromosome number of these two aneuploids is the same, it is sometimes easy to determine which condition is present by an analysis of meiotic chromosome pairing relations. When normal chromosome

TABLE X. Protocol for Estimating Pollen Fertility

1. *Squash a mature undehisced anther in a drop of aceto-carmine. Remove debris and gently place a coverslip over stain. Alternately, dust pollen from a dehisced anther in a drop of stain and apply coverslip.*
2. *Heat slide gently over flame of an alcohol burner or allow unheated slide to sit for 15-30 minutes before observing.*
3. *Observe slide under low magnification of a light microscope. Using a dual entry desk recorder, tabulate the number of stained and unstained pollen grains. Estimate pollen fertility by determining percent of stained pollen grains.*

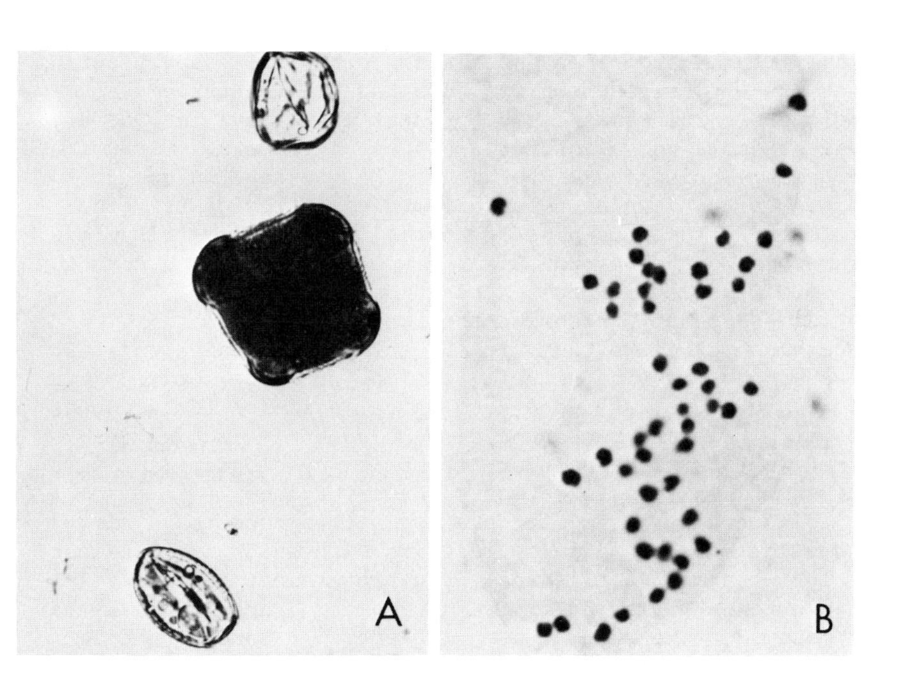

FIGURE 2. Meiotic analysis of in vitro derived interspecific hybrids. (A) Sample of partially fertile pollen from 4x (Nicotiana stocktonii x N. tabacum). (B) Metaphase I pollen mother cell of N. stocktonii x N. tabacum (2x = 48).

pairing relations are not disturbed by the aneuploidy, the PMCs of the double monosomic will contain two univalents, while PMCs of the nullisomic will contain only bivalents (Table IX). However, as with monosomy, nullisomy can upset meiotic pairing and result in the presence of two or more univalents at metaphase I. In these cases, further analysis of the aneuploid condition requires progeny analysis.

A trisomic is a plant which has one more than the normal complement of chromosomes. The extra chromosome can be one of the normal chromosomes of the complement (primary trisomy), an isochromosome for one of the chromosomes arms of the complement (secondary trisomy), or a chromosome with two arms of two nonhomologous chromosomes (tertiary trisomy). Other types of trisomic conditions have been described (43); however, these will not be discussed as their identification is limited to those few species where detailed pachytene analysis can be performed.

In a primary trisomic, the extra chromosome usually pairs with homologous chromosomes at pachytene to form a trivalent. However, this chromosome association often falls apart during later meiotic stages. During metaphase I, the three homologous chromosomes can be present as a trivalent, as a bivalent and a univalent, or as three univalents. Each of the primary trisomics of tomato and *Datura* have unique phenotypes that can be used for comparison to identify the particular chromosome present in the unidentified trisomic. Chromosome morphology can also be used in tomato and corn trisomics to identify the extra chromosome.

In a secondary trisomic, the two ends of the isochromosome will often pair and form a ring univalent. The presence of this ring univalent, along with the somatic chromosome count, are good cytological evidence for secondary trisomy. However, because a ring univalent is not always formed in PMCs of a secondary trisomic, cytological analysis will sometimes fail to detect secondary trisomy.

In a tertiary trisomic, the extra chromosome may pair with either or both chromosome pairs with which it contains homologous arms. A trivalent or a pentavalent is expected in pachytene PMCs of a tertiary trisomic. Rarely, the extra chromosome may be present as univalent. As some of the chromosome associations dissolve at metaphase I, various configurations can be expected (Table IX). The presence of a pentavalent or a quadrivalent with a univalent in a plant with a somatic chromosome number of $2n + 1$ is a good indication that the plant in question is a tertiary trisomic.

Changes in chromosome structure are probably more commonly encountered in regenerated plants than are changes in chromosome number. Analysis of these structural changes is greatly facilitated by the presence of highly distinct pachytene chromosomes. In species where detailed pachytene analysis can be performed, the location and extent of the change can be detected by observing pachytene configurations of structural hybrids. Structural hybrids are individuals which are heterozygous for any chromosomal change. Where a deletion or duplication has occurred, a buckle or hump will be observed. An inversion will be visible as a loop, whereas a cross-shaped configuration will be observed if a translocation has occurred. Cytological analysis of many species, however, is limited to metaphase I and later meiotic stages. In these cases, crosses to known linkage and marker stocks can be used to obtain a more complete analysis of the structural changes than can be obtained with meiotic analysis alone. A complete discussion of the chromosome configurations of structural hybrids, along with details of genetic methods for analyzing structural changes, is given in Burnham (9).

In general, the location and number of cross-overs will affect the metaphase I configurations of structural hybrids. Depending on the location and order of the inserted segment in a duplication, recombination can result in the presence of an acentric chromosome and a ring chromosome, or an acentric and a dicentric chromosome at metaphase I. At later stages the acentric chromosome will appear as a fragment, while an anaphase bridge will be formed as a dicentric chromosome is pulled to both poles. Similar configurations will be observed in PMCs with pericentric inversions. Crossing-over in chromosomes with pericentric inversions results in an acentric and a dicentric chromosome at metaphase I. One or more bridges can be present at either or both anaphase divisions, while fragments may be observed at anaphase I. Recombination does not alter the number of centromeres in chromosomes carrying paracentric inversion. Therefore, paracentric inversions can be cytologically detected only at pachytene.

The cross-shaped configuration of translocation chromosomes will frequently appear at metaphase I as a ring or chain of chromosomes. If two pairs of chromosomes are involved in the translocation, a ring or chain of four will be observed. As the number of chromosomes with translocated regions increases, the number of chromosomes in the ring or chain will also increase. Semi-sterility and altered linkage groups are frequently encountered in the progeny of a translocation heterozygote and should be tested for if a translocation is suspected to have occurred.

Recently, several interspecific hybrids have been produced through use of various *in vitro* techniques. These hybrids, many of which cannot be obtained using conventional breeding techniques, have been produced through embryo culture (72), ovule culture (73) and somatic cell hybridization (95). As with the more conventionally derived hybrids, there is usually an interest in the meiotic behavior of the cell culture derived hybrids. By studying meiotic chromosome associations, an estimate of homology between the chromosomes of the parental species can be obtained.

When the chromosomes of the hybrid are present at the 2x ploidy level, the occurrence of univalents is usually of most interest. If no homology exists between the chromosomes of the parental species, only univalents will be observed (Fig. 2B). In closely related species, where a great deal of homology still exists, a large number of bivalents will be formed. Somatic hybrids contain two copies of both parental genomes. In these hybrids, the occurrence of multivalents will signal the presence of homology between the chromosomes of the parental species.

IV. CONCLUDING REMARKS

Basic cytogenetic protocols have been described for cultured cells and plants regenerated from cell cultures. These techniques are useful in analysis of plant tissue cultures as chromosome instability has been observed in numerous cell lines. The instability is most often reported as aneuploidy and polyploidy. The frequency of both aneuploidy and polyploidy is less in plants than in the cultures used for regeneration studies. As structural changes have also been observed in chromosomes of plants regenerated from cell cultures, karyotypic analysis of chromosomes would be quite useful. Nonetheless, karyotypic studies on cell cultures or regenerated plants are scarce. In addition, there are only a few reports on meiotic analysis of regenerated plants. In each case, results suggest interesting chromosome pairing relationships in aneuploids. It is hoped that cytogenetic analysis will be applied more widely to analysis of interesting genetic material derived from plant cell cultures.

REFERENCES

1. Ahloowalia, B.S., *In* "Current Chromosome Research" (K. Jones and P.E. Brandham, eds.), p. 115. Elsevier/North Holland, Amsterdam, Netherlands, (1976).

2. Balzan, R., *Caryologia 31*, 75 (1978).
3. Bayliss, M.W., *Chromosoma 51*, 401 (1975).
4. Bayliss, M.W., *Protoplasma 92*, 109 (1977).
5. Bayliss, M.W., and Gould, A.R., *J. Expt. Bot. 25*, 772 (1974).
6. Bennici, A., *Z. Pflanzenzucht. 72*, 199 (1974).
7. Bennici, A., and D'Amato, F., *Z. Pflanzenzucht. 81*, 305 (1978).
8. Berlyn, G.P., and Miksche, J.P., "Botanical Microtechnique and Cytochemistry", Iowa State Univ. Press, Ames, Iowa, (1976).
9. Burnham, C.R., "Discussions in Cytogenetics", Univ. of Minnesota, St. Paul, (1962).
10. Burns, J.A., *Tobacco Sci. 8*, 22 (1964).
11. Chandra Roy, S., *Protoplasma 102*, 171 (1980).
12. Chu, Y.E., and Lark, K.G., *Planta 132*, 259 (1976).
13. Chupeau, Y., Missonier, C., Hommel, M.C., and Goujand, J., *Molec. Gen. Genet. 165*, 239 (1978).
14. Collins, G.B., *In "Nicotiana* Procedures for Experimental Use" (R.D. Durbin, ed.), p. 17. U.S.D.A. Technical Bulletin 1586, (1979).
15. Cooper, L.S., Cooper, D.C., Hildebrandt, A.C., and Riker, A.J., *Amer. J. Bot. 51*, 284 (1964).
16. D'Amato, F., *In* "Plant Cell, Tissue and Organ Culture" (J. Reinert and Y.P.S. Bajaj, eds.), p. 343. Springer-Verlag, Berlin. (1977).
17. D'Amato, F., *In* "Frontiers of Plant Tissue Culture 1978". (T.A. Thorpe, ed.), p. 287. Univ. of Calgary Press, Calgary (1978).
18. Darlington, C.D., and La Cour, L.F., *In* "The Handling of Chromosomes" George Allen and Unwin, Sixth Edition, London, (1976).
19. Demoise, C.F., and Partanen, C.R., *Amer. J. Bot. 56*, 147 (1969).
20. Devreux, M., Saccardo, F., and Brunori, A., *Caryologia 24*, 141 (1971).
21. Dobel, P., Rieger, R., and Michaelis, A., *Chromosoma 43*, 409 (1973).
22. Dudits, D., Hadladzky, GY., Levi, E., Fejer, O., and Lazar, G., *Theoret. Appl. Genet. 51*, 127 (1977).
23. Evans, D.A., *Z. Pflanzenphysiol. 95*, 459 (1980).
24. Evans, D.A., and Gamborg, O.L., *Plant Sci. Lett.* (submitted) (1980).
25. Evans, D.A., Jensen, R.A., and Flick, C.E., *Science* (in press) (1980).
26. Evans, D.A., Wetter, L.R., and Gamborg, O.L., *Physiol. Plant 48*, 225 (1980).
27. Filion, W.G., *Chromosoma 49*, 51 (1974).

28. Furner, I.J., King, J., and Gamborg, O.L., *Plant Sci. Lett. 11*, 169 (1978).
29. Gamborg, O.L., Shyluk, J.P., Fowke, L.C., Wetter, L.R., and Evans, D.A., *Z. Pflanzenphysiol. 95*, 255 (1979).
30. Geile, W., and Wagner, E., *Plant, Cell and Environ. 3*, 141 (1980).
31. Gill , B.S., and Kimber, G., *Proc. Nat. Acad. Sci. 71*, 1247 (1974).
32. Gleba, Y.Y., and Hoffmann, F., *Naturwissen. 66*, 547 (1979).
33. Greilhuber, J., *Theoret. Appl. Genet. 50*, 121 (1977).
34. Guo, C.-L., *In Vitro 7*, 381 (1972).
35. Harn, C., *Caryologia 25*, 429 (1972).
36. Heinz, D.J., and Mee, G.W.P., *Amer. J. Bot. 56*, 450 (1971).
37. Horak, J., *Biol. Plant. 14*, 423 (1972).
38. Iordansky, A.B., Zurabishvilli, T.B., and Badaer, N.S., *Theoret. Appl. Genet. 51*, 145 (1978).
39. Kao, K.N., *In* "Plant Tissue Culture Methods" (O.L. Gamborg and L.R. Wetter, eds.), p. 60. National Res. Council, Saskatoon, Canada, (1975).
40. Kao, K.N., *In* "Cell Genetics in Higher Plants". (D. Dudits, G.L. Farkas, and P. Maliga, eds.), p. 233. Akademiai Kiado, Budapest, (1976).
41. Kao, K.N., *Molec. Gen. Genet. 150*, 225 (1977).
42. Kao, K.N., Miller, R.A., Gamborg, O.L., and Harvey, B.L., *Can. J. Genet. Cytol. 12*, 297 (1970).
43. Khush, G.S., *In* "Cytogenetics of Aneuploids". Academic Press, New York, (1973).
44. Kihlman, B.A., and Kronborg, D., *Chromosoma 51*, 1 (1975).
45. King, P.J., Potrykus, I., and Thomas, E., *Physiol. Veg. 16*, 381 (1978).
46. Kool, A.J., and Pelcher, L.E., *Protoplasma 97*, 71 (1978).
47. Krumbiegel, G., and Schieder, O., *Planta 145*, 371 (1979).
48. Lamm, R., *Hereditas 84*, 235 (1976).
49. Maliga, P., Kiss, Z.R., Nagy, A.H., and Lazar, G., *Molec. Gen. Genet. 163*, 145 (1978).
50. McClintock, B., *Stain Tech. 4*, 53 (1929).
51. Mehra, P.N., and Anand, M., *Physiol. Plant. 45*, 127 (1979).
52. Mehra, A., and Mehra, P.N., *Bot. Gaz. 135*, 61 (1974).
53. Melchers, G., Sacristan, M.D., and Holder, A.A., *Carlsberg Res. Commun. 43*, 203 (1978).
54. Merritt, J.F., and Burns, J.A., *J. Hered. 65*, 101 (1974).
55. Mitra, J., Mapes, M.O., and Steward, F.C., *Am. J. Bot. 47*, 357 (1960).
56. Murashige, T., and Nakano, R., *Amer. J. Bot. 54*, 963 (1967).

57. Nagao, T., *Jap. J. Crop Sci. 47*, 491 (1978).
58. Nagao, T., *Jap. J. Crop Sci. 48*, 385 (1979).
59. Newcomer, E.H., *Science 118*, 161 (1953).
60. Nishi, T., and Mitsuoka, S., *Jap. J. Genet. 44*, 341 (1969).
61. Nishiyama, I., and Taira, T., *Jap. J. Genet. 41*, 357 (1966).
62. Noda, K., and Kasha, K.J., *Stain Tech. 53*, 155 (1978).
63. Novak, F.J., *Caryologia 27*, 45 (1974).
64. Novak, F.J., Ohnoutkova, L., and Kubalakova, M., *Cereal Res. Comm. 6*, 135 (1978).
65. Ogihara, Y., and Tsunewaki, K., *Jap. J. Genet. 54*, 271 (1979).
66. Ogura, H., *Jap. J. Genet. 51*, 161 (1976).
67. Orton, T.J., *Theoret. Appl. Genet. 56*, 101 (1980).
68. Papes, D., Jelaska, S., Tomaseo, M., and Devide, Z., *Experientia 34*, 1016 (1978).
69. Partanen, C.R., *In* "Tissue Culture Methods and Applications" (P.F. Kruse and M.K. Patterson, eds.), p. 791. Academic Press, New York, (1973).
70. Power, J.B., Frearson, E.M., Hayward, C., George, D., Evans, P.K., Berry, S.F., and Cocking, E.C., *Nature 263*, 500 (1976).
71. Power, J.B., Berry, S.F., Chapman, J.V., and Cocking, E.C., *Theoret. Appl. Genet. 57*, 1 (1980).
72. Raghavan, V., *In* "Experimental Embryogenesis in Vascular Plants" (V. Raghavan, ed.), p. 319. Academic Press, New York, (1976).
73. Reed, S.M., and Collins, G.B., *J. Hered. 69*, 311 (1978).
74. Roper, W., *Z. Pflanzenphysiol. 93*, 245 (1979).
75. Sacristan, M.D., *Chromosoma 33*, 273 (1971).
76. Sacristan, M.D., and Melchers, G., *Molec. gen. Genet. 105*, 317 (1969).
77. Schieder, O., *Molec. Gen. Genet. 162*, 113 (1978).
78. Schubert, I., Anastassova-Kristeva, M., and Rieger, R., *Experiment. Cell. Res. 120*, 433 (1979).
79. Schweizer, D., *Chromosoma 40*, 307 (1973).
80. Schweizer, D., *In* "The Plant Genome" (D.R. Davies and D.A. Hopwood, eds.), p. 61. John Innes Charity, Norwich, England (1980).
81. Schweizer, D., and Ambros, P., *Plant Systemat. Evol. 132*, 27 (1979).
82. Sharma, A.K., and Sharma, A., "Chromosome Techniques". Butterworths, London, (1972).
83. Sheridan, W.F., *In* "Genetic Mechanisms with Plant Materials" (L. Ledoux, ed.), p. 263. Plenum Press, London (1975).

84. Singh, B.D., and Harvey, B.L., *Cytologia 40,* 347 (1975).
85. Singh, B.D., Harvey, B.L., Kao, K.N., and Miller, R.A., *Can. J. Genet. Cytol. 14,* 65 (1972).
86. Smith, H.H., *In* "*Nicotiana* Procedures for Experimental Use" (R.D. Durbin, ed.), p. 1. U.S.D.A. Technical Bulletin 1586, (1979).
87. Smith, H.H., Kao, K.N., and Combatti, N.C., *J. Hered. 67,* 123 (1976).
88. Smith, S.M., and Street, H.E., *Ann. Bot. 38,* 223 (1974).
89. Snow, R., *Stain Tech. 38,* 9 (1963).
90. Stringham, G.R., *Can. J. Bot. 48,* 1134 (1970).
91. Sunderland, N., *In* "Plant Tissue and Cell Culture" (H.E. Street, ed.), p. 177. Univ. of Calif. Press, Berkeley, (1977).
92. Takatori, F.H., Murashige, T., and Stillman, J.E., *Hort. Sci. 3,* 20 (1968).
93. Torrey, J.G., *Physiol. Plant. 20,* 265 (1967).
94. Van't Hof, J., *In* "Methods in Cell Physiology" Vol. III. (D.M. Prescott, ed.), p. 114. Academic Press, New York, (1968).
95. Vasil, I.K., Vasil, V., White, D.W.B., and Berg, H.R., *In* "Plant Regulation and World Agriculture" (T.K. Scott, ed.), p. 63. Plenum Press, New York, (1979).
96. Zenk, M.H., *In* "Frontiers of Plant Tissue Culture 1978" (T.A. Thorpe, ed.), p. 1. Univ. of Calgary Press, Calgary (1978).

PRODUCTION OF ISOGENIC LINES: BASIC TECHNICAL ASPECTS OF ANDROGENESIS

Colette Nitsch

Genetique et Physiologie du Developpement des Plantes
Centre National de la Recherche Scientifique
Gif-sur-Yvette, France

I. INTRODUCTION

Androgenesis is a relatively new *in vitro* culture technique which allows for the production of haploid and doubled haploid plants originating from very young pollen cells. Since 1967, when J.P. Bourgin and J.P. Nitsch obtained fully grown haploid plants from *Nicotiana sylvestris* and *N. tabacum* anthers for the first time (5), geneticists and plant breeders have become more and more interested in the numerous possibilities of research offered by this development.

The literature contains a great deal of information on the subject. Recent reviews can be found in Reinert and Bajaj (31), Thorpe (37), Sharp *et al.* (32), Hughes *et al.* (14) and Vasil (38, 39); as well as in the Proceedings of the Symposium on Plant Tissue Culture (Science Press, Peking, 1978). The purpose of this article is to convey the basic ideas which can serve as a guideline to allow use of the technique on any plant of choice. To help clarify the explanation, examples will be taken from two different families of plants in particular: one a dicotyledon in the Solanaceae (some *Nicotianas*) and secondly a monocotyledon in the cereals (*Zea mays*).

Research over the past decade has clearly indicated that success of the technique involves two important considerations, viz:

(1) the physiological state of the donor plant, and

ISBN 0-12-690680-7

(2) androgenesis, i.e., the aptitude and the conditions necessary to modify the normal development of the pollen and to force it towards embryogenesis.

In addition, a third practical consideration has spurred interest in the method, viz: its usefulness for agriculture. However, the method has some deficiencies. All these will be discussed in this chapter.

II. THE DONOR PLANT

In order to allow for the *in vitro* development of pollen into an adult plant it is very important to start with healthy pollen cells. Plant breeders know from practice that one pollen cell is not the same as another, good pollen comes from good healthy plants. It is even more so for *in vitro* culture than for plant breeding. Therefore it is only from plants grown under the best environmental conditions that one can obtain suitable pollen (24).

a) The donor plant should be taken care of from the time of flower induction to the sampling of pollen. They should be well fed and placed in an optimum light regime with high light intensity. Unfortunately, artificial light cannot replace sunshine. It is therefore suggested that if the plant has to be grown indoors, one should use greenhouses with natural daylight supplemented with artificial light, if necessary, rather than growth chambers with only artificial light.

b) To prevent any stress during pollen development it is advisable to repot the plant just before flower induction and to give optimum nutrition over the period of pollen development. In other words to water the donor plant with nutrient solution for a few days before removal of the pollen sample.

c) The use of any kind of pesticide whether externally applied or systemic should be avoided for the three to four weeks preceding sampling. When placed in *in vitro* culture, i.e., taken away from its natural protective devices, the tissue is more easily killed by chemicals and usually does not recover from such treatments.

d) The last consideration about the donor plant concerns the pollen itself. The state of pollen development when the flower is cut from the donor plant is critical for success in this work. Success is heavily dependent on the variety. Therefore, it is absolutely necessary before starting any *in vitro* culture to follow the development of the pollen from the tetrad state to the binucleate state by staining, preferably following the Feulgen technique. It is a good method which differentially stains the two nuclei: the vegetative nucleus stains pale pink while the generative nucleus stains dark red (see details in Section V *D.1.a*).

As there is variation in each cultivar, the state of pollen development suitable for androgenesis goes from just before the first pollen mitosis until just after, however always before the appearance of starch grains. For example; in the cereals (wheat, rice, maize) when the microspore is formed, its nucleus sits next to the pore (2, 8, 29, 30, 35, 36, 44). Just before pollen mitosis, the nucleus migrates to the opposite side from the pore and mitosis occurs. Next the vegetative nucleus migrates back towards the pore and the generative cell is at the opposite side from the pore. This latter cell then divides to form two gametes. Haploid plants can be obtained if the pollen is put in culture when the nucleus is in the center of the grain, i.e., halfway in its migration. On the other hand, in the Solanacae, depending on the variety, the pollen must be taken either before mitosis (*N. alata*), during mitosis (*N. tabacum*) or just after mitosis (*N. sylvestris*).

III. ANDROGENESIS

The underlying principle of androgenesis is to stop the development of the pollen cell whose fate is normally to become a gamete, i.e., a sexual cell, and to force its development directly into a plant, as is done with somatic cells. This abnormal pathway is possible if the pollen cell is taken away from its normal environment in the living plant and placed in other specific conditions.

a) This induction toward androgenesis is enhanced by keeping the anthers (either in the flower bud or on the culture medium) at low temperature. Besides the synchronizing effect of the cold, it stops existing metabolism and if the cold period is long enough, it will allow the resting pollen cell to start out on a new metabolic pathway when placed in *in vitro* culture on inductive conditions for vegetative cell division (27). This effect of low temperature can be compared to a sort of induced dormancy or vernalization. The degree of cold which should be given is dependent on the species. This requirement can also be replaced by heat in the case of rape (18, 20). For the Solanacae and wheat the cold treatment given varies between 3° and $6^{o}C$ for 3 to 15 days, but maize responds better to a temperature of $14^{o}C$. The length of time in the cold apparently is not critical, as long as it is more than three days, and it can last for up to two weeks without injury to the pollen. The maximum effect is reached around the third day. After the pollen is put in culture, a stepwise gradual increase in the temperature from 14° to 18° to $25^{o}C$; keeping the culture 5 to 7 days at each temperature, enhances the production of embryos.

b) The culture medium used for anther or pollen culture is of importance, but the important constituents apparently are not those which one would usually assume. This view, which comes from examining many reports (e.g., 1-4, 6-21), will be discussed by looking at the different media components needed for successful culture. The major elements most commonly used are those in Murashige and Skoog's medium (23), which has been modified for some species, especially in respect to the NH_4 concentration (8, 9). The minor elements and vitamins do not appear to be necessary for the induction of androgenesis. However, these become more important during the development of the young embryo into a plant. Chelated iron salt has been shown to play an important role for the differentiation of globular embryos into heart shape embryos and further onto complete plants.

Sugar is a constituent which needs to be studied in detail in the different families. For example, the success of the method can depend upon the concentration of sucrose. 2 or 3% sucrose is optimum for the Solanacae, 6% for wheat or rice and up to 12% is necessary for corn. Glucose has been shown to have a positive effect in some cases. However, it can generally be replaced by sucrose at the right concentration.

Growth substances are usually key media addenda for successful *in vitro* culture. Androgenesis is one of the rare cases using this technique where exogenous growth substances have no effect or even can be detrimental. Such a statement requires some explanation. As mentioned earlier, androgenesis is a mechanism by which pollen cell development is modified and forced toward embryogenesis, thus avoiding the sexual pathway. Embryogenesis is inhibited by growth substances as shown by Halperin and Wetherell (11). Enhancing cell division of the vegetative pollen cell, while it is still enclosed in the exine, gives a good start towards embryogenesis. However, the presence of growth substances, especially auxins and cytokinins in the culture medium, increases the cell division to such an extent that the pollen cell loses its character and becomes disorganized forming callus.

Therefore, we have come to the conclusion that substances favoring cell division have to be avoided, if possible, in androgenesis. In the case of Solanacae, it has been shown that growth substances are not necessary, and it is the same for maize where embryos and haploid plants are obtained without any growth substances in the medium. Moreover, the presence of an antiauxin such as Triiodobenzoic acid (TIBA) enhances embryogenesis (unpublished data). Whenever growth substances are used in the culture medium (e.g., in wheat, rice), the process of obtaining plants is complicated by the fact that the pollen becomes a callus from which bud formation has to be induced. For some species this is a major undertaking in itself. It could even be a real barrier for the general use of the approach, since the induction of buds from callus is not possible in all plant species, in our present state of knowledge. In short, we suggest, that to obtain haploid plants via androgenesis, one should adapt the culture medium by modification of sugar or the amino acid composition rather than by use of growth substances; although, it has been shown that a low amount of cytokinin for a short time at the beginning of the induction period might increase the number of embryoids. In such cases, the tissue should be transferred to a medium lacking cytokinin soon after its effect on induction has been achieved; otherwise, it may prevent the differentiation of the root of the embryo and no plants will be produced.

Nitrogen metabolism is quite an important feature of the *in vitro* culture technique, especially when the technique is used for developmental purposes. The presence of nitrate or of ammonium salts in the medium cannot, in some instances, replace the need for amino acids which appear to play a very specific role at different stages of the developmental process. When the technique is used to achieve the complete cycle of development from a single cell to a complete plant, it is important not to overlook these substances. Here again no general rule can be given, and empirical studies have to be carried out with each plant species. However, if the emphasis should be put on only one amino acid, we have concluded, by studying the effect on several species, that glutamine is probably beneficial for most plants as an aid to achieving the *in vitro* differentiation of a cell to a complete plant.

In many instances, authors have mentioned the use of casein hydrolysate in the culture medium. This indeed is one way of doing it, i.e., by placing in the medium a complete amino acid mixture, with the hope that the cell will take from the mixture what it needs. However, knowing feed-back inhibition effects and the interaction between different amino acids on metabolism, I suggest that a study be made of the *in vivo* metabolism of these substances in relation to the developmental process which is to be reproduced *in vitro*, and also to make an analysis of the different amino acids in the pool at the different stages of development being studied. Such an approach will give a guideline for the composition of the optimum culture medium. For example, a complete analysis of the amino acids present in the tissue at several steps of the *in vivo* development of the zygotic embryo has shown the influence of some specific amino acids; thus in *Nicotiana* the need for glutamine and serine has been shown, and in *Zea mays* it was that of proline. In both cases the presence of these amino acids in the culture medium enhanced the growth of the embryos originating from pollen.

c) Besides the induction of androgenesis and the effect of the culture medium on the growth of the induced pollen, another factor has to be mentioned. The environmental conditions in which the cultures are to be placed can enhance the differentiation of the globular embryos into plantlets with cotyledons and roots. More physiological studies need to be done for a complete understanding of this mechanism. However, on the basis of present knowledge, we recommend that pollen induced to undergo androgenesis should be placed in liquid medium, as this favors the further development of the embryo. This can be done independently of the medium on which induction has been done, whether it is on solid or liquid medium.

A second important environmental factor is light. We have shown in the work done on *N. sylvestris* that high intensity (5000 lux) white light is inhibitory for androgenesis and that darkness or blue light are not. However, low intensity (500 lux) white light or even better red light markedly enhanced the number of plants obtained, and furthermore, red light shortened the time necessary to produce plants from pollen by about 20%. In other words, 15 days under red light (660 Å) at 25°C is enough to obtain 500 plantlets from a pollen suspension of one flower bud, whereas under low intensity (500 lux) of white light, 20 days are necessary and the number of plantlets produced is slightly less (450).

IV. USEFULNESS OF THE METHOD

In a plant breeding program of selection, the time necessary to have an isogenic line is reduced from 5 or 6 (minimum) generations to 1 or 2 using the technique of androgenesis. If the doubling of the chromosome is achieved by treating the anthers with colchicine at the time of the first pollen mitosis just before the *in vitro* culture, the homozygous line is obtained from the first generation. When the culture is done with the untreated microspore, the plant produced is haploid, and therefore sterile and needs to be doubled. Two generations of plants at the maximum are therefore necessary to produce an isogenic line (3, 5, 7, 13, 38).

The production of haploids by *in vitro* culture is essentially unlimited when the technique is properly adapted to the species. Moreover such methods as nursing allows the rescue of tissues that could not grow in nature. It means that not only the number of combinations of genomes which are available for selection can be increased, but the nursing effect makes available genotypic combinations which have never been or will never be obtained by sexual crosses (17).

Having a fixed line in one generation is also a main advantage for breeders, as it allows multiplication from seeds and thus increases the sanitary aspect of a population. Indeed, one limit of vegetative propagation is that after a few generations the plants become very sensitive to viruses (22), which then need to be eliminated (see Chapter 6). Reproduction by seeds reduces the multiplication of virus. Anyhow, in the worse cases, it is easier to eliminate the virus from seeds, than to clean the young plant by meristem culture. In other words, less virus is introduced by seed propagation than by cuttings.

The limit of the method is that the technique has to be adapted to the plant for each species before the breeder can use it. This work should better be done by a physiologist. Therefore, to really obtain the best results from this method for plant improvement, plant physiologists and breeders should work together on the same program (19).

The second difficulty in this work lies in the ease with which the haploid plant can be doubled (16). Haploids double spontaneously in some species or can be easily achieved in other species. However, it may be more difficult for others, as in the case for *Zea mays*.

V. PROTOCOL FOR POLLEN CULTURE

The requirements for successful culture for *Nicotiana* and cereals is as follows (after Nitsch 25, 26).

A. *Materials*

Place to keep the buds at low temperature from 5 to 10°C
Place to perform sterile culture

Refrigerated centrifuge
Conical centrifuge tubes with closures (to keep them sterile)
Syringes and filters to sterilize the culture medium by filtration
Screen with pore diameter adjusted to the size of the pollen to be put in culture
Small pyrex petri dishes from 3 to 5 cm in diameter
Parafilm[R] to seal the petri dishes
Inverted microscope for *in vivo* observation
Stains and high magnification microscope for observation
Incubator with temperatures set between 25° and 30°C

B. *Culture Media for Solanaceae*

(1) macroelements (1 liter stock solution)

KNO_3	950 mg
NH_4NO_3	720 mg
$MgSO_4 \cdot 7H_2O$	185 mg
$CaCl_2$	166 mg
KH_2PO_4	68 mg

(2) stock solution FeEDTA: for 100 ml

Na_2EDTA	745 mg
$FeSO_4 \cdot 7H_2O$	557 mg

Dissolve the two salts separately, mix the solutions and bring to a boil before adjusting the volume. Use 5 ml per liter of medium.

(3) medium for isolated pollen culture

stock minerals	100 ml
FeEDTA	0.5 ml
sucrose	2 g
myo-inositol	500 mg
glutamine	50 mg
serine	3 mg

Adjust pH to 5.8

C. *Reagents*

(1) Acetocarmine
Boil with refluxing for one hour, 5 g of carmine in 100 ml of 45% acetic acid.
Cool, filter and store in a brown bottle.

(2) Toluidine blue (for 100 ml)
1% solution in 0.05% citrate-phosphate buffer pH (4.0).

buffer:	Citric acid	(0.1 M)	15.4 ml
	+ Na_2HPO_4	(0.2 M)	9.6 ml

(3) Schiff's reagent (for 100 ml)

	Fuschin base	1 g
+	Potassium metabisulfite	2 g
+	1N HCl	15 ml
+	H_2O	85 ml

Stir for two hours, then add 1 g activated charcoal and stir another 5 minutes Let stand overnight in the cold. Filter and keep in a brown bottle in the cold.

D. *Methods*

1. *Induction.* The state of the pollen being precisely determined by Feulgen staining (see below), the flower buds are collected at the time of the first pollen mitosis. The unopened buds are sterilized by immersing for 2 min in freshly prepared 7% calcium hypochlorite solution, followed by rinsing twice with sterile water.

a. *Feulgen staining.* Fix anthers for 4 to 24 hours in ethanol:acetic acid (3:1). After removal from the fixative, hydrolyze with 1N HCl for 8 minutes at 60°C. Promptly dip the vial in ice water to immediately stop the effect of HCl. Remove the HCl and replace with Schiff's reagent. Keep overnight in the refrigerator and observe the next day directly in the stain or keep at room temperature in the dark for 3 to 4 hours before observation.

b. *Production of homozygous diploids.* The flower buds are placed into a 0.04% colchicine solution with 2% dimethyl sulfoxide added as a carrier, for 4-12 hours in the cold under vacuum. The colchicine solution is then washed out and the buds kept in a humid condition at low temperature (5 to 10°C).

c. *Further induction period.* For *Nicotiana*, but not the cereals, an additional induction period of three days is necessary. The anthers are dissected out and placed into the medium made of the macroelements plus 2% sucrose and 5 ml per liter of FeEDTA at pH 5.8. 50 anthers are floated onto 5 ml of medium into a 50 ml Erlenmeyer flask for a minimum of 72 hours. At the end of this inductive period microspores with two identical nuclei can be seen.

2. *Isolation of the Pollen.* In the case of large anthers, as in *Nicotiana*, about 50 anthers are placed in a small beaker containing 20 ml of basal medium. The pollen grains are then squeezed out of the anthers by pressing them against the side of the beaker with a glass rod or the piston of a syringe. Anther tissue debris is removed by filtering the suspension thus obtained through a nylon sieve having a pore diameter which is slightly wider than the diameter of the pollen (e.g., 40 microns for *Nicotiana*, 100 microns for maize, etc.).

This pollen suspension is then centrifuged down at low speed (500-800 rpm for 5 min), the supernatant containing fine debris is discarded and the pellet of pollen is resuspended in fresh medium and washed twice. Rinsing is particularly important if the anther tissue contains inhibitory substances, which could prevent growth of the microspores.

3. *Culture of Microspores.* The pollen obtained above is mixed with an appropriate culture medium (see V *B* 3) at a density of 10^3 to 10^4 grains per ml. The final suspension is pipetted into 5 cm wide pyrex petri dishes at 2.5 ml per dish.

To ensure good aeration, the layer of liquid in the dish should be as thin as possible. Each dish is then sealed with Parafilm^R to avoid dehydration, and 14 dishes are placed together in a 20 cm-wide dish. These are incubated at 25°C for *Nicotiana*.

In species in which the flower buds are very small, as for example with fruit trees or cereals, it is difficult to remove the anthers. In these cases we separate each floret from the main axis, cutting off the bud at the level of the receptacle and treat the whole bud in the same manner as for the anthers of *Nicotiana*. A pollen suspension can be just as easily prepared using the whole buds as by using anthers only.

The microspores are best grown in liquid medium. The cultures are placed at 25°C under low intensity white light (500 lux). Young embryos can be observed after 30 days under such conditions. Pro-embryos can be observed after staining two week-old cultures. For this early observation, the two week-old pollen suspension is centrifuged down, the pollen is hydrolyzed with 5N HCl for 18 minutes, then washed with distilled water. One more centrifugation is necessary to eliminate the water. A few drops of Schiff's reagent are added onto the pollen, which is kept in a closed tube in the dark for one hour. Microscopic observation will show pro-embryos of about 10 cells.

REFERENCES

1. Anagnostakis, S.L., *Planta 115*, 281 (1974).
2. Anonymous, *Acad. Sinica Acta Gen. 2*, 302 (1975).
3. Behnke, M., *Z. Pflanzenzuchtg. 75*, 262 (1975).
4. Bosemark, N.O., *Hereditas 69*, 193 (1971).
5. Bourgin, J.P., and Nitsch, J.P., *Ann. Physiol. Vég. 9*, 377 (1967).
6. Breukelen, E.W.M., Ramanna, M.S., and Hermsen, J.G.T., *Euphytica 24*, 567 (1975).
7. Chen, Y., and Li, L.T., *In* "Proc. Symp. on Plant Tissue Culture", p. 199. Science Press, Peking (1978).
8. Chu, C.-c., *In* "Proc. Symp. on Plant Tissue Culture", p. 43. Science Press, Peking (1978).
9. Chuang, C.-c., Duyang, T.-w., Chia, H., Chou, S.-m., and Ching, C.-k., *In* "Proc. Symp. on Plant Tissue Culture", p. 51, Science Press, Peking (1978).
10. Foroughi-Wehr, B., Wilson, H.M., Mix, G., and Gaul, H., *Euphytica 26*, 361 (1977).
11. Halperin, W., and Wetherell, D.F., *Am. J. Bot. 51*, 274 (1964).
12. Heberle, E., and Reinert, J., *Naturwissenschaften 64*, 100 (1977).
13. Hondelmann, W., and Wilberg, B., *Z. Pflanzenzuchtg. 69*, 19 (1973).
14. Hughes, K.W., Henke, R., and Constantin, M., (eds.), "Propagation of Higher Plants through Tissue Culture", Conf. 7804111, U.S. Tech. Inf. Serv., Springfield, VA. (1979).
15. Irikura, Y., *Res. Bull. Hokkaido Nat. Agric. Exp. Station 112*, 1 (1975).
16. Jacobsen, E., *Inaugural Dissertation, Bonn* (1978).
17. Jacobsen, E., and Sopory, S.K., *Theor. Appl. Genet. 52*, 119 (1978).
18. Keller, W.A., and Armstrong, K.C., *Z. Pflanzenzuchtg. 80*, 100 (1978).
19. Keller, W.A., and Stringam, G.R., *In* "Frontiers of Plant Tissue Culture 1978", (T.A. Thorpe, ed.), p. 113. University of Calgary Press, Calgary, (1978).
20. Keller, W.A., Rajhathy, T., and Lacapra, J., *Can. J. Genet. Cytol. 17*, 655 (1975).
21. Kohlenbach, H.W., and Wernicke, W., *Z. Pflanzenphysiol. 86*, 463 (1978).
22. Morel, G., and Martin, C., *C.R. Acad. Sci. Paris 235D*, 1324 (1952).
23. Murashige, T., and Skoog, F., *Physiol. Plant. 15*, 473 (1962).

24. Nitsch, C., *In* "Genetic Manipulation with Plant Material", (L. Ledoux, ed.), p. 297. Plenum Press, New York, (1975).
25. Nitsch, C., *In* "Cell Genetics in Higher Plants", (D. Dudits, G.L. Farkas and P. Maliga, eds.), pp. 221, 225. Akadémiai Kadó, Budapest, (1976).
26. Nitsch, C., *In* "Plant Cell, Tissue and Organ Culture", (J. Reinert and Y.P.S. Bajaj, eds.), p. 268. Springer-Verlag, Berlin, (1977).
27. Nitsch, C., and Norreel, B., *C.R. Acad. Sci. Paris 276D*, 303 (1973).
28. Nitzsche, W., and Wenzel, G., *Adv. Plant Breeding 8*, 1 (1977).
29. Oono, K., *Bull. Nat. Inst. Agric. Science, Tokyo, No. 26*, (1975).
30. Picard, E., and de Buyser, J., *C.R. Acad. Sci. Paris 281D*, 127 (1975).
31. Reinert, J., and Bajaj, Y.P.S., (eds.), "Plant Cell, Tissue, and Organ Culture", Springer-Verlag, Berlin, (1977).
32. Sharp, W.R., Larsen, P.O., Paddock, E.F., and Raghavan, V., (eds.), "Plant Cell and Tissue Culture", Ohio State Univ. Press, Columbus (1979).
33. Sopory, S.K., Jacobsen, E., and Wenzel, G., *Plant Sci. Lett. 12*, 47 (1978).
34. Sunderland, N., and Robert, M., *Nature 270*, 236 (1975).
35. Thomas, E., and Wenzel, G., *Naturwissenschaften 62*, 40 (1975).
36. Thomas, E., Hoffman, F., and Wenzel, G., *Z. Pflanzenzuchtg. 75*, 106 (1975).
37. Thorpe, T.A., (ed.), "Frontiers of Plant Tissue Culture 1978", Univ. of Calgary Press, Calgary, (1978).
38. Vasil, I.K., (ed.), "Perspectives in Plant Cell and Tissue Culture", *Int. Rev. Cytol.* Suppl. 11A. Academic Press, New York, (1980).
39. Vasil, I.K. (ed.), "Perspectives in Plant Cell and Tissue Culture", *Int. Rev. Cytol.* Suppl. 11B, Academic Press, New York, (1980).
40. Wenzel, G., *Kartoffelbau 30*, 126 (1979).
41. Wenzel, G., Hoffman, F., and Thomas, E., *Theor. Appl. Genet. 48*, 205 (1976).
42. Wenzel, G., Hoffman, F., and Thomas, E., *Z. Pflanzenzuchtg. 78*, 149 (1977).
43. Wernicke, W., Harms, C.T., Lorz, H., and Thomas, E., *Naturwissenschaften 65*, 540 (1978).
44. Wilson, H.M., *Plant Sci. Lett. 9*, 233 (1977).

IN VITRO FERTILIZATION AND EMBRYO CULTURE

Edward C. Yeung
Trevor A. Thorpe

Department of Biology
University of Calgary
Calgary, Alberta, Canada

C. John Jensen

Agricultural Research Department
Risø National Laboratory
Roskilde, Denmark

I. INTRODUCTION

Pollination followed by fertilization normally leads to the production of an embryo, which in the intact plant is linked with normal seed development. Most angiosperms are outbreeders, and thus self pollination is limited. Furthermore interspecific and intergeneric hybridizations are also rare in nature. In plant breeding, however, selfing and hybridization are methods commonly used to obtain desirable crosses. *In vitro* pollination and fertilization is one experimental approach which aids in achieving the above. Since many of the embryos thus obtained will not survive, they must often be rescued. Thus zygotic embryo culture methods become important. These twin aspects form the basis of this chapter. The basic principles and applications of these techniques will be outlined. This will be followed by a protocol for embryo culture.

ISBN 0-12-690680-7

II. *IN VITRO* POLLINATION AND FERTILIZATION

The technique of *in vitro* pollination and fertilization was pioneered by scientists at the University of Delhi, India, in the early 1960's using *Papaver somniferum* (8). This was followed by success with *Argemone mexicana*, *Eschscholzia californica*, *Dicranostigma franchetianum*, *Nicotiana tabacum*, *N. rustica*, *Petunia violacea*, and *Antirrhinum majus* (26). These topics have been discussed in detail by Rangaswamy (26) and Zenkteller (36).

A. Principles

Success in this technique depends upon two basic considerations (26). These are (1) using pollen grains and ovules at the proper stage, and (2) obtaining nutrient media that will support pollen germination, pollen tube growth and embryo development. Thus many aspects of the floral biology of the test species must be understood. These include anthesis, dehiscence of anthers, pollination, pollen germination, pollen tube growth, ovule penetration, fertilization, and embryo and endosperm development.

Pollen is normally collected by aseptically excising ready-to-dehisce anthers from flower buds or flowers. The anthers are then allowed to dehisce on a suitable sterilized substratum, e.g., filter paper.

To prevent *in vivo* pollination, flower buds are emasculated (in bisexual species) and bagged prior to anthesis. Controlled pollination can be carried out on (1) the stigma, (2) the placenta, or (3) the ovule. For stigmatic pollination, the bagged flower buds are collected just before or at the day of anthesis. After aseptically removing all floral parts except the pistil, the stigma is pollinated aseptically, and the pistil is then implanted in sterilized agar nutrient medium. Pistils excised before anthesis are kept in sterile culture until pollinated. In carrying out placental or ovular pollination, the organ involved is excised aseptically and pollen is deposited on it and then it is placed in culture. Fertilization takes place in 1 to 3 days, and the fertilized ovules are usually transferred to a new medium to allow for embryo development.

B. Applications and Problems

1. Production of Hybrids. Work by Zenkteller and his colleagues showed that the technique could be used to produce interspecific and intergeneric hybrids, which would otherwise not be formed in nature (37). Using *Melandrium album* and *M. rubrum* as ovule parents, they were able to obtain development with 15 different species in four families as pollen parents. Thus the technique opens up the possibility of obtaining novel, but useful hybrids.

2. *Overcoming Sexual Self-Incompatibility*. This is a physiological barrier which prevents fusion of sexually different gametes, which are otherwise fertile, produced by the same individual of a heterosporous species. The phenomenon of incompatibility is discussed by Nettancourt and Devreux (17). Most success has been achieved with placental pollination (27). In the process, the entire ovule mass of an ovary intact on the placenta is cultured after pollination. This allows the original *in vivo* arrangement of the ovules to be retained.

3. *Induction of Haploid Plants*. Techniques such as delayed pollination, distant hybridization, pollination with abortive or irradiated pollen, and physical and chemical treatments to the ovary (14) have been used in attempts to induce haploidy. More recently anther (microspore) culture has been tried more successfully (see Chapter 8). An example of the use of *in vitro* pollination and fertilization is the work of Hess and Wagner (5). They obtained about 1% haploids of *Mimulus luteus*, when the ovule mass was pollinated *in vitro* with pollen from *Torenia fournieri* (both species from the family, Scrophulariaceae).

4. *Studies on Pollen Physiology and Fertilization*. The technique is potentially useful in studying aspects of pollen physiology and fertilization. As an example, Balatkova and Tupy were able to determine that pollen tubes of *Nicotiana tabacum* could affect fertilization even after gamete formation in pollen cultures (1).

As Rangaswamy (26) has pointed out the extent to which the technique can be put to practical use is governed by (1) the number and nature of the parameters that affect the operation, and (2) the ease with which the parameters can be controlled. The measure of success of test tube fertilization is the production of viable seed. This is dependent upon many factors, most of which have nothing to do directly with the technique. Such factors as (1) age of the explants, particularly the ovules, (2) adequate pollen germination, (3) proper growth of pollen tubes and microgametogenesis, (4) pollen tube entry into ovules, and (5) the degree of fertilization, all influence the development of the seed.

Little is known about the effects of physical factors such as light, temperature and humidity on *in vitro* pollination and fertilization. Furthermore such factors as the location of placement of the pollen, and keeping the ovule and placenta surfaces free of water play a role in achieving success in *in vitro* pollination (26).

In addition to the above factors relating to fertilization, success is also influenced by factors involved in the development of the fertilized ovule into seeds and of the zygote into mature viable embryos (26). The nutrient medium is obviously very important. One component that has had a beneficial affect in some cases has been the addition of casein hydrolysate (500 mg/1) (2, 13). Thus, it is possible for normal development of viable seeds to occur. However, in some cases embryo degeneration and other abnormal events take place. Under these conditions embryos may be excised and grown to maturity in culture.

III. EMBRYO CULTURE

Plant embryo culture has its origins at the beginning of this century in the independent pioneering studies of Hännig and Brown (19). Among the outstanding early researchers were LaRue (12) and Tukey (32) who encountered problems of precocious germination in their studies. A major breakthrough was the discovery by van Overbeek *et al.* (34) that very small (less than 200 μm) hybrid embryos of *Datura* could be grown in culture in media containing coconut milk. As such, plant embryo culture is one of the earliest branches of *in vitro* culture to be applied to practical problems, such as with orchid seeds. This topic of embryo culture has been discussed recently by Raghavan (21-23), Monnier (15, 16) and Norstog (19). A list of plants in which embryo culture techniques have been applied can be found in Raghavan (21). That reference also indicates the variety of media that have been used successfully.

A. *Principles*

The underlying principle of the method is the aseptic excision of the embryo and its transfer to a suitable medium for development under optimum culture conditions. Some of the methods used in the excision and culture of embryos of vascular plants have been described by Sanders and Ziebur (29), Raghavan (20) and Torrey (31). In general, it is relatively easy to obtain pathogen-free embryos, since they are within the sterile environment of the ovule. Thus entire ovules, seeds or capsules containing ovules are sterilized and the embryos aseptically separated from the surrounding tissue (22).

Seeds with hard seed coats are generally surface sterilized and then soaked in water for a few hours to a few days. In the latter case, they are surface sterilized again before embryo excision. Splitting the seeds under aseptic conditions and transferring the embryos directly to a nutrient agar medium is the simplest technique used. With smaller embryos it is important that they be removed uninjured. This often requires the use of a dissecting microscope. In the case of orchids the entire ovule is cultured, since there are no functional storage tissues and the seed coat is reduced to a membranous structure (24). It is also important that the excised embryo not become desiccated during the above operations.

The most important aspect of embryo culture work is the selection of the medium necessary to sustain continued growth of the embryo. The nutrient media formulations used over the years vary tremendously, and many of them have not been rigorously determined. Nevertheless it is possible to make certain generalizations, i.e., the younger the embryo, the more complex is its nutrient requirements. Thus, while relatively mature embryos can be grown in an inorganic salt medium supplemented with a carbon energy source such as sucrose, relatively young embryos require, in addition, different combinations of vitamins, amino acids, growth hormones and in some cases natural endosperm extracts such as coconut milk. Since embryos are often embedded in the ovular sap under considerable osmotic pressure, culture in the presence of an osmoticum, such as mannitol, is also recommended (22).

The changing nutritional requirement for successful embryo culture has often meant transferring the embryo from one medium to another for optimum growth. However, Monnier devised a culture method, illustrated in Figure 1, which allowed for the uninterrupted growth of globular stage (ca. 50 μm) *Capsella* embryos to maturity (15, 16). The composition of the medium used in each part of the culture dish is shown in Table I. He stressed the importance of obtaining uninjured embryos with their suspensors intact for successful culture. The importance of the suspensor for the growth of young embryos has been confirmed by Yeung and Sussex in their studies on plantlet formation from cultured embryos of *Phaseolus coccineus* (Table II) (35). Embryos selected at later stages did not require the attached suspensor for normal development.

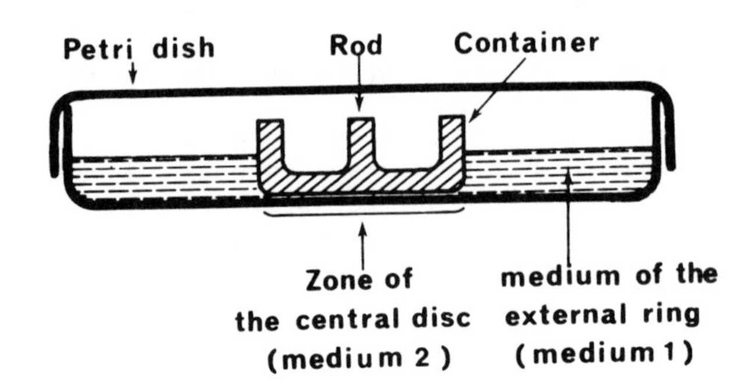

FIGURE 1. Device allowing the juxtaposition of two media with different compositions (Table I). The first agar medium is liquified by heating and then poured around the small central glass container. This medium will give the external ring. After cooling and solidification of the agar medium, the central container is removed. In the hole the second medium of a different composition is poured. The embryos are cultivated on this second medium, in the central part of the petri dish, and, as a result of diffusion, they are subjected to the action of a variable medium with time (15, 16).

After the embryos have grown into plantlets *in vitro*, they are generally transferred to sterile soil or vermiculite and grown to maturity in the greenhouse.

B. *Applications*

1. *Overcoming Embryo Inviability.* Embryo abortion occurs quite frequently as a result of unsuccessful crosses in breeding. Although in most of these cases successful fertilization and early embryo development occurs, a number of irregular events subsequently takes place, resulting in the eventual death of the embryo and collapse of the seeds (22). A major cause of early embryo abortion is the failure of the endosperm to develop properly. Aseptically culturing the embryo in a nutrient medium has often overcome this problem. Successful cases have included embryos arising from *interspecific hybrids* in cotton, barley, tomato, rice, and jute.

TABLE I. Composition of the Two Media used in Different Parts of the Culture Dish (Figure 1) to obtain Uninterrupted Growth of Globular Stage Capsella Bursa-pastoris Embryos to Maturity (15)

Chemical	*Medium 2 (Central zone) mg/l*	*Medium 1 (External ring) mg/l*
KNO_3	1900	1900
$CaCl_2 \cdot 2H_2O$	1320	484
NH_4NO_3	825	990
$MgSO_4 \cdot 7H_2O$	370	407
KCl	350	420
KH_2PO_4	170	187
Na_2EDTA	0	37.3
$FeSO_4 \cdot 7H_2O$	0	27.8
H_3BO_3	12.4	12.4
$MnSO_4 \cdot H_2O$	33.6	33.6
$ZnSO_4 \cdot 7H_2O$	21	21
KI	1.66	1.66
$Na_2MoO_4 \cdot 2H_2O$	0.5	0.5
$CuSO_4 \cdot 5H_2O$	0.05	0.05
$CoCl_2 \cdot 6H_2O$	0.05	0.05
Glutamine	600	0
$B_1 = B_6$	0.1	0.1
Sucrose	180,000	0
Agar (Difco)	7,000	7,000

Success has also been achieved with *intergeneric hybrids*, e.g., of *Hordeum* and *Secale*, *Hordeum* and *Hordelymus*, *Triticum* and *Secale*, and *Tripsacum* and *Zea* (see ref. 22 for details).

2. *Monoploid Production in Barley.* One novel use of embryo culture has been in the production of monoploids and doubled monoploids of barley. The cultivated barley (*Hordeum vulgare*) is a diploid species ($2n = 2x = 14$). It is one of the oldest cereals known to man and ranks fourth in the world's cereal production (only surpassed by wheat, rice and maize). Barley is a self-pollinating annual and is found as either two- or six-rowed, spring or winter types. Traditionally, the barley grain (a caryopsis) has chiefly been used in human diet, animal feed, and in producing alcoholic beverages. Its conspicuous chromosomes, its morphological characters, and

TABLE II. The Effect of the Suspensor on Growth and Formation of Plantlets of Phaseolus coccineus Embryos Cultured in vitro[a]

Treatment	*Fresh weight ± standard error*	*No. of plantlets/ total no. cultured (% precocious germination)*
Embryo-proper only	*3.19 ± 0.51*	*37/89 (42)*
Embryo-proper with suspensor attached	*8.91 ± 1.16*[b]	*84/95 (88)*
Embryo-proper with detached suspensor in direct contact	*6.22 ± 0.78*[b]	*37/51 (73)*
Embryo-proper with heat-killed detached suspensor in direct contact	*4.10 ± 0.43*	*16/43 (37)*

[a]*Embryos were cultured at the heart stage and had an average fresh weight of 0.87 ± 0.02 mg at the time of culture. Growth measurements were made after 10 days in culture, whereas the results on plantlet formation were taken after 8 weeks in culture. For details see Yeung and Sussex (35).*
[b]*significance at the 1% level.*

ease of handling have made barley valuable for mutation studies. Genetically it is among the best known higher plants. It is estimated that there are about 15,000 to 20,000 cultivars. Jensen listed many of the advantages of monoploids as tools in plant breeding or genetics (7). Monoploids in barley are sporophytes with the basic chromosome number of 2n = x = 7. The frequency of spontaneous monoploids is extremely low and there are thus two artificial routes for their production (6). One route is via anther or microspore culture (see Chapter 8), which has given very limited success to date. The alternative method involves the female gametes (megaspores) in hybridization, followed by somatic chromosome elimination. The first report of this latter approach was by Kao and Kasha (10), and the principles are discussed by Kasha (9).

In essence, the technique is based on an interspecific cross with *Hordeum vulgare* as the female and *H. bulbosum* as the male. The scheme is illustrated in Figure 2. Fertilization proceeds readily and zygote induction is high. However, the chromosomes of *H. bulbosum* are rapidly eliminated from the cells of the developing embryo. The endosperm develops for 2-5 days and then disintegrates. In the developing monoploid embryo cells the division and development is slower

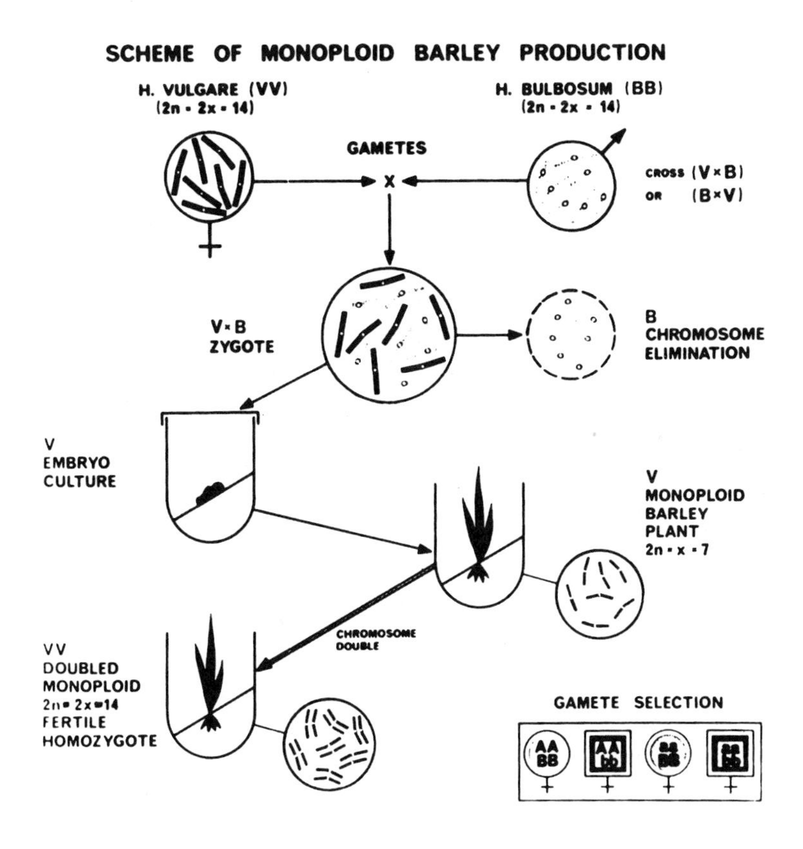

FIGURE 2. Scheme of monoploid and doubled monoploid barley production via the interspecific hybridization method followed by elimination of H. bulbosum chromosomes (6, 7).

than the diploid cells. This slower growth together with the disintegration of the endosperm leads to the formation of small embryos which have to be dissected from the fruits and cultured in order to complete development and germinate. Following embryo culture, the developing plantlets are raised under normal greenhouse conditions and chromosome doubling induced on established plants. The method has the advantage that very high frequencies of female gametes are induced to form embryos, from which the *H. bulbosum* chromosomes are lost. This approach also works with hexaploid wheat (3), and is thus potentially a very useful method in plant improvement.

3. *Overcoming Seed Dormancy and Related Problems*. The causes of seed dormancy are varied and include the presence of endogenous inhibitors, specific light, temperature, and storage requirements, and immaturity of the embryo. As examples, embryo culture has been successfully used to overcome dormancy due to inhibitors in *Iris* seeds (25); while Knudson (11) succeeded in germinating immature orchid embryos in a nutrient agar medium containing sugar. This latter study led ultimately to the commercial application of embryo culture for orchid propagation.

Physiologically related to the problem of dormancy is the requirement for specific signals from the host for initiating germination in seeds of obligate parasites. This problem has been overcome by embryo culture. One apparent necessary component in the medium is casein hydrolysate, indicating that specific amino acids may well be required from the host for germination.

Embryo culture has also proven useful in reducing the breeding cycle of new varieties in cases where long dormancy and/or slow growth of the seedlings cause long breeding seasons, as in roses and weeping crab apple. A potential use of the technique is in the production of seedlings from seed of naturally vegetatively-propagated plants such as bananas and *Colocasia*, whose seeds do not germinate in nature.

Finally, the use of embryo culture, in seed testing trials, has proven to be very useful in determining the viability of particular seed lots. This use arose out of early findings by Tukey that there was a good correspondence between the growth of excised embryos of non-after-ripened peach seeds and germination of the after-ripened seeds (33).

4. *In Studies in Experimental Embryogenesis*. In addition to the applied uses of embryo culture the technique has proven useful in studies of the growth requirements of embryos, the effects of phytohormones and environmental conditions on zygotic embryogenesis, the nutrition and metabolism of

embryos, etc. These topics and others have been reviewed recently by Norstog (18) and Raghavan (23). One major problem in embryo culture is precocious germination. In this phenomenon, excised immature embryos of some species do not continue normal embryo development in culture, but grow directly into small, weak plantlets. It appears that an interplay of inhibitors, high osmotic pressure and possibly low O_2 tension in the ovule may be involved in the *in vivo* regulation of normal embryo development.

IV. PROTOCOL FOR EMBRYO CULTURE

The following protocol for embryo culture is based on the methods used by Jensen (6, 7) for barley at Risø. With modification this basic protocol should be applicable to embryo culture in general. In any case empirical studies are needed with new test material.

A. *Material*

1. *Growing Spring Barley for Donor Plants: Selfed Embryos, Haploid and Hybrid Production.* Use a standard soil to plant the seed. Water them and place pots in a lighted greenhouse on rockwool (glass-wool) mats. These mats are watered and fertilized automatically via pump and plastic trickle tubes. The greenhouse is lit by 400 W electric lamps (HQIL-Osram) giving a light intensity of about 65 x 10^3 lux (3 lamps per m^2), for 18 hrs. per day. The temperature regime is 18^oC day, and 13^oC night. The time from planting the seeds to flowering is approximately 6 weeks. About two weeks after the emergence of the plants, the soil is treated with a systematic fungicide and insecticide.

2. *Cut-Shoots of Barley* (from field grown barley). This technique provides for controlled nutrient feeding to the spikes (30). The cut material is placed in 20 l containers with a modified Hoagland solution. Shoots are cut just before the emasculation stage and placed in a fresh water culture solution which is aerated via compressed air (reduction valve). Containers are placed in a growth room. These can be used for controlled pollination or other experimental manipulations (6).

Ovaries are ready for excision for culture from about 3-5 days (very small) onward. Ovaries 18-26 days old are large enough for ease in handling. Obviously the age selected will be dependent on the objective of the work.

B. Procedure

1. Harvesting and Sterilization of Ovaries (Fruits on Spikes).

(1) Remove fruits carefully from spike and 'dehusk' (remove lemma and palea) and place ovary ("fruit") in a clean glass petri dish (5 cm). Gather ovaries from one spike into one dish and label.
(2) Add clean water, to which a few drops of Tween 80^R or Tween 20^R have been added.
(3) Prepare a 5% solution of calcium hypochlorite, $Ca(ClO)_2$. To make 300 ml sterilizing solution weigh out 15 g $Ca(ClO)_2$ (*using a face mask or bind a cloth to protect mouth and nose*). Add water to make 300 ml. Stir and filter. Add 3 ml Tween 80^R to the filtered solution (solution I).
(4) Cover the ovaries in the petri dishes with *solution I*, shake and discard the solution, i.e., via hypodermic syringe, or apply vacuum suction or decant through a net.
(5) Pour fresh *solution I* and cover ovaries, leave for 15 min. Shake occasionally.
(6) Discard *solution I* and rinse with sterile water.
(7) Repeat four times, adding sterile water, shaking and discarding. Do this operation in a sterile environment, i.e., laminar flow bench.
(8) Add last sterile water solution to just cover the ovaries. Store if necessary at 5°C in the dark (can be kept for one week).

2. Excision and Culture.

(1) Clean the laminar air-flow bench, dissecting utensils and dissecting microscope with 70% ethyl alcohol
(2) Organize the working area inside sterile flow bench:
   i) Dissecting sterile petri dishes
   ii) 5 to 6 cm diameter sterile plastic petri dishes (treated for all cultures).
   iii) Heavy glass container with lid to hold approximately 50 ml 96% ethyl alcohol (for dipping and flaming dissecting instruments).

iv) Heavy glass container with sterile, distilled water for cooling flamed instruments.

v) Gas Bunsen burner on solid footing.

vi) Sterile paper towels.

vii) Dissecting microscope (approximately from 5x to 100x magnification); sliding stage; glass fiber (cool) light source.

viii) Dissecting instruments:

a) Dissecting needles.

b) Surgical blade and handle (preferably 'Swan-Morton' blade No. 11).

c) Pointed forceps (Kosmo no. 5, Fontax-Soloox, Junker S. A. No. 5, or Crem 5 SA - (anti-acid). Dumont forceps are also suitable).

d) Micro-dissecting kit (see below).

Micro-dissecting equipment. When excising embryos smaller than 100 μ, special fine pointed needles and scalpels are needed. It is cheapest to make these yourself from tungsten (W) wire. Commercial micro-dissecting utensils are available from: ROBOZ, Special Instrument Co., Inc. (can be obtained via Fisher Scientific Company, International Division, 52 Fadem Rd., Springfield, N.J. 07081, U.S.A.). The following are of advantage to use:

Micro knife (Catalog No. RS-375)
Dissecting needles (tungsten, 6 microm.) (No. RS-378)
Dissecting scissors (type: Vannes, No. RS-620)

However, these instruments are fairly costly. To prepare your own from tungsten wire refer to Cutter (4). Also refer to Romberger (28). In addition, suitable micro-dissecting instruments may be obtained from Medical Supply Companies, (e.g., Fine Science Tools Ltd., 1687 Arborlynn Drive, North Vancouver, B.C., Canada V7J 2V7).

ix) Plastic gloves (e.g., Kimguard disposable gloves).

x) Plastic (polyethylene) wash bottle (preferably 250 ml) with flexible walls to contain 70% ethyl alcohol.

xi) Marker (water and alcohol resistant speed marker), marking tape.

xii) Support, sterilized, to place dissecting instruments upon.

xiii) Sterile glass container for waste water.

xiv) Parafilm^R strips.
xv) Culture media.
xvi) Sterile water.
xvii) Sterile Pasteur pipets with stopper.

(3) Sterilize your hands (gloved) and instruments and excise embryos:

i) Place sterile dissecting petri dish-lid (or -bottom) on microscope stage.

ii) Take one ovary at a time from the 'stored', prepared dish onto the dissecting dish, view under magnification and locate the basal end holding the embryo.

iii) Using one hand, hold the ovary lightly in the middle with a pair of flamed forceps. With the other hand make an excision (with a flamed, *sharp* scalpel) across the ovary at about 2/3 of its length from the base (the micropylar end) and push the chalazal end away. Flame and cut carefully from the cut surface of the ovary along one side. Flame and very carefully cut again along the other side. Now flame and cool in sterile water and flip the ovary wall open to expose the embryo sac or embryo. Carefully push your *flamed* and *cooled* needle or scalpel underneath the embryo (or sac) lift it out and push it into the washing solution (liquid culture medium). Washing the embryo helps to remove inhibiting substances from the endosperm.

iv) Make a note on the embryo size and development.

v) Quickly transfer the embryo to the prepared dish with the culture medium (2 ml per dish), close the lid after each transfer. Care must be taken not to damage the embryo and to place it with the ventral surface on the medium.

vi) Seal the dish with Parafilm^R, label and place at the appropriate temperature and light conditions. Very young embryos are cultured initially in the dark for 1-2 weeks at 18°C, as light encourages precocious germination. When properly differentiated, the embryos are moved to 12 hr light regime (500-1000 lux) at a temperature of 20° to 22°C. With further differentiation, the embryos are transferred to a solid medium in culture tubes or vials, where development into plants with good roots occur within 1-2 weeks. Established plants are then transferred to higher light intensities (40-50 x 10^3 lux).

(4) Terminate embryo culture and start plant culture.
   i) Prepare a light soil mixture using ca. 1/2 standard soil to 1/2 sand (by volume).
   ii) Place pots in a flat; plant the plantlets from the *in vitro* sterile dishes directly into the soil which has been watered lightly.
   iii) Place a plastic tent over the flat containing the potted plantlets and place in the greenhouse.
   iv) After ca. 1 week, remove the top of the plastic tent and harden the plantlets to the drier atmosphere of the greenhouse.
   v) After another week, remove the plastic tent. Water more frequently when growth is resumed.
   vi) Plant into bigger pots using the standard soil.

C. Culture Media

1. Chemicals. It is advisable to buy analytical grade chemicals. Store chemicals in the dark in the dry state. Vitamins, hormones and amino acids should be stored at the recommended temperatures, i.e., store dry at $-18^{o}C$ and $+5^{o}C$, as required.

2. Preparation of Stock Media. Stock solution should be prepared as 10x or 100x strength of the final concentration and stored cold ($5^{o}C$). Mineral stock solutions should be prepared in separate groups as follows: nitrates, sulfates, halides, irons, and P&Mo. All organic stocks should be stored sterile and cold (preferably frozen at $-18^{o}C$).

Another method is to make up a 10x strength stock solution, complete for all salts, and store it as a ready-mix solution in the frozen state ($-18^{o}C$) in plastic bags. Sugars and other organic compounds are added when the solution is finally made up to normal strength.

3. Preparation of Liquid Media. Following the formulas in Table III for media, C-17 and C-21, the stock solutions are mixed, water added, pH adjusted and sterilized via a sterile membrane filter unit (e.g., Millipore filter-type: GS 0.22 μ, filterholder type: XX430700). The medium is forced through the filter by means of pressurized air of about 2.5 p.s.i. (The macro- and micronutrients may be autoclaved).

The sterility of the medium is checked by testing it for contamination before use. This test is simple: pour 5 ml into a sterile petri dish, let it stand for 4 days at $28^{o}C$ and screen for contamination.

TABLE III. Composition of Media used in Embryo Culture of Barley[a]

Chemical	B-II[b] mg/l	C-17[c] mg/l	C-21[d] mg/l	C-45[e] mg/l
MACRONUTRIENTS				
KNO_3		300	300	900
$CaCl_2 \cdot 2H_2O$	740	250		400
$MgSO_4 \cdot 7H_2O$	750	325	300	300
$(NH_4)_2SO_4$				60
$NaH_2PO_4 \cdot H_2O$		100		75
KCl	750	150	300	
$KH_2PO_4 \cdot H_2O$	910	150	500	170
$Ca(NO_3)_2$			500	300
NH_4NO_3		200		500
MICRONUTRIENTS				
KI		0.10		
H_3BO_3	0.5	0.5	15.0	1.0
$MnSO_4 \cdot 4H_2O$	3.0	0.5		5.0
$ZnSO_4 \cdot 7H_2O$	0.5	0.25		5.0
$Na_2MoO_4 \cdot 2H_2O$	0.025	0.012		0.25
$CuSO_4 \cdot 5H_2O$	0.025	0.012		0.012
$CoCl_2 \cdot 6H_2O$	0.025	0.012		0.012
Na_2EDTA				
$FeSO_4 \cdot 7H_2O$				
Ferric citrate	10.0	3.0	20.0	20.0
Fe EDTA		17.5	10.0	28.0
VITAMINS				
Nicotinamide				1.0
Thiamine HCl	0.25	0.25	10.0	10.0
Pyridoxine HCl	0.25	0.25		1.0
Inositol	50.0	50.0	150	100.0
Ca-pantothenate	0.25	0.25		
Glycine		0.75		
L-ascorbic acid		0.50		1.0
AMINO ACIDS				
L-glutamine	400			600.0
L-glutamic acid		150	300	
L-alanine	50	30		100.0
L-cysteine	20			
L-arginine	10	20	50	
L-leucine	10	10		
L-phenylalanine	10	20		
L-tyrosine	10			
L-tryptophan	10			

TABLE III. (continued)

Chemical	B^{II} [b] mg/l	C-17 [c] mg/l	C-21 [d] mg/l	C-45 [e] mg/l
AMINO ACIDS (cont'd)				
L-aspartic acid		30	100	100.0
L-proline		50	50	
L-valine		10		
L-serine		25	25	50.0
L-threonine		10		100.0
SUCROSE	34,000	60,000	45,000	45,000
AGAR (DIFCO)	6,000*	-	-	-
*Purified				
pH	5.0	5.5	5.5	5.8

[a] In addition: Media B^{II}, C-17, C-21, and C-45 have the following additives per l medium:

B^{II}: Malic acid -1 g dissolved in 50 ml H_2O, pH brought to 5.0 with NH_4OH.

C-17: Citric acid, 500 mg in 50 ml H_2O, pH adjusted to 5.3 with NH_4OH. Tri-potassium citrate, 300 mg, added to final medium, pH of medium adjusted with KOH. Filter sterilize.

C-21: Citric acid, 50 mg in 50 ml H_2O, and tri-potassium citrate, 250 mg, pH adjusted to 5.0 with NH_4OH. Added to medium and final pH brought to pH 5.5 with KOH. Filter sterilize.

C-45: Malic acid - 0.3 g dissolved with citric acid - 0.3 g, in 50 ml water, pH brought up to 5.0 with NH_4OH.

[b] B^{II} = Medium described by Norstog (18).

[c] C-17 = Medium used by Jensen based partially on B^{II} and other media used as liquid culture on small and non-uniform monoploid embryos (6, 7).

[d] C-21=Medium used by Jensen on uniform, well developed embryos (6, 7).

[e] C-45 = Medium used by Jensen on embryos of barley, from one to two weeks old (6, 7).

Before using a medium on a large scale it is best to check its accuracy by testing it against a standard culture performance.

Use 2 ml of the sterile medium for sterile plastic petri dishes (6 cm). In the liquid culture method, a sterile Millipore[R] (MF, 47 mm 0.45 μm) filter is placed on 8 ml of medium in a 5-6 cm petri dish and the embryos placed thereon.

REFERENCES

1. Balatkova, V., and Tupy, J., *Biol. Plant. Acad. Sci. Bohemoslov. 10*, 266 (1968).
2. Balatkova, V., and Tupy, J., *Biol. Plant. Acad. Sci. Bohemoslov. 14*, 82 (1972).
3. Barclay, I.R., *Nature 256*, 410 (1975).
4. Cutter, E.G., *In* "Methods in Developmental Biology", (F.W. Wilt and N.K. Wessels, eds.), p. 623. T.Y. Crowell Co., N.Y., (1967).
5. Hess, D., and Wagner, G., *Z. Pflanzenphysiol. 72*, 466 (1974).
6. Jensen, C.J., *In* "Barley Genetics III", (H. Gaul, ed.), p. 316. Verlag Karl Thiemig, München, (1976).
7. Jensen, C.J., *In* "Plant Cell, Tissue and Organ Culture", (J. Reinert and Y.P.S. Bajaj, eds.), p. 299. Springer-Verlag, Berlin, (1977).
8. Kanta, K., Rangaswamy, N.S., and Maheshwari, P., *Nature 194*, 1214 (1962).
9. Kasha, K.J., *In* "Haploids in Higher Plants", (K.J. Kasha, ed.), p. 67. University of Guelph Press, Guelph, (1974).
10. Kao, K.N., and Kasha, K.J., *In* "Barley Genetics II", (R.A. Nilan, ed.), p. 82. Washington State Univ. Press, Pullman, (1969).
11. Knudson, L., *Bot. Gaz. 73*, 1 (1922).
12. LaRue, C.D., *Bull. Torrey Bot. Club 63*, 365 (1936).
13. Maheshwari, P., and Kanta, K., *In* "Pollen Physiology and Fertilization", (H.F. Linskens, ed.), p. 187. North-Holland, Amsterdam, (1964).
14. Maheshwari, P., and Rangaswamy, N.S., *Adv. Bot. Res. 2*, 219 (1965).
15. Monnier, M., *Rev. Cytol. et Biol. vég. 39*, 1 (1976).
16. Monnier, M., *In* "Frontiers of Plant Tissue Culture 1978", (T.A. Thorpe, ed.), p. 277. University of Calgary Press, Calgary, (1978).
17. Nettancourt, D. de, and Devreux, M., *In* "Plant Cell, Tissue and Organ Culture", (J. Reinert and Y.P.S. Bajaj, eds.), p. 426. Springer-Verlag, Berlin, (1977).

18. Norstog, K., *In Vitro 8*, 307 (1973).
19. Norstog, K., *In* "Plant Cell and Tissue Culture", (W.R. Sharp, P.O. Larsen, E.F. Paddock and V. Raghavan, eds.), p. 179. Ohio State University Press, Columbus, (1979).
20. Raghavan, V., *In* "Methods in Developmental Biology", (F.H. Wilt and N.K. Wessels, eds.), p. 413. T.Y. Crowell, New York, (1967).
21. Raghavan, V., "Experimental Embryogenesis in Vascular Plants". Academic Press, New York, (1976).
22. Raghavan, V., *In* "Plant Cell, Tissue and Organ Culture", (J. Reinert and Y.P.S. Bajaj, eds.), p. 375. Springer-Verlag, Berlin, (1977).
23. Raghavan, V., *Int. Rev. Cytol. Suppl. 11B*, 209 (1980).
24. Raghavan, V., and Torrey, J.G., *Am. J. Bot. 51*, 264 (1964).
25. Randolph, L.F., and Cox, L.G., *Proc. Am. Soc. Hort. Sci. 43*, 284 (1943).
26. Rangaswamy, N.S., *In* "Plant Cell, Tissue and Organ Culture", (J. Reinert and Y.P.S. Bajaj, eds.), p. 412. Springer-Verlag, Berlin, (1977).
27. Rangaswamy, N.S., and Shivanna, K.R., *Nature 16*, 937 (1967).
28. Romberger, J.A., *BioScience 16*, 373 (1966).
29. Sanders, M.E., and Ziebur, N.K., *In* "Recent Advances in the Embryology of Angiosperms", (P. Maheshwari, ed.), p. 297. Proc. Int. Soc. Plant Morphology. Delhi, (1963).
30. Subrahmanyam, N.C., and Kasha, K.J., *Barley Genet. Newsl. 3*, 62 (1973).
31. Torrey, J.G., *In* "Tissue Culture - Methods and Applications", (P.K. Kruse, Jr. and M.K. Patterson, Jr., eds.), p. 166. Academic Press, New York, (1973).
32. Tukey, H.B., *J. Hered. 24*, 7 (1933).
33. Tukey, H.B., *Proc. Am. Soc. Hort. Sci. 45*, 211 (1944).
34. van Overbeek, J., Conklin, M.E., and Blakeslee, A.F., *Am. J. Bot. 29*, 472 (1942).
35. Yeung, E.C., and Sussex, I.M., *Z. Pflanzenphysiol. 91*, 423 (1979).
36. Zenkteller, M., *Int. Rev. Cytol. Suppl. 11B*, 137 (1980).
37. Zenkteller, M., Misiura, E., and Guzowska, I., *In* "Form, Structure and Function in Plants", (B.M. Johri Commemoration Volume), p. 180. Sarati Prakashan, Meerut, (1975).

IN VITRO METHODS APPLIED TO RICE

Kiyoharu Oono

Division of Genetics
National Institute of Agricultural Sciences
Yatabe, Tsukuba, Ibaraki, Japan

I. INTRODUCTION

Since the 1930's, plant tissue culture has progressed mostly on dicotyledonous plants, and studies on monocotyledonous plants have lagged far behind. The *in vitro* culture of rice started with organ culture, when Fujiwara and Ojima cultured exised roots (9). The culture conditions for various stages of immature embryos were examined by Amemiya *et al.*, and five-day-old embryos were found to be the youngest for successful germination (1, 2). Nakajima and Morishima obtained F_1 hybrids, *Oryza sativa* Japonica 4x x *O. minuta* or *O.* sp (Paraguay) 2x through embryo culture; seeds of these hybrids had embryos capable of germination, but they lacked the necessary endosperm to support germination (29).

Successful callus culture of rice was achieved first by Furuhashi and Yatazawa, who cultured the nodes of young seedlings on Heller's medium with vitamins and 2,4-D (2ppm) (10). Regeneration of plants was done from callus of various origin by Tamura (seed callus) (48), Kawata and Ishihara (root callus) (19), Niizeki and Oono (pollen) (31), Nishi *et al.* (seed callus) (34), and Maeda (seed callus) (22) in 1968.

Studies on metabolism in rice callus were carried out by several workers. Furuhashi and Yatazawa found that the growth of rice callus tissue was reduced when methionine was excluded from the complete mixture of amino acid medium (11). Nishi traced the changes in the uptake capacities of phosphorus and calcium during callus induction in rice as an approach to the mechanism of callus induction, and found a

ISBN 0-12-690680-7

decrease in the capacity of phosphorus uptake and an increase in that of calcium (32). Ohira *et al.* studied the nutrient requirements of rice in liquid culture, and found that the following major nutrients (at the indicated concentrations) were beneficial: NO_3-N (40 mM), NH_4-N (5.0 mM), P (2.0 mM) and K (40 mM) (35). Cobalt, iodine, pyridoxine, nicotinic acid and m-inositol were found not to be essential.

II. METHODS OF RICE TISSUE CULTURE

Callus tissue can be induced from seed, root, leaf, leaf sheath, endosperm, etc., in rice. The callus inducation method from these tissues is essentially the same as conventional procedure, viz: sterilization of tissues (e.g., 20 minutes in the 10% w/v calcium hypochlorite solution after dipping into 70% ethanol for a few seconds), rinsing with sterile water and inoculation of the explants to the media. Other procedures, such as regeneration of plants and transfer of plants to soil are the same as the anther culture methods. In rice anther culture, haploid plants are usually obtained through two procedural steps, i.e., callus induction from the pollen and regeneration of plants from pollen callus.

A. *Equipment*

The following equipment is necessary:

(1) Laminar air flow cabinet or sterile room.
(2) Culture room.
(3) Microscope.
(4) Two pairs of watch-maker's forceps.
(5) A pair of small scissors.
(6) A scalpel made of nichrome wire (1.4 mm ϕ) mounted on a steel or wooden handle.
(7) Tissue paper.
(8) Test tubes (18 x 180 mm).
(9) Aluminum foil (60 x 60 mm) and paper or cotton plugs.
(10) Glass slides and cover slips.

B. *Plant Materials*

Anthers at the uninucleate pollen stage are collected: panicles 2-4 days prior to heading usually contain suitable anthers in some portion of the panicle. Plants raised in the greenhouse or in the field can be used. However, anthers tend not to be suitable when plants are grown under short-day conditions.

C. *Reagents*

(1) 70% ethanol.
(2) Acetic-alcohol (1:3).
(3) Ferric chloride.
(4) Ferric ammonium sulfate.
(5) Aceto-carmine stain.

D. *Media*

1. *Callus Induction Medium.* Miller's (25) or Chu *et al.*'s (7) basic medium with 2,4-dichlorophenoxyacetic acid (2,4-D) (10^{-5}M), sucrose (30 g/l), agar (10 g/l) and adjusted to pH 5.8. Composition of basic media is shown in Table I. 10 ml of medium is then measured into each test tube, capped with a paper plug, and sterilized by autoclaving at 120°C for 20 minutes. The medium is then placed on a slant to cool and solidify.

2. *Differentiation Medium.* Murashige and Skoog's (26) basic medium with naphthaleneacetic acid (NAA) (10^{-6} M), yeast extract (3 g/l), casein hydrolysate (2 g/l), sucrose (70 g/l), agar (10 g/l), and ajusted to pH 5.8. 6-benzyl-aminopurine (5 x 10^{-5} M) may use instead of yeast extract and casein hydrolysate. The rest of the procedure is identical to that of the callus induction medium.

E. *Procedures*

As the inside of rice spikelets are usually free from pathogens, the purpose of inoculation procedures is to avoid the contamination during the initial manipulation.

(1) Select the exact panicles at the uninucleate pollen stage. In order to observe the pollen stages, the anthers are fixed with acetic-alcohol containing 4% ferric chloride and squashed with aceto-carmine stain. Approximate uninucleate pollen stages are observed by squashing the fresh anther with water.
(2) Sterilize the panicle by dipping in 70% ethanol for 3-5 seconds.

TABLE I. Composition of Basic Media for Rice Anther Culture

Compound	Miller (25)	Chu (7)	MS (26)
	mg/l	mg/l	mg/l
NH_4NO_3	1000	-	1650
$(NH_4)_2SO_4$	-	463	-
KNO_3	1000	2830	1900
$Ca(NO_3)_2 \cdot 4H_2O$	347	-	-
$CaCl_2 \cdot 2H_2O$	-	166	440
KH_2PO_4	300	400	170
$MgSO_4 \cdot 7H_2O$	35	185	370
$MnSO_4 \cdot 4H_2O$	4.4	4.4	22.3
$ZnSO_4 \cdot 7H_2O$	1.5	1.5	10.6
H_3BO_3	1.6	1.6	6.2
KI	0.8	0.8	0.83
$CuSO_4 \cdot 5H_2O$	-	-	0.025
$Na_2MoO_4 \cdot 2H_2O$	-	-	0.25
$CoCl_2 \cdot 6H_2O$	-	-	0.025
KCl	65	-	-
Na-Fe-EDTA	32	-	-
$FeSO_4 \cdot 7H_2O$	-	27.85	27.85
Na_2-EDTA	-	37.25	37.25
Inositol	-	-	100
Glycine	2	2	2
Nicotinic acid	0.5	0.5	0.5
Pyridoxine·HCl	0.1	0.5	0.5
Thiamine·HCl	0.1	1	0.1

(3) Absorb the excess ethanol with tissue paper.

(4) Cut the lemma of spikelets in half to thirds using scissors sterilized by flame and ethanol.

(5) Pick out the anthers from spikelets using forceps and inoculate the surface of agar medium slants. Sterilize the forceps by flame and use two pairs of cooled forceps alternately. Inoculate 30-40 anthers per test tube.

(6) Cap the test tube with aluminum foil instead of a paper plug in order to prevent the medium from drying out during long periods of culture.

(7) Incubate the tubes containing anthers in the culture room in the dark at 28°C.

(8) When calluses are formed and developed to 50-300 mg, generally after 5-8 weeks, transfer ca. 50 mg of callus to

the differentiation medium.

(9) When plants are regenerated and grown 10-15 cm in height, remove the aluminum foil. After 20-40 hours, cut off ½ of the leaf blades to protect the plants from excessive evapotranspiration and plant them into pots containing sandy soil.

(10) Shoots redifferentiated without roots by medium with 6-benzylaminopurine are transferred to the new medium for rooting; Murashige and Skoog's basic medium with NAA (10^{-6} M), yeast extract (1 g/l), sucrose (30 g/l) and agar (10 g/l) adjusted to pH 5.8.

F. *Chromosome Analysis*

(1) Fix vigorously growing calluses or root tips of plant in acetic-alcohol for 24-72 hours.

(2) Treat with a 4% solution of ferric ammonium sulfate for 24 hours as a mordant.

(3) Stain with aceto-carmine at 90°C for 1-2 minutes.

(4) Squash the tissue and observe the chromosomes.

G. *Collection and Analysis of Data*

In rice anther culture, the following analyses yield the basic information:

(1) The frequency of callus formation:

$$\frac{\text{No. of anthers which formed callus}}{\text{No. of anthers inoculated}}$$

(2) No. of calluses formed in an anther.

(3) The frequency of shoot redifferentiation:

$$\frac{\text{No. of pollen calluses redifferentiated into shoots}}{\text{No. of pollen calluses inoculated}}$$

$$\frac{\text{No. of plants regenerated from callus}}{\text{No. of calluses inoculated (Count the divided calluses)}}$$

(4) The frequency of variant:

$$\frac{\text{No. of variants regenerated (e.g., \textit{albino}, etc.)}}{\text{No. of regenerated plantlets}}$$

(5) No. of matured plants from redifferentiated shoots.

(6) Chromosomal variation in callus and plants.

(7) Effects of the modification of the cultural conditions to callus formation and regeneration (medium, temperature, light, etc.).

(8) Difference between the genotypes on callus formation and regeneration.

(9) Changes in totipotency by prolonged culture.

(10) Observation of the phenotypic differences on matured plants (A_1) (haploid, diploid, triploid, etc.).

(11) Field tests on agronomic characters of the progenies (A_2, A_3, ...) on fixation, mutation, inbreeding depression, etc.

H. Comments

(1) The above procedure makes it possible to inoculate 2500-3000 anthers within 6 hours.

(2) Plants are obtained without transfer of callus to differentiation medium when NAA (10^{-5} M) was used instead of 2,4-D in callus induction medium.

(3) Pollen calluses are observed in 3 weeks to 4 months after inoculation of anthers; ca. 13% of calluses were observed in the first month, 46% in the second month, 33% in the third month and 8% in the fourth month.

(4) Pollen callus induced by anther culture tends to show decreased totipotency with long periods of culture or subculture. To obtain high frequency of plant regeneration from callus, it is necessary to transfer the callus to differentiation medium immediately.

(5) Plants are regenerated from callus in 3 to 6 weeks after subculture of callus to differentiation medium and it is possible to transfer them to soil after 3 to 4 weeks of shoot redifferentiation.

III. RICE TISSUE CULTURE FOR GENETIC STUDIES

A. Single Cell Systems

For critical studies in genetics or breeding, it is convenient to have single cell systems established. An individual cell with an intact set of genes can be considered totipotent and thus the regeneration ability of such single cells is the desired goal of such studies. It is through plant cell culture that this practical application is achieved. Suspension culture, pollen culture and protoplast culture are the original sources of cells for single cell systems.

1. *Suspension Culture*. Maeda isolated single cells through suspension culture of rice callus (21). Rice calluses usually develop as friable small clumps on agar, but a small number of single cells can be obtained in suspension culture. Iwanaga and Kawai (unpublished) used suspension culture and filtration by mesh to obtain single cells of rice callus. In some cases, they used a brittle culm variety. Through their method, they obtained mainly (80-100%) single cells of 20-43 μ but some cell clumps (2-3 mm) still remained. Consequently, they failed to establish complete single-cell strains.

2. *Pollen Culture*. Kuo *et al.* successfully induced calluses from isolated pollen grains of which anthers had been cultured 4 days (20). Suspended pollen grains on agar media formed callus after 38-48 days. These calluses were pale yellow and subculturing was possible.

3. *Protoplast Culture*. The isolation of rice protoplast was achieved from callus cells (16, 53, 54) and from leaf blades (52), and the fusion of protoplasts has been achieved (16). Deka and Sen obtained viable protoplasts from mesophyll cells (50-60% yield) and from callus cells (60-70% yield) (8). To successfully isolate mesophyll protoplasts, the age of plants, cell turgidity and the nature of the plant parts used were important factors. Plating efficiencies on an average of 30% were achieved in cultures of callus and mesophyll protoplasts and differentiation of roots occurred after about 2 weeks. Cai *et al.* also successfully induced some callus from the protoplasts isolated from callus (3). If synchronous growth regulation, control of spontaneous mutation, and maintenance of totipotency are achieved, the value of single cell systems will be greatly enhanced.

B. *Marker Genes*

When genetic studies are intended *in vitro*, it is essential to have various markers. However, the expression of several characters, such as chlorophyll or pigment synthesis and morphological traits, is often suppressed *in vitro*. However, biochemical mutants, physiological mutants such as temperature sensitive mutants *(ts, cs)*, respiratory deficient mutants, isozymes, etc., can be utilized as the markers.

Nakagahara *et al.* investigated isozyme polymorphism of esterase in rice with leaves of about 800 Asian varieties and observed 14 anodic bands (9 of them easily distinguishable) and 27 different zymograms with these (28). The slowest isozyme band, 1A, was specified by a dominant allele at the esterase locus Est_1. The second band group, 6A (Est_2^s) and 7A (Est_2^f) and the third band group 12A and 13A were both controlled by codominant alleles (27).

Esterase isozymes as markers for *in vitro* rice callus genetics were examined by Oono *et al.* (39). Fig. 1 shows the electrophoresis of leaf blade and callus in seven varieties, Chinsurah Boro II, Dao ren qiao, Qui lu ai 3, Norin 8, Belle patna, North rose and Geraldine. Leaf blades were low in enzyme activities at bands 1C, 6A, 7A, but high at bands 1A, 11A, 12A, 13A. Electrophoretic patterns of calluses on Miller's medium were generally stronger than those grown on MS medium. Some variation of activity at band 1C, 6A, 7A were observed during callus induction and regeneration. At an early stage of callus induction, an endosperm specific band (E) appeared, and disappeared in

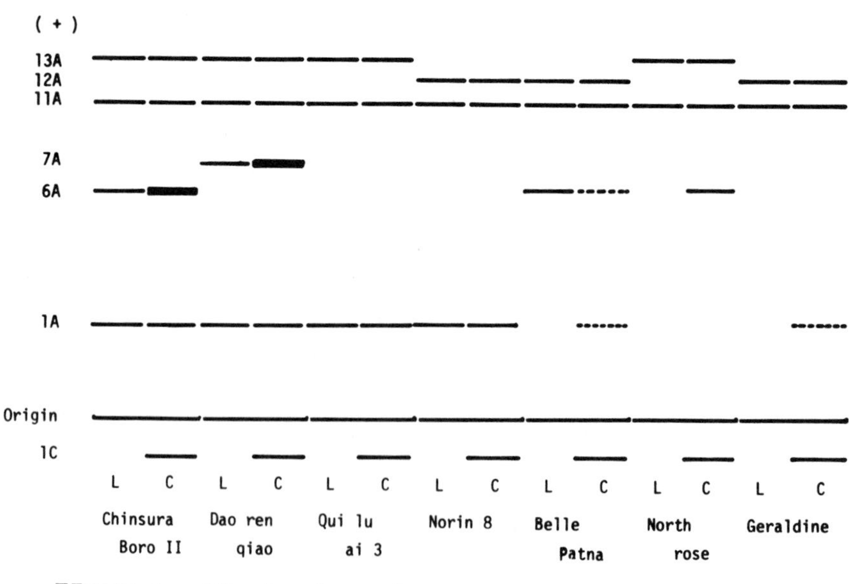

FIGURE 1. Electrophoretic pattern of esterase isozyme of rice in leaf blade (L) and callus (C), (after Oono et al., 39).

subculture, near band 6A in Norin 8, Qui lu ai 3 and Geraldine. It was concluded that band 6A (Est_2^s), 7A (Est_2^f) and 1A (Est_1) could be used as markers for rice *in vitro* genetics.

As electrophoretic patterns of esterases in calluses showed specific bands in pimiento, soybean, sweet potato, etc., it seems possible to use esterase as a marker. Isozyme banding of α-amylase in callus tissues of three normal varieties (Ginbozu, Aichiasahi, Te-tep) and three dwarf varieties (Tan ginbozu, Waisei shinriki, Kotake tamanishiki) were found to be different from each other according to Saka and Maeda (42). One of the dwarf variety showed low α-amylase activity, but had as many as ten isozymes. Therefore, the dwarfism could not be correlated with changes in isozyme banding patterns.

IV. APPLICATION TO PLANT BREEDING

As shown in Table II, *in vitro* methods in plant breeding can be classified into 5 processes, viz: broadening the genetic variability, fixation, selection, propagation and gene preservation. These will now be discussed.

A. *Fixation Through Anther Culture*

1. Induction of Haploid Callus and Plants. The procedure of rice anther culture includes callus induction from pollen grains and regeneration of plants. The most suitable developmental stage of pollen for anther culture is the uninucleate stage; Wang *et al.* found the late uninucleate stage most adequate (56). Frequencies of callus development varied from genotype to genotype, and were influenced by such factors as pollen vigor, seed fertility, shedding of pollen at flowering and changes occurring in the cultured anthers. Genotype differences were noted also in the requirement of growth substances (14).

The medium and conditions for anther culture are about the same as for ordinary diploid tissues. Auxin (2,4-D or NAA) is required for callus induction. The callus induction rate with Miller or MS basic medium was 2.1% (an average from 25 varieties and hybrids). Only slight modifications on the induction rate were observed by Oono, as a result of making modifications in the concentrations of auxin, cytokinin, sugar, pH, etc. (36). On the other hand, induction rates of more than 10% were not rare, such as 44.7% in the Japonica

TABLE II. Methods in Test Tube Breeding[a]

		Breeding process				
		Variation	Fixation	Selection	Propagation	Gene preservation
Intact Plant		Mutation	Pedigree	Genotype	Seed	Low temperature storage
		Hybridization	Bulk	Phenotype	Vegetative propagation	Conservative cultivation
		Recombination				
		Ploidy				
	Organ	Embryo culture		Meristem culture (virus free)	Clonal multiplication	
		Test tube fertilization				
	Tissue and Cell	Mutation	Anther and pollen culture	Cultural environment Genotype	Clonal multiplication	Freeze storage in super-low temperature
		Somatic hybridization				
Test Tube	Organelle	Transformation Mutation		Phenotype	Organelle culture	

	Breeding process				
	Variation	Fixation	Selection	Propagation	Gene preservation
	Recombination (Uptaking)				
Molecule	Recombinant DNA molecule Hybrid plasmid	Gene cloning	Colony hybridization Hybridization	Gene amplification	Freezing Lyophilizing bacteria

[a] After Oono (37).

variety H-124 on MS medium observed by Chen and Lin (6). Chu *et al.* found that the most appropriate N concentration to increase the induction rate is $(NH_4)_2 \cdot SO_4$ (3.5 mM) with KNO_3 (28 mM) as a nitrogen source (designated as N_6 medium) (7). Potato extract medium was proposed by the 302 Research Group of China for the same purpose; their medium with 30% extract of potato tubers instead of all of Miller's nutrient components usually showed higher induction rates (50).

2. *Redifferentiation.* In the early days of anther culture studies in rice, the approach to find the optimum medium conditions for regeneration of plants involved using various combinations of IAA and kinetin with the basic medium or the auxin-free medium. Intensive research followed on the combination of auxin and cytokinin, amino acids, base analogues, growth retardants, and sugars to improve the regeneration efficiency. Three medium requirements for regeneration in rice were established, viz:

(1) High sucrose concentration,
(2) either yeast extract and casein hydrolysate or 6-benzyladenine, and
(3) low concentration of auxin.

With such media, 50-80% regeneration frequencies were obtained; and in some cases scores of shoots were formed from one callus piece. However, overall the totipotency of pollen calluses was extremely low (56). The totipotency of pollen calluses was lost more rapidly than diploid seed callus. This phenomenon may be explained on the basis of genetic or physiological modification in callus. Due to this unfavorable phenomenon immediate regeneration from callus is required in rice.

3. *Characteristics of Regenerated Plants.* Chromosomal observations of regenerated plants from rice anther culture are shown in Table III. 40% were found in haploid and 48.5% in diploid state. The 302 Research Group also obtained the same kind of results, and 60% of about 1000 pollen plants obtained from F_1 anther culture were found to be diploid, 35% haploid and 5% triploid or aneuploid (49).

Chlorophyll deficient plants, especially albino, were frequently regenerated from pollen callus. Albinos appeared not only in haploid but also in diploid and triploid plants. Sun *et al.* found in rice albino plants, through electron-microscopy, that proplastids with primary lamellae were present in the leaf cells, but failed to develop into normal chloroplasts with lamellae (44). DNA-like fibrils were found

Table III. Chromosome Numbers of Plant Regenerated from Pollen Calluses (x=12)[a]

Variety and hybrid	*x*	*2x*	*2x+1*	*3x*	*4x*
Norin 8	6	1			
Norin 20	4	2			
Toride 2		4			
Minehikari		12			
F_1, Norin 8 x Akamochi	1	1		1	
F_1, Muyozetsuto x (Kinmaze x Morak Sepilai x Kinmaze)	1	13			
F_1, (Kaltha x Norin 25) x Norin 25				2	
F_1, Murasaki-ine x (Norin 25 x Compena x Norin 25)					1
F_1, Akamochi x Muyozetsuto	16	1			
F_1, LT-36 x Murasaki-ine			2		
Oryza perennis				2	
Total	28	34	2	5	1

[a]*After Oono (36).*

in albino plastids but no ribosomes were found. They considered that the albinism was directly caused by the lack of ribosomes. They further found the difference in electrophoretic pattern of soluble proteins between the albino and the green plants using polyacrylamide gel; the albino plants lacked one of the major bands (46). The same authors also discovered that albino plants lacked 23s and 16s rRNAs (45).

Spontaneous diploidization of haploid plants was observed at the frequencies of 2.5% (Norin 20, one in 40) to ca. 2% (pooled data, 510 plants) (49). Relatively long duration of callus culture increased the frequencies of diploid regeneration (59). As shown before, 50% of regenerated plants were diploid. But, with the transplantation of callus at the age of 30-50 days for differentiation, the rates of fertile plants increased to 80%. Treatment with colchicine also induces diploidization; e.g., the 302 Research Group obtained 40.4% diploids from 587 plants (49).

Through the examination of the spontaneous diploid plants from two varieties, Minehikari and Toride 2 at the second (A_2) and the third generation (A_3), no inbreeding depression was observed. Generation after generation of regenerated plants from Indica-Japonica hybrids revealed that the phenotypes of A_1 varied when pollen origin was different. However, each of the various A_1 phenotypes was judged homozygous, since no segregation was found in their A_2 and A_3 generations (Table IV). According to the 302 Research Group, 90 percent of 187 pollen plant lines obtained from 31 hybrid F_1 showed uniform offspring without any segregation (49). The recombination and segregation of four characters, plant height, heading date, shattering habit and apiculus color have been compared between the 49 pollen plants obtained from various hybrid F_1 plants and the F_1 hybrids, and it was found that both were alike. This result suggested that there was no difference in the frequency of producing pollen plants among the microspores of various genotypes. No obvious competition between different genotypes were observed among the anther cultures. However, the Group of Genetics and Plant Breeding *et al.* observed that the anthers of the F_1 hybrid, Indica-Japonica, responded differently to various media (13). When anthers are cultured on N_6 medium to which Japonica rice is adapted, the majority of the plants produced resembled the Japonica type or intermediate type, different from the usual F_2 segregation ratios between Indica and Japonica types. This suggests that the culture medium has a certain degree of screening effect on the pollen plant (A_1).

Mutant phenotypes in pollen size, seed fertility, plant height, etc., appeared among the regenerated plants of Minehikari, Toride 2 and so on. These mutant characters appeared in both homozygous and heterozygous states among the regenerated plants and their descendants. This means that the mutations were induced both in the haploid phase and at the time of chromosome doubling. The 302 Research Group also reported that about 10% of the pollen plants showed segregation of characters in the progenies (49).

Not many studies on wild rice species have been done as yet but some triploids (36) and haploids (55) have been obtained in *O. perennis* and *O. spontanea*.

4. Breeding of New Varieties by Anther Culture. The model system for the practical application of anther culture in rice breeding is as follows:

1st year	Sowing of parents and obtaining F_1 by crossing Anther culture of F_1 Obtaining A_1 (x, $2x$)
2nd year	Individual selection, line selection Diploidization

3rd and 4th years	Performance test Test for physiological characters Test for adaptation

Several improved varieties have already been obtained through anther culture (41, 59). A comparison of the first varieties obtained through anther culture, Hua Yü 1 and 2, and Nipponbare obtained by a bulk method is given in Fig. 2. Anther culture efficiently reduces the breeding period.

For actual breeding practice, the following should be considered:

(1) When we subject n-pairs of genes in breeding of a certain combination, 2^n kinds of haploid plants should theoretically appear through anther culture of F_1 mother plants. In a practical breeding situation, we usually take account of intrachromosomal recombinants obtained by crossing over, 3-4 times of the safety rates of the manipulating population, and the number of genes for interchromosomal recombination. Although the necessary number of plants may vary in each case, large number of plants should be examined therefore in breeding through anther culture.

(2) Adaptability to the local conditions of environment and year-to-year deviation are usually taken into account in the selection of strains in diploid methods. Need of selections against different environmental conditions must be considered.

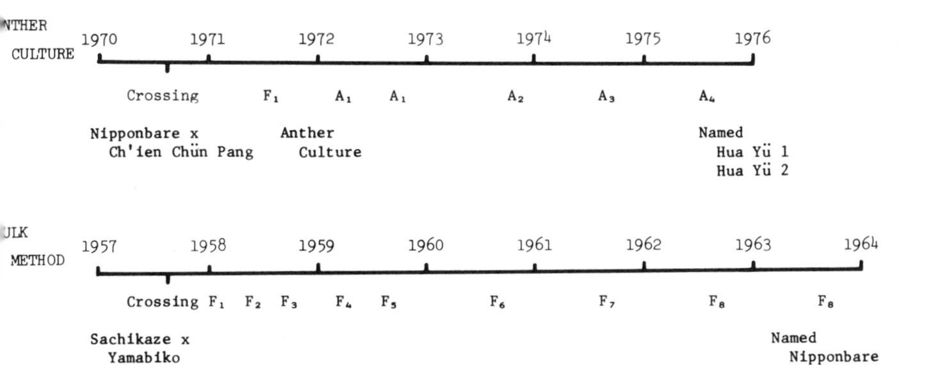

FIGURE 2. Breeding of new varieties by anther culture and bulk method (after Oono, 37).

TABLE IV. *Comparison of Some Characters among Progenies of Pollen Plants, Two Parental Varieties and A_2 and A_3 Strains*[a]

Parent and progeny	Generation	No. of plants	Heading date	Plant height (cm)	Panicle length (cm)	No. of effective tillers per plant	Seed fertility (%)
Muyozetsuto (female parent)		39	Sept. 13± 1	127.4± 5.2	22.8±1.4	14.5±3.6	83.4
B_1F_2 (F_2 of male parent)		39	Oct. 8±13	110.3±15.3	24.4±2.3	15.5±4.3	28.8
F_2 of hybrid		38	Sept. 20± 6	129.1±16.7	20.3±4.0	14.9±4.0	79.2
AG-1-2 (maculata)	A_2 (1970)	2	Sept. 10	73.0± 0	15.0±0	11.0±0	-
AG-1-15		16	Sept. 3± 1	106.9± 6.2	22.2±1.5	14.7±5.7	87.3
AG-1-22 (normal)		27	Sept. 2± 1	108.6± 4.3	22.4±1.6	13.0±3.5	88.3
AG-1-22 (short culm)		10	Sept. 2± 1	72.0± 4.1	17.6±2.4	9.2±3.9	68.8
AG-1-24		39	Sept. 4± 1	113.1± 6.4	22.9±1.5	13.2±4.4	86.1
AG-1-2 (maculata)	A_3 (1971)	50	Sept. 8	58.3± 4.7	16.7±1.0	15.3±6.2	-
AG-1-22 (normal)		56	Sept. 8	94.4± 6.3	22.6±1.7	13.1±4.6	-
AG-1-22 (short culm)		82	Sept. 8	67.2± 4.9	20.2±2.0	10.1±2.8	-

Parent and progeny	Generation	No. of plants	Heading date	Plant height (cm)	Panicle length (cm)	No. of effective tillers per plant	Seed fertility (%)
AG-1-24		54	Sept. 8	98.9± 3.9	23.5±1.3	11.3±2.7	-

Material: Muyozetsuto x (Kinmaze x Moak Sepilai) x Kinmaze B_1F_1)
Seeding: May 22, 1970 Transplanting: June 23, 1970
: June 1, 1971 : June 23, 1971

[a]After Oono (36).

B. Broadening the Genetic Variability by Mutation

Research on protoplast fusion and genetic engineering has progressed recently, and the application of these techniques is for the purpose of broadening genetic variability, as the source of genes for breeding. However, at this moment, using mutation is one of the easiest ways of broadening the genetic variability. As shown in Table V, many different types of mutations occur in tissue culture, e.g., genomic, chromosomal, gene and cytoplasmic. These results suggest the possibility of broadening the genetic variability through tissue culture. Therefore, quantitative and qualitative analyses of mutations in regenerated plants should be made.

1. Chromosomal Variation in Rice Pollen Callus. Rice callus tissues are unstable at the chromosomal level as shown in Fig. 3. Chromosome numbers in 50-55 day-old calluses of anther culture from the F_1 plants of a cross Arborio x Loctjan were examined; these calluses were induced on 2,4-D or NAA (5×10^{-6} M each) media. The calluses examined varied in size (weight) from less than 10 mg to 1000 mg, and they showed very wide variation in chromosome number. The frequencies of haploid cells were 34.5% in the calluses induced on 2,4-D medium and 31.1% on the NAA medium. The frequencies of diploid cells were 42.5% and 25.9% in the two different media respectively. As shown in the figure, the largest chromosome number observed was 114. The size of several observed calluses were less than 1 mm in diameter,

TABLE V. Mutations in Plant Tissue Culture Observed in Regenerated Plants

Type of mutation		*Crop (Reference)*
Genome mutation	*Polyploid*	*Rice (68)*
	Aneuploid	*Solanum (15)*
		Rice (36)
	Haploid	*Oryza punctata (30)*
Chromosome mutation	*Deletion*	*Haworthia setata (58)*
	Translocation	*Rice (half sterility) (36)*
Gene mutation		*Rice (36)*
Cytoplasmic mutation		*Tobacco (24)*
		Maize (12)

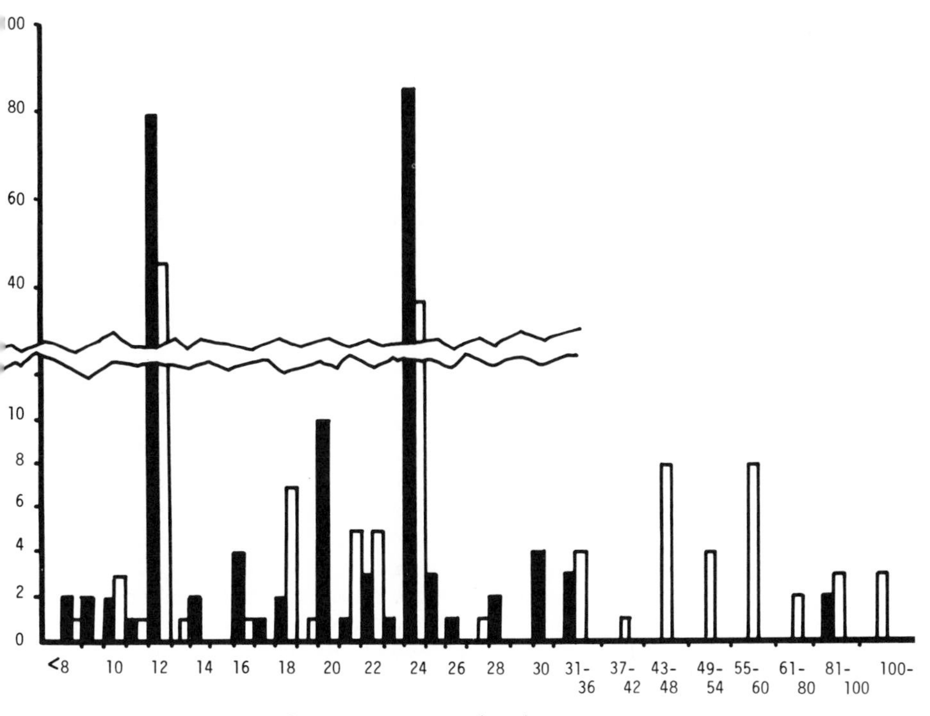

FIGURE 3. Chromosome numbers of cells of pollen calluses. Materials: Calluses obtained after 50-55 days incubation of anthers of F_1 plant in Arborio x Loctjan. Total number of calluses observed = 22. Total number of cells observed = 343. ■ *: Calluses obtained on medium of Miller + 2,4-D (5×10^{-6} M).* □ *: Calluses obtained on medium of Miller + NAA (5×10^{-6} M).*
After Oono (36).

and using the size of callus, the time needed for a cell division can be estimated from the following equation:

$$2^x = \text{number of cells in callus} \doteqdot \frac{\text{callus size}}{\text{cell size}} = \frac{(1\ \text{mm})^3}{(20\ \mu)^3}$$

$$x \doteqdot 17$$

In one callus 8 different chromosome numbers between 10 and 36 appeared, and in another callus 15 between 12 and 144 were found. These results suggest that cell divisions in pollen calluses are not always normal and that polyploidy and aneuploidy is common. It appears that the culture conditions do not exert strong negative selection pressures on the abnormal cells or chromosomal mutants.

2. *Mutation in Rice Diploid Callus.* Gene mutations appear in rice anther culture (36). We studied spontaneous mutations in callus culture. Mutation analysis was applied to 1121 plants (D_1) regenerated from 75 calluses which originated from the rice seeds of a genetically pure line (progeny of a doubled haploid from a spontaneous haploid of *Oryza sativa L.*, var. Norin 8), and their progenies (D_2, D_3). Eighty three plants (7.4%) were albino. Seed fertilities in 489 D_1 plants showed wide variation; 58.7% of them were less than 80% fertile and 18.2% of them were less than 40% fertile. Heading dates of D_2 plants (6382 plants of 762 D_2 lines) ranged between Aug. 19 and Oct. 7 (average, Sept. 24 $\pm$ 2.9, in contrast to Sept. 17 $\pm$ 1.7 for the control) and their heights ranged between 29 and 135 cm (average, 107 $\pm$ 11.2 cm, in contrast to 116.9 $\pm$ 6.8 cm for the control). Plants less than 50 cm in height were mostly malformed and/or completely sterile. Low seed fertility was also observed in D_2 plants. Frequencies of D_2 plants with seed fertility less than 80% or 40% were respectively 19.5% and 5.0%. Mutations which occurred in the D_1 generation were observed through segregation in the D_2 generation. As shown in Table VI, an examination of five marker characters (seed fertility, plant height, heading date, morphology, and chlorophyll content) showed that only 28.1% of the D_2 lines were normal in observed characters and 28.0% carried two or more mutated characters. Among 6382 D_2 plants, 1491 D_2 plants (23.4%) were considered mutants of recessive homozygotes or dominant heterozygotes including chromosomal aberration.

104 D_2 mutants and 5 normal D_2 plants were further investigated in D_3 generation to confirm their breeding behavior. Five normal D_2 lines and four lines from the mutant D_2 (regarded as mutants from the decreased seed fertility) were found as normal in all five characters in D_3. The remaining 100 D_3 lines either showed segregation or appeared as true breeders for any of the five mutant characters. Mutants for chlorophyll characters, plant height, heading date, etc., were obtained as fixed D_3 lines.

C. *Selection of New Genetic Characters*

Since it is easy to induce mutations in cultures by irradiation, chemical mutagens and spontaneous mutations

TABLE VI. Changes of Characters of Mutants in D_2 Lines Derived from Panicles of D_1 Plants Regenerated from Rice Calluses[a]

Mutated characters[b]	*No. of lines*	*%*
Normal	214	28.1
Ploidy (4x)	12	1.6
Fer.	273	35.8
Ht.	19	2.5
Hd.	3	0.4
Mor.	1	0.1
Ch.	27	3.5
Fer. & Ht.	74	9.7
Fer. & Hd.	14	1.8
Fer. & Mor.	7	0.9
Fer. & Ch.	59	7.7
Ht. & Ch.	3	0.4
Hd. & Ch.	1	0.1
Mor. & Ch.	1	0.1
Fer., Ht. & Hd.	11	1.4
Fer., Ht. & Mor.	2	0.3
Fer., Ht. & Ch.	27	3.5
Fer., Hd. & Ch.	2	0.3
Fer., Mor. & Ch.	1	0.1
Fer., Ht., Hd. & Mor.	1	0.1
Fer., Ht., Hd. & Ch.	9	1.2
Fer., Ht., Mor. & Ch.	1	0.1
Total	762	100

[a] *After Oono et al. (40).*

[b] *Abbreviations for the mutated characters are as follows: Fer; seed fertility, Ht.; plant height, Hd.; heading date, Mor.; morphological traits, Ch.; chlorophyll deficiency.*

(see Chapter 5), introduction of new breeds could be done if a proper selective maneuver is possible with the change of culture conditions. Very few of the selection studies reviewed (e.g., 23, 51, 57), deal with the inheritance of regenerated plants. For agricultural application, efficient methods for mutant induction and selection are desired.

1. *Improvement of the Grain Quality.* Chaleff and Carlson tried to obtain high lysine mutants through *in vitro*

selection to improve food quality (5). In higher plants, lysine as well as methionine (both essential amino acids), are derived from asparate biosynthetically. The activity of the first enzyme appears to be regulated through feedback inhibition primarily by lysine and to a lesser degree by threonine. Aspartokinase activity in cell free extracts of rice shoots is also sensitive to feedback inhibition by the lysing analogue s-(β-aminoethyl)-cysteine (SAEC). Chaleff and Carlson considered that resistance to SAEC could be utilized to select for rice mutants in which lysine biosynthesis is released from regulatory control. Calluses from rice seeds were treated with 1% ethyl methanesulfonate for 1 hour, and cell lines resistant to 2 mM SAEC and with increased amount of total protein and amino acid content were selected. Whether these changes in the quality and quantity of protein can be reflected in the seed endosperm remains to be determined.

2. *Disease Resistance*. Blast is a major disease of rice and, therefore, the breeding of resistant varieties is important in Japan. Ishii tried to use the formation of aerial hyphae and hyphal growth conspicuous on callus tissue induced from embryos as the signs of a strain's susceptibility (17). However, it was concluded that blast resistant genes did not manifest themselves in the callus tissue since various rice varieties did not show any differences. In testing by the smear inoculation method on plantlets at an early stage of redifferentiation from embryo-callus, no difference between resistance and susceptibility was observed in various combinations of rice varieties and fungus strains. Carlson obtained plants which were less susceptible to *Pseudomonas tabaci* by selection against methionine sulfoximine (4). Therefore, resistant mutant selection by fungal toxins seems to be a more probable route for success. Takahashi *et al.* reviewed the toxins of *Pyricularia orizae* as α-picolic acid, piricularin, pyriculol, tenuazonic acid and 3,4-dihydro-3,4,8-trihydroxy-1(2H)-napthalenone (47). Since these toxins have inhibitory effect on rice seedlings, it is worth determining if the rice strains which are resistant to these toxins are actually resistant against the fungus.

3. *Saline Resistance*. Saline resistance is a desired characteristic of rice in Southern Asia. Many varieties were studied for the screening of saline resistant genes. Several tolerant varieties were selected but they were found not to be sufficiently resistant. We examined the feasibility of obtaining saline resistant mutants by

spontaneous mutation in callus culture of diploid rice (38). Some callus lines resistant against 1% NaCl were isolated after 6-12 months of subculture of Norin 8 calluses; 72 D_1 plants were regenerated from three resistant callus clones, and 19 D_2 plants from lines with normal seed fertility were examined with respect to their germination on 1% NaCl medium. Three out of nine D_2 lines regenerated from a resistant callus clone reproduced resistant D_2 plants in segregation. Furthermore this resistance of the seedling was inherited in D_3's also.

D. *Preservation of Rice Tissue*

Preservation of germplasm is one of the important areas in plant breeding. Seed preservation in low temperatures is already established and is being applied practically (18). Although it is possible to preserve stocks of aneuploid or mutants in rice, it is costly and laborious to keep them free from disease or insects. Sala *et al.* successfully preserved rice callus tissue (43), and probably there will be possibilities to establish effective methods of single cell-callus-plant level preservation in rice.

V. SUMMARY

Contributions on tissue culture of rice for agricultural purposes were reviewed from the view points of genetics and breeding. After anther culture techniques were established in rice, it became possible to apply them to rice breeding. Therefore, a brief historical review was carried out on tissue culture, single cell culture and genetic markers in culture.

For the application of anther culture in rice breeding, the laboratory protocol and the characteristics and problems were described, and a comparison with conventional breeding systems was made. A model system of breeding was also given. Efficiency improvement and prevention of albino regeneration are the major problems in the process of rice breeding through anther culture.

Chromosomal and genetic mutations occur frequently in rice tissue culture; systematic analysis on the regenerated plants and their pedigrees indicated the repeated occurrence of mutations with the frequencies as high as ones obtained via mutagens. Therefore, mutations occurring during callus differentiation can be considered as an additional source of new characters for breeding and selection. By this means, saline resistant mutants have been selected.

REFERENCES

1. Amemiya, A., Akemine, H., and Toriyama, K., *Bull. Natl. Inst. Agric. Sci. D6,* 1 (1956).
2. Amemiya, A., Akemine, H., and Toriyama, K., *Bull. Natl. Inst. Agric. Sci. D6,* 41 (1956).
3. Cai, Q., Qian, Y., Zhou, Y.,and Wu, S., *Acta Botanica Sinica 20,* 97 (1978).
4. Carlson, P.S., *Science 180,* 1366 (1973).
5. Chaleff, R.S., and Carlson, P.S., *In* "Genetic Manipulations with Plant Material" (L. Ledoux, ed.), p. 351. Plenum, New York (1975).
6. Chen, C., and Lin, M., *Bot. Bull. Academia Sinica 17,* 18 (1976).
7. Chu, C., Wang, C., and Sun, C., *Scientia Sinica 18,* 659 (1975).
8. Deka, P.C., and Sen, S.K., *Molec. gen. Genet. 145,* 239 (1976).
9. Fujiwara, A., and Ojima, K., *J. Sci. Soil Manure, Japan 28,* 9 (1955).
10. Furuhashi, K., and Yatazawa, M., *Kagaku (Japan) 34,* 623 (1964).
11. Furuhashi, K., and Yatazawa, M., *Plant Cell Physiol. 11,* 569 (1970).
12. Gengenbach, B.G., Green, C.E., and Donovan, C.M., *Proc. N.A.S. (U.S.) 74,* 5113 (1977).
13. Group of Genetics and Plant Breeding and Group of Botany, Chekiang Agricultural University and Agricultural Research Institute, Chuchi, Chekiang, *Acta Genetica 5,* 153 (1978).
14. Guha-Mukherjee, S., *Nature 204,* 497 (1973).
15. Harn, C., *SABRAO Newsletter 3,* 39 (1971).
16. Harn, C., *SABRAO Newsletter 5,* 107 (1973).
17. Ishii, K., *Doctorate thesis of Nihon University* (1978).
18. Ito, H., *Bull. Natl. Inst. Agric. Sci. D13,* 163 (1965).
19. Kawata, S., and Ishihara, A., *Proc. Japan Acad. 44,* 549 (1968).
20. Kuo, M., Cheng, W., Hwang, J., and Kuan, Y., *Acta Genetica Sinica 4,* 333 (1977).
21. Maeda, E., *Proc. Crop Sci. Soc. Japan 34,* 139 (1965).
22. Maeda, E., *Proc. Crop Sci. Soc. Japan 37,* 51 (1968).

23. Maliga, P., *In* "Frontiers of Plant Tissue Culture 1978" (T.A. Thorpe, ed.), p. 381. University of Calgary Press, Calgary (1978).
24. Maliga, P., Sz-Breznobits, A., Marton, L., and Joo, F., *Nature 225*, 401 (1975).
25. Miller, C.O., *Morderne Methoden der Pflanzenanalyse, Vol. 6*, 194 (1963).
26. Murashige, T., and Skoog, F., *Physiol. Plant. 15*, 473 (1962).
27. Nakagahara, M., *Japan. J. Breed. 27*, 141 (1977).
28. Nakagahara, M., Akihama, T., and Hayashi, K., *Japan J. Genet. 50*, 373 (1975).
29. Nakajima, T., and Morishima, H., *Japan. J. Breed. 8*, 105 (1958).
30. Niizeki, H., *Japan. J. Breed. 27, Suppl. 1*, 124 (1977).
31. Niizeki, H., and Oono, K., *Proc. Japan Acad. 44*, 554 (1968).
32. Nishi, T., *Physiol. Plant. 23*, 561 (1970).
33. Nishi, T., and Mitsuoka, S., *Japan. J. Genet. 44*, 341 (1969).
34. Nishi, T., Yamada, Y., and Takahashi, E., *Nature 219*, 508 (1968).
35. Ohira, K., Ojima, K., and Fujiwara, A., *Plant Cell Physiol. 14*, 1113 (1973).
36. Oono, K., *Bull. Natl. Inst. Agric. Sci. D26*, 139 (1975).
37. Oono, K., *Tropical Agriculture Research Series 11*, 109 (1978).
38. Oono, K., and Sakaguchi, S., *Japan. J. Breed. 28, Suppl. 2*, 124 (1978).
39. Oono, K., Nakagahara, M., and Hayashi, K., *In* "Studies of Single Cell Culture on Higher Plants" (Y. Watanabe, ed.), Series 109, p. 58. Research Council, Minist. Agri. Forest. Fish., Tokyo (1978).
40. Oono, K., Okuno, K., and Kawai, T., *In* "Studies of Single Cell Culture on Higher Plants" (Y. Watanabe, ed.), Series 109, p. 51. Research Council, Minist. Agri. Forest. Fish., Tokyo (1978).
41. Rice Research Laboratory, Tiensin Agricultural Research Institute, 302 Research Group, Institute of Genetics, Academia Sinica, *Acta Genetica Sinica 3*, 19 (1976).
42. Saka, H., and Maeda, E., *Proc. Crop. Sci. Soc. Japan 42*, 307 (1973).
43. Sala, F., Cella, R., and Rollo, F., (in press) (1979).
44. Sun, C., Wang, C., and Chu, C., *Sci. Sinica 17*, 793 (1974).
45. Sun, C., Wu, S., and Wang, C., *In* "Proceeding of Symposium on Plant Tissue Culture", p. 151. Science Press, Peking (1977).

46. Sun, C., Wu, S., Chu, C., and Wang, C., *Acta Genetica Sinica 4,* 359 (1977).
47. Takahashi, N., Marumo, S., and Otake, N., "Tennenbutsu Kagaku (Chemistry of Physiologically Active Natural Products)", p. 1. University of Tokyo Press, Tokyo (1973).
48. Tamura, S., *Proc. Japan. Acad. 44,* 544 (1968).
49. 302 Research Group, Institute of Genetics, Academia Sinica, *Acta Genetica Sinica 3,* 277 (1976).
50. 302 Research Group, Institute of Genetics, Academia Sinica, and Hokiang Rice Institute, Heilungkiang, *Acta Gentica Sinica 4,* 302 (1977).
51. Thomas, E., King, P.J., and Potrykus, I., *Z. Pflanzenzüchtg. 82,* 1 (1979).
52. Tseng, T., and Shiao, S., *Bot. Bull. Academia Sinica 17,* 63 (1976).
53. Tseng, T., Liu, D., and Shiao, S., *Bot. Bull. Academia Sinica 16,* 55 (1975).
54. Wakasa, K., *Japan. J. Genet. 48,* 279 (1973).
55. Wakasa, K., and Watanabe, Y., *Japan. J. Breed.* (in press) (1979).
56. Wang, C., Sun, C., and Chu, Z., *Acta Botanica Sinica 16,* 43 (1974).
57. Widholm, J.M., *In* "Plant Tissue Culture and Its Bio-technological Application" (W. Barz, E. Reinhard and M.H. Zenk, eds.), p. 112. Springer-Verlag, Berlin, (1977).
58. Yamabe, M., and Yamada, T., *La Kromosomo 94,* 2923 (1973).
59. Yin, K., Hsu, C., Chu, C., and Pi, F., *Sci. Sinica 19,* 227 (1976).

IN VITRO METHODS APPLIED TO SUGAR CANE IMPROVEMENT

Ming-Chin Liu

Department of Plant Breeding
Taiwan Sugar Research Institute
Tainan, Taiwan, China

I. INTRODUCTION

Research on sugar cane (*Saccharum* species hybrids) tissue and cell cultures was started in Hawaii in 1961 by Nickell (37), and its history has been reviewed by Heinz *et al.* (12). Taiwan Sugar Research Institute (TSRI) has been carrying out a program of application of tissue culture techniques to further sugar cane improvement since 1970 (21). At a later date, similar studies were also begun in Fiji (19). Other institutes in Florida, Louisiana, Maryland and France have also been carrying out some tissue culture studies.

II. LITERATURE REVIEW

A. *Tissues*

Callus can be initiated from almost any sugar cane tissue, e.g., shoot and root apical meristems, young leaves, root band tissue of a young node, immature inflorescences (callus comes mostly from the rachis) and pith parenchyma (29, Liu and Chen, unpublished data).

ISBN 0-12-690680-7

B. Media

Modified Murashige and Skoog (MS) medium (33) containing 3 mg/l 2,4-dichlorophenoxyacetic acid (2,4-D) has been used for callus induction (13, 29) (Table I). The addition of coconut milk and myo-inositol to the medium are essential for efficient stimulation of callus formation. White's (51) medium is less efficient than MS medium in supporting cell growth of both callus and suspension cultures (3, 29). Modified MS medium supplemented with 1 mg/l kinetin, 1 mg/l α-naphthaleneacetic acid (NAA) and 400 mg/l casein hydrolysate has proven excellent in inducing plant differentiation from callus (Table I). Modified Schenk and Hildebrandt (SH) medium (43) has worked very well in the rooting of shoots (Table I) (Liu and Chen, unpublished data).

C. Suspension Culture

The growth and degree of dispersion of sugar cane cells in liquid modified MS medium is superior to results obtained in modified White's medium (39). Suspended cells grown in the former were particularly sensitive to subculturing time at each growth cycle. The cultures usually turned brown and cells were lyzed if subcultured at the stationary phase. Suspended cells grew slowly in the medium without meso-inositol (3).

D. Plant and Root Differentiation

In general, sugar cane callus can easily differentiate into shoots. However, a marked influence of the donor's genotype on the initiation of callus and regeneration of plants has been observed. Old cultivars such as Badila, POJ2878 and POJ2883, etc., regenerate much less frequently than recently released ones, such as F160, F170, and F177, etc. (29, Liu and Chen, unpublished data). Tissue sources also affect differentiation rate. Unemerged inflorescences have the greatest capacity to differentiate into plantlets, and the genetic constitution of the plants derived from them remain unchanged. Therefore, they have seldom been used for selection of variants. No significant difference in differentiation capacity has been detected between apical meristems and young leaf sources (30).

The Hawaiian researchers have been confronted with difficulties in rooting the differentiated shoots. They suggest

TABLE I. Media used by the Taiwan Sugar Research Institute for Callus Induction, and Shoot and Root Differentiation

	Ingredient	Callus formation MS-C (33) (mg/l)	Shoot differentiation MS-D (33) (mg/l)	Root promotion Modified SH (43) (mg/l)
(a)	NH_4NO_3	1650	1650	-
	KNO_3	1900	1900	2500
	$NH_4H_2PO_4$	-	-	300
	KH_2PO_4	170	170	-
	$CaCl_2 \cdot 2H_2O$	440	440	200
	$MgSO_4 \cdot 7H_2O$	370	370	400
(b)	H_3BO_3	6.2	6.2	5.0
	$MnSO_4 \cdot 4H_2O$	22.3	22.3	-
	$MnSO_4 \cdot H_2O$	-	-	10.0
	$ZnSO_4 \cdot 7H_2O$	8.6	8.6	1.0
	KI	0.83	0.83	1.0
	$Na_2MoO_4 \cdot 2H_2O$	0.25	0.25	0.1
	$CuSO_4 \cdot 5H_2O$	0.025	0.025	0.2
	$CoCl_2 \cdot 6H_2O$	0.025	0.025	0.1
(c)	Na_2EDTA	37.3	37.3	37.3
	$FeSO_4 \cdot 7H_2O$	27.8	27.8	27.8
(d)	Myo-inositol	100	100	1000
	Thiamine·HCl	1.0	1.0	5.0
	Pyridoxine·HCl	-	-	0.5
	Nicotinic acid	-	-	5.0
	Casein hydrolysate	-	400	-
	Coconut milk	10% (v/v)	10% (v/v)	10% (v/v)
(e)	2,4-D	3.0	-	-
	α-Naphthaleneacetic acid			
	Kinetin	-	1.0	-
(f)	Sucrose	2000	2000	2000
(g)	Agar	9000	8000	8000
	pH (adjusted with 1 NaOH)	5.7	5.7	5.7

several methods for encouraging root production, such as culturing the shoots in water, storing them at 15°C, trimming the leaves and adding NAA (5 mg/l) to the medium, etc. (34). All of these measures would retard the shoot growth and are not in the interest of mass production if one wants to raise plantlets on a large scale. Addition of dalapon to the medium as reported previously (38), is not effective either. After much effort we have found that the modified SH medium (Table I) is effective in promoting root production provided that the following two requirements are met, viz; (1) there is a sufficient quantity of medium (usually 60 ml contained in a 250 ml Erlenmeyer flask works well), and (2) culture takes place under diffused sunlight conditions.

E. *Histology*

Histological examination has revealed that callus initially arises from the phloem cells of a young vascular bundle. This phenomenon applies to both apical meristems and young leaf tissue sources (31).

Early organogenesis is characterized by the development of a small green nodule of about 1.0-1.5 mm on the surface of a callus. Anatomical examination shows that the green nodule is virtually an apical meristem flanked with leaf primordia, and has the same shape as the shoot apex in a field-grown sugar cane plant. No radicle or coleorhiza is associated with the callus-derived shoot apex, unlike that in a sugar cane seed embryo which does have these two tissues (2). Cells near the shoot apex differentiate into tracheary elements with secondary wall thickenings within 30 days after subculturing (22).

A hypothesis of single cell origin for the callus-derived shoot apex has been suggested (22), but the evidence is not entirely conclusive at the present time (35).

F. *Genetic Variability*

Investigation on genetic variability occurring in cultured cells or differentiated plantlets has concentrated on the analysis of chromosome number, morphological characters and isoenzyme patterns. The results obtained are as follows:

1. Suspension Culture. Chromosome numbers varied from 2n=92 to 2n=191 with a mean of 111 in the suspension cells of cv. F164 as compared with the donor chromosome number 2n=108 (30).

2. *Regenerated Plantlets.*

a. Morphological character. The most conspicuous change in morphological character in plants derived from callus occurred in the auricle length which accounted for 8.6% of the change as compared to donor clones. Dewlap shape ranked next (6.5%), followed by hair group (6.2%) and form of top leaves (1.9%). The frequencies of morphological changes occurring in callus derivatives were different depending upon their origin. For example, the callus derivatives from F146 had 1.8% changes. Others had frequencies as follows: N:Co310, 2.4%; F161, 4.8%; F164, 6.7%; F162, 7.4%; 56-2080, 15.8%; F170, 27.6%; and F156, 34.0% (24).

b. Chromosome number. Chromosome number varied from 2n=86 to 2n=126 in the F156's callus-derived plants as compared with their donor number of 114, whereas those of F164's callus derivatives varied from 2n=88 to 2n=108 as compared with their donor number of 108. Those of F146's cell population generally centred around the original number, 2n=110, and it was considered to be the most genetically stable cultivar (24).

c. Isoenzyme patterns. The isoenzyme patterns fractioned by polyacrylamide disc electrophoresis have been used to study the genetic constitution of regenerated plants (49). The callus-derived line, 70-6132, from cv. F164 was short two bands in the esterase zymogram as compared with the donor, whereas 70-6132 had the same band number as its donor, F164 (25). Three variant plants C_1, C_2, and C_3, derived from 61-1248 (a breeding line) had two bands, three bands, and one band less, respectively, in the esterase zymogram than those of their donor which had 11 bands (30).

G. *Disease Resistance*

Tissue culture is a good tool for studying host-parasite relationships and offers a new approach in the production of disease resistant plants from susceptible donors. A system for studying *Phytophthora parasitica* in tobacco callus has been described by Helgeson *et al.* (15). Callus cultures derived from resistant plants would not support growth of the fungus, unlike cultures from sensitive plants which do support its growth. However, in our study, using the same approach, we found that callus cultures derived from either resistant or susceptible sugar cane varieties could nurse the growth of the downy mildew fungus (*Sclerospora sacchari* Miyake) (7). Therefore, the usefulness of tissue culture methods for studying disease resistance remains debatable.

III. LABORATORY PROTOCOL

A. *Materials and Methods*

The basic equipment and procedures for conducting tissue culture research work are described in detail in Chapters 1 and 2 in this book. The following methods and procedures have been applied to sugar cane.

1. Preparation of Medium. The medium for sugar cane, as for other plant species, is composed of basic major groups of ingredients: inorganic nutrients (a, b, c in Table I), vitamins and organic supplements (d in Table I), growth regulators (e in Table I), and a carbon source (f in Table I). Media for callus formation and shoot differentiation are those modified from Murashige and Skoog (MS) medium (33), and referred to as MS-C and MS-D, respectively; that for root promotion is a variation of Schenk and Hildebrandt (SH) medium (43) (Table I).

2. Tissue Explants. Subapical meristematic regions including the 1st to 8th internode (node designation is according to Artschwager, 2) are excised from the stem tip; the second innermost roll of young leaves right above the shoot apex is taken in lengths of 2 - 3 cm; parts of unemerged inflorescences are picked up with forceps from a sugar cane arrow still enclosed by the leaf sheath.

3. Excision Procedures. Excision tools, such as knives, forceps and petri dishes are thoroughly sterilized. Because the apical and the rolled young leaf tissues are wrapped deep in the leaf sheaths, they are generally pathogen-free, and it is not necessary to disinfect them. The outer leaf sheaths are stripped off to approximately the 10th node (Fig. 1). The other tender sheaths are twisted carefully until they are broken off at the layer of insertion of the leaf sheath on the nodes until only the first and second innermost leaf sheaths remain attached to the shoot apex. The apical dome (Fig. 2) and the 2 cm portion of the leaf sheath (Fig. 3) near the apex are speparately cut off and immediately inoculated onto the MS-C medium.

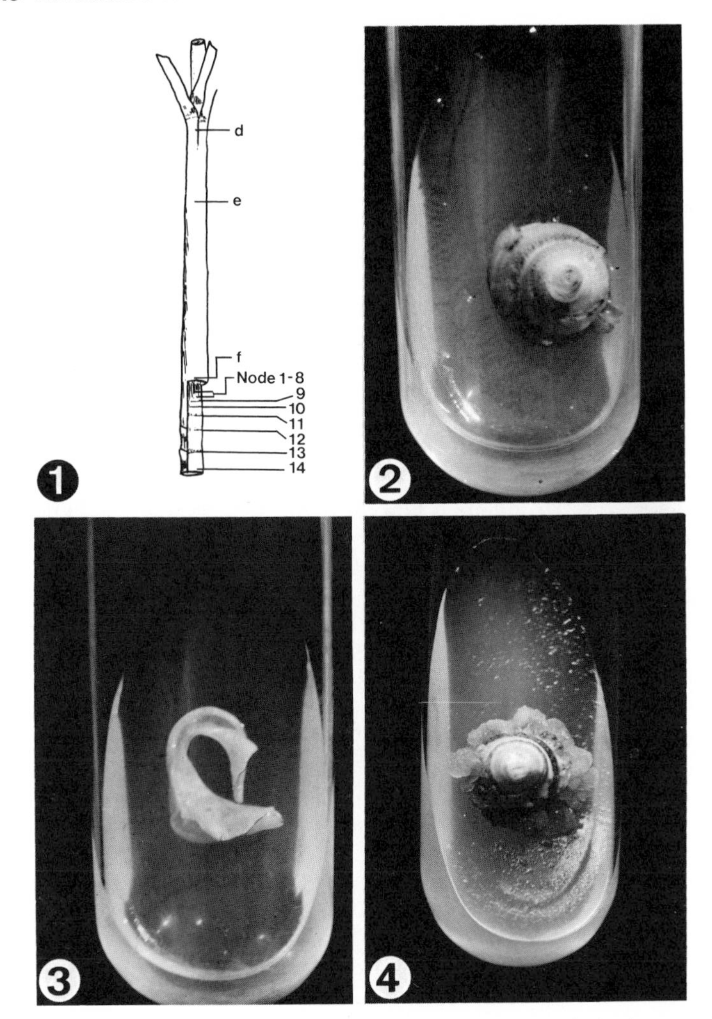

FIGURE 1. Top part of sugar cane plant with leaf sheaths removed and stem tip split open to expose growing point and leaf portion to be excised for explantation. Nodes 14, 13, and 12 have mature leaf sheaths, d indicates the position of the termination of the leaf sheath of node 11, e of node 10, and f of node 9.

FIGURE 2. An excised growing point. One day after explantation. (1.0 X).

FIGURE 3. A piece of a meristematic leaf 2 cm above the growing point. One day after explantation. (1.0 X).

FIGURE 4. Callus proliferating from exposed wounded surface on a growing point. Six weeks after explantation. (0.8 X).

There is a major problem with excised sugar cane meristematic tissue in that it secretes a brown exudate into the medium. This is often so great that cultures do not survive. The reason for this is that the meristematic tissue contains a high concentration of polyphenols, which are oxidized to form quinones and water catalyzed by the enzyme, phenoloxidase, when exposed to oxygen. Quinones are well known to be toxic to microorganisms and inhibitory to plant cellular growth. In order to reduce the brown exudate, the most important measure is to excise the tissues quickly. The tender leaf sheaths are twisted carefully until they are broken off along the layers of leaf traces. A leaf trace is a vascular bundle extending between its connection with a leaf and that of another vascular unit in the stem. Therefore, if the twisting operation is well done, the leaf sheath will be separated smoothly without injury to the meristematic cells at the node (Fig. 2). Thus, there is less chance for the phenolic compounds to be exposed to the air. This operation is rather difficult and needs repeated practice. Even the most skillful technician can attain only 60% success, i.e., only 60% of the meristematic tissue excised can be kept white.

In addition to skillful excision, the following two measures are helpful in reducing the oxidation of the polyphenols.

(1) Use of cysteine. Right after excision, the apical dome (Fig. 2) is immersed in sterilized water containing 100 mg/l L-cysteine·HCl for a few minutes before inoculating it onto the medium, which is supplemented with 50 mg/l L-cysteine·HCl.
(2) Favorable physical conditions. After explantation, the apical domes are immediately placed in the dark for one week, and then exposed to light. It is not necessary to incubate young inflorescences and leaf tissues in the dark.

4. *Culture Conditions*. The inoculated tissues are incubated in a culture room kept at 26 - 28°C under cool white fluorescent light of an intensity of about 3,870 lux, at a day/night cycle of 12/12 hours.

5. *Callus Formation*. After 5 - 7 days in culture, callus cells start to form from the broken regions of the explants. Six weeks later a considerable mass of callus will be formed as shown in Fig. 4 (apical dome), Fig. 5 (young leaf) and Fig. 6 (young inflorescence). Unemerged inflorescence tissue has the greatest capacity to form callus, from which numerous shoots can be regenerated directly, without having to be transferred to MS-D medium for differentiation.

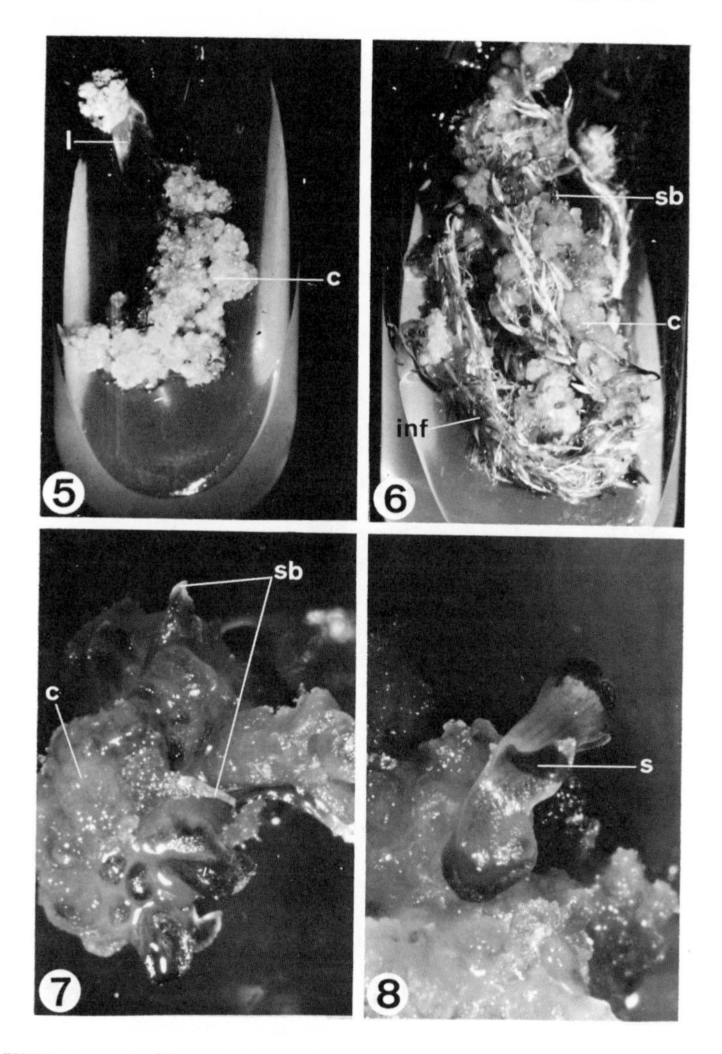

FIGURE 5. Callus (c) arising from cut surface of a piece of meristematic leaf (l). Six weeks after explantation. (1.0 X).

FIGURE 6. Callus mass (c) proliferating vigorously from young inflorescence tissue (rachis) (inf). Note numerous shoot buds (sb) differentiated directly from the callus. Five weeks after explantation. (1.25 X).

FIGURE 7. Callus mass (c) with numerous green nodules (shoot buds (sb). (1.75 X).

FIGURE 8. The green nodules gradually develop into a leafy and succulent shoot (s). (2.3 X).

6. *Transfer and Maintenance of Callus Culture.* After 4 to 5 weeks of growth, the callus mass can be subcultured onto MS-C medium for callus proliferation, or onto MS-D medium for plant regeneration. This two direction approach can be carried out every 4 to 5 weeks for both maintenance of callus and to produce plantlets continuously.

7. *Shoot Differentiation from Callus.* Ten to fourteen days after transferring to a fresh MS-D medium, numerous green bumps can be seen on the callus surface. These are the growing points and leaf primordia (Fig. 7) and they gradually develop into shoots. Most of them exhibit a leafy and succulent appearance (Fig. 8). Three to four weeks later the entire surface is covered with healthy green shoots (Fig. 9).

8. *Mutation Induction Procedures.*

a. *Ethyl methane sulphonate (EMS) treatment.* The best time for applying chemical mutagens is when the callus surface shows the dense green nodules (Fig. 7). The callus is immersed in a 0.02 or 0.04 M millipore-filtered EMS solution for 24 to 48 hours at 20°C, then washed with sterile water and transferred to fresh MS-C medium for growth. The green nodule is equivalent to the bud primordium stage *in vivo* and is apparently most vulnerable to the chemical.

b. *Gamma ray irradiation.* Again, the best time for treatment is when the callus exhibits the dense green bumps. It can be treated with 2, 3, 4, and 5 kR of gamma rays in a gamma chamber with a ^{60}Co source. Right after the irradiation, the treated callus is cut into 0.5 cm square pieces and each is placed at a spacing of 2 x 2 cm on a MS-C medium in a 90 mm petri dish. Each piece of callus will be covered with approximately 10 - 15 green nodules (shoot apices). Five weeks later, the callus pieces which have grown into shoots 2 - 4 mm high are transferred into a 30 x 150 mm tube containing MS-C medium. Plantlets are grown to maturity according to the procedures described below.

9. *Rooting of Shoots.* There is no true embryoid stage in callus to give rise to a complete plantlet. Instead, the shoots and roots develop independently. Some varieties of cane develop shoots only from their callus. In such cases, a special effort has to be made to stimulate root production. We have been successful in rooting such shoots by transferring them onto a modified SH medium (Fig. 10).

FIGURE 9. Vigorously growing shoots derived from a single piece of callus. (1.5 X).
FIGURE 11. Plantlets growing in pots of vermiculite. (0.1 X).

10. Successful Rearing of Plantlets. When the plantlets have grown 6 to 9 cm in height in flasks (Fig. 10), they are transferred to vermiculite-containing pots (Fig. 11), and watered with Hoagland solution every two days. When they are 15 to 18 cm, they can be transferred to an intermediate-bed filled with soil. Four to six weeks later, they are ready to be transplanted into the field within a plot size of 6 x 1.37 m for field trials, with their donors planted by their sides (Fig. 12).

B. *Selection Methods for Callus Derivatives and Analysis of Data*

1. Chromosome Number Determination. To count the chromosomes in pollen mother cells, young spikelets from callus-derived plants are fixed in Newcomer's solution (36), and squashed in 4% iron aceto-carmine. Somatic cells are examined by the leaf squash method of Price (42). Five to ten cells per plant should be examined.

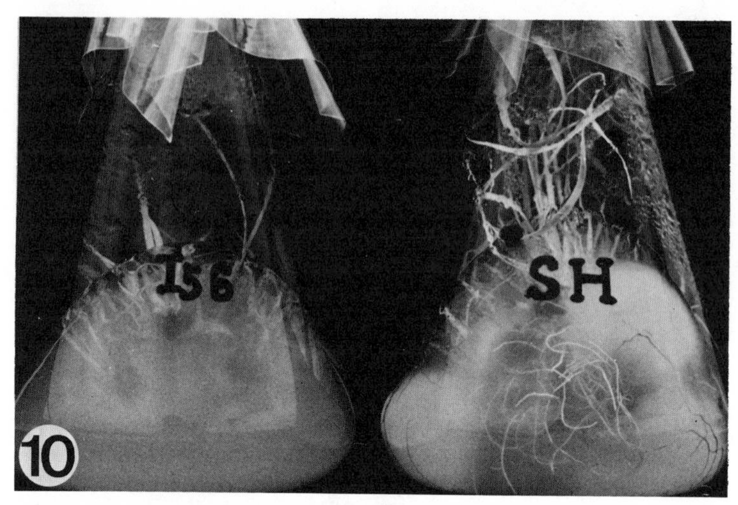

FIGURE 10. I_{56} test medium for shoot differentiation in a series with modified MS media. I_{56} is identical to MS-D medium in Table I. SH medium is effective in stimulating rooting of shoots.

FIGURE 12. Callus-derived plantlets undergoing field testing along with their donor clones.

2. *Yield Test.* Two-node cuttings excised from the second year asexually propagated callus derivatives and their donors are planted (0.4 m apart), in autumn in 8-row plots,

each row having a size of 6 x 1.37 m. The experimental array is a randomized complete block design with 2 - 3 replications. Canes are harvested when they are mature, i.e., about 18 months after planting. Data are collected during harvesting time and subjected to an analysis of variance according to the field design (46).

3. *Sucrose Analysis*. Every promising callus-derived plant is first checked for their field Brix by a hand refractometer. If their values are high enough, they then undergo chemical analyses according to the procedures described in the Cane Sugar Handbook, edited by the Taiwan Sugar Corporation (47). Since sugar content is greatly influenced by environmental factors such as temperature, soil fertility and water, and sampling technique, derivatives and their donors are analyzed monthly from October to February each year. For the purpose of reliable comparison, callus derivatives and their donors are always grown at the same site and their sugar data, either field Brix or percent sucrose, are investigated at the same time. In order to test how much the average sucrose content of callus derivatives differs from that of the donor clone, a paired t test is run for each high-sucrose line (46).

4. *Disease Resistance Test*. At the present time, disease resistance studies have concentrated on smut disease. The reaction of *in vitro* regenerated plantlets from a susceptible clone is tested by the wound-paste method described by Leu and Teng (20). A chi-square test is run to determine if there is a difference in smut infection between callus-derived lines and their donor clone.

IV. THE POTENTIAL OF USING TISSUE CULTURE METHODS FOR SUGAR CANE IMPROVEMENT

A. *Introduction*

Sugar cane is a major economic crop in tropical and subtropical regions. Almost 100 million tons of sugar were produced worldwide during the period of 1979-1980. Of this total 61.2% was produced from sugar cane.

The major problems involved in sugar cane production are to (1) increase cane yield, (2) raise sucrose content in cane and (3) obtain disease-resistant varieties. These goals have been achieved by the conventional breeding method -- sexual hybridization in the past 50 years. However, the genetic variability created through sexual crossing will likely be exhausted some time in the future, thus requiring the exploitation of new methods. Cell and tissue culture techniques offer such a new approach to the broadening of genetic bases in higher plants (44). This is based on the phenomenon that callus or isolated cells maintained *in vitro* over a long period are usually cytogenetically unstable and give rise, after regeneration, to plantlets which are often chromosomally and morphologically different from their donor plants. This capacity to obtain genetic variability by tissue culture methods constitutes the theoretical basis for selecting desirable mutants for agricultural use.

B. *Approaches to Sugar Cane Improvement Through Tissue Culture Methods*

1. Selection of High-Yielding Callus-Derived Lines. For screening high-yielding lines, replicated field tests for the selected callus derivatives have been held each year since 1974. Results obtained in the years 1974/1975, 1977/1978, and 1978/1979 are presented in Table II.

The results obtained in the year 1974/1975 indicated that 70-6132 was 32, 34, and 6% higher in cane yield, sugar yield and stalk number, respectively, than its donor, F164. The differences between 70-6132 and its donor for cane and sugar yield were significant at the 5% probability level, revealing that their differences were real. Of the 4 agronomic characters measured, stalk number was the most important character that influenced the cane yield of 70-6132 (Table II) (25).

In the F160 derivatives, 74-3216 was 5, 2, and 14% higher in cane yield, sugar yield and stalk number, respectively, than its donor. Again, among the agronomic characters measured, stalk number was the most important one influencing cane yield. Because of its good performance, 74-3216 has been released to the regional field trail conducted at various locations (Table II) (26). Another F160 derivative, 75-3070, was 8% and 1.5% higher in cane yield and sugar yield, respectively, than its donor (Table II) (27). Although these two derivatives (74-3216 and 75-3070) were only 5 - 8% higher in cane yield than their donor, they are very valuable since they surpass the No. 1 commercial variety used in Taiwan's sugar cane growing regions.

TABLE II. Means and Comparisons Between Callus Derivatives and Their Donors for Four Characters

Derivative and variety	Cane yield		Sucrose	Sugar yield		Stalk number	
	$\bar{X}$	CD/D[b]		$\bar{X}$	CD/D	$\bar{X}$	CD/D
	ton/ha	%	%		%	1000/ha	%
F164[c]-70[d]-6132[e]	130.8[a]	132	13.53	17.7	134[a]	62.7	106
F160[c]-74[d]-3216[f]	155.7	105	12.52	19.5	102	64.0	114
F160[c]-75[d]-3070[g]	167.0	108	12.65[a]	21.1	115	58.0	97
F164 (donor)[e]	99.1		13.42	13.3		59.2	
F160 (donor)[f]	148.9		12.89	19.2		56.0	
F160 (donor)[g]	154.0		11.88	18.3		60.0	

[a] The difference between callus derivative and its donor is significant at the 5% level. [b] CD/D = callus derivative/donor. [c] The donor. [d] Year in which the callus-derivatives were transplanted into the field from a sterilized medium. [e] Callus derivative (assigned by a serial number) and its donor tested in 1974/1975 (25). [f] Callus derivative and its donor tested in 1977/1978 (26). [g] Callus derivative and its donor tested in 1978/1979 (27).

2. *Selection of Callus-Derived Lines Having High Sucrose Content.* Since 1950, the average sugar-producing rates in the factories of Taiwan Sugar Corporation (TSC) have been decreasing in each decade, i.e., from 12.32% in the 1950s to 11.75% in the 1960s, and finally it dropped to 10.51% during the period of 1970-1975 (48). This constitutes a major problem for TSC. Currently, TSC is making every effort to raise the sugar production rate. Improved manipulation in field culti-

breed high-sugar varieties.

The sucrose content of selected high-sugar callus derivatives and their donors are presented in paired pattern at each date analyzed (Table III). Among these, 71-4829 is a striking example of a line whose percent sucrose has been consistently higher than its donor, H37-1933, since its establishment. The difference was significant at the 0.01 probability level, indicating that the increase in 71-4829's sucrose is genetically controlled. The same explanation can be applied to the pair 75-3070 and F160. Unfortunately, 75-3070 has sclerotic disease symptom, and therefore has not entered into regional field testing program. Derivatives 76-3035 and 76-3303 were slightly higher in sucrose content than their donors (Table III) (26).

3. Screening of Callus-Derived Lines for Smut (Ustilago scitaminea) Resistance. Smut has been one of the major diseases in Taiwan since 1964. Because it has plagued Taiwan's sugar cane fields very seriously, screening cane varieties resistant to this fungus before releasing them to the regional yield-trials has become routine.

Cultivar F177 is Taiwan's No. 2 variety, occupying 10.8% of the total sugar cane acreage during the crop year 1978/1979. However, its growing area has been limited owing to its high susceptibility to smut disease. In order to get rid of this defect, selection of disease-resistant lines has been conducted since 1976. Among 60 callus-derived lines of F177, nine of them were selected at the Field Test I stage based on their better agronomic performance. These selected lines were propagated in the field. When they grew up, 40 cuttings from each line were sent to the Smut Resistance Test Nursery, Department of Plant Protection, TSRI, for disease-resistance testing. Diseased stools were investigated and recorded each month from December 1978 to June 1979 (Table IV). Among the 9 lines tested, 76-5530 was much lower in infection rate (43.8%) as compared with its donor (F177)'s 88.2%. A chi-square test was run between each derivative and the donor by a contingent table method (46). Data in Table IV show that 3 out of 9 derivatives had a significant chi-square value, i.e., there is a difference between the derivatives and their donor in response to smut infection (27). Derivative 76-5530 is now under field trial to evaluate its other agronomic characters.

4. Induced Variations by the Use of Chemical and/or Physical Mutagens. One of the objectives in using tissue culture techniques for sugar cane improvement is to facilitate

TABLE III. High-Sucrose Lines Selected from Callus Derivatives

Date of analysis	Available sucrose (%)[a]							
	71-4829	Donor (H37-1933)	75-3070	Donor (F160)	76-3035	Donor (F160)	76-3303	Donor (F162)
Oct.	11.12	9.12	9.50	8.92	9.26	9.20	10.17	10.69
Nov.	13.11	11.18	11.96	10.44	13.52	13.19	-	-
Dec.	13.25	10.92	13.16	11.88	-	-	14.66	13.80
Jan.	15.33	11.99	14.16	13.89	15.05	14.41	-	-
Feb.	14.64	12.51	14.46	14.20	14.83	13.45	16.25	16.00
Average	13.49	11.14	12.65	11.87	13.17	12.56	13.69	13.49
Mean difference	2.35[c]		0.78[b]		0.61[d]		0.20[d]	

[a] Average of data collected from 3-4 consecutive years.

[b,c] Significant at the 0.05 and 0.01 probability levels according to paired t-tests.

[d] Not significant.

TABLE IV. *Smut-Resistant Plants Selected*[a] *from the Callus Derivatives of a Susceptible Clone, F177*

Derivative	Survivors following inoculation	No. of infected stools	No. of healthy stools	Infection rate %	Chi-square
76-5502	28	28	0	100.0	3.14[c]
76-5512	44	39	5	88.6	0.0[c]
76-5517	33	33	0	100.0	3.87[b2]
76-5519	33	31	2	93.9	0.43[c]
76-5530	16	7	9	43.8	7.48[b2]
76-5531	32	32	0	100.0	3.87[b1]
76-5539	22	19	3	86.4	0.04[c]
76-5544	41	40	1	97.6	2.03[c]
76-5550	13	13	0	100.0	1.75[c]
F177 (donor)	17	15	2	88.2	

[a]*Method according to Leu and Teng (20).* [b1,b2]*Significant at the 0.05 and 0.01 levels of probability, respectively.* [c]*Not significant.*

artificial induction procedures. The following three mutagens have been used for the purpose of increasing the mutation frequency for a given character.

a. Colchicine. By treating cv. F164's suspension cells with 100 mg/l colchicine, the chromosome numbers increased to as many as 309, almost tripling its donor number of 108. Three variant plants, 61-1248-C_1, -C_2 and -C_3 have been regenerated from the cell culture of 61-1248, a breeding line which was treated with colchicine. Their chromosome numbers were found to increase to 2n = 156, which was aneuploid compared with their donor's number of 104. Plants having such high chromosome numbers were stunted and lacked vigor (30).

b. EMS. Three derivatives have been obtained through the treatment of callus with EMS solutions, viz., 76-3035, from cv. F160, by using 0.02 M for one day; 77-3178 and 77-3179, both from cv. F177, by 0.04 M for two days. All the derived lines were similar to their donors without noticeable changes in agronomic characters (Liu and Chen, unpublished data).

c. Gamma ray. A variant from cv. F164 induced by 2.0 kR demonstrated conspicuous alterations in agronomic characters. It became shorter and thinner as compared with its donor (Liu and Chen, unpublished data).

With respect to the practical use of ionizing irradiation in sugar cane, detection of mutated plants is tedious and often much more difficult than the induction procedure *per se*. Chimera formation and diplontic selection are two particular problems in mutation breeding of vegetatively propagated crops. The importance of elimination of chimeras and decrease diplontic selection has been discussed previously (16,17). Elimination of chimeras can apparently be achieved by careful single stalk selection in the first vegetative mutation generation (vM_1).

5. *The Role of Protoplast Culture in Crop Improvement.*

a. Prospects. Since the development of an enzymatic procedure to isolate protoplasts from plants by Cocking (8), plant protoplasts have become an important experimental tool in the study of physiological, biochemical, and genetical problems. Protoplasts in culture can regenerate cell walls and can divide into cell clumps. Under appropriate conditions they can fuse and the fused products can be cultured; even the regeneration of intergeneric somatic hybrids has been recently reported (32). Protoplasts can also take up microorganisms and organelles (9), as well as isolated DNA molecules (41). Thus, protoplast culture can provide the opportunity to (1) combine by fusion different plant species which are sexually incompatible, and (2) introduce foreign genetic materials such as organelles or DNA into the genome for genetic transformation studies (see also Chapter 4, this volume).

b. Approaches. Several approaches to this problem of crop improvement are possible (see 9, 10, 45). However, to date, we have approached it in the following ways:

(1) From sugar cane leaf tissue. Since 1974, we have been working on sugar cane protoplasts with the intention of modifying the cell contents and with a final goal of achieving somatic hybridization between sugar cane and a legume crop, e.g., soybean (*Glycine max*). A large number of protoplasts

have been readily isolated from meristematic leaf tissues (4). About 7% of the freshly isolated protoplasts have more than one nucleus. These multinucleates apparently arise through spontaneous fusion during protoplast isolation. Immediately after isolation, the viable protoplasts have no walls, but after culturing for 24 hours, new cell walls are formed, as determined by fluorescence microscopy (5). The protoplasts undergo their first cell divisions with 2 to 5 days if cultured by the hanging drop method. A cell cluster is formed after 3 weeks in culture. Protoplasts from the same source are induced to fuse by treatment with polyethylene glycol (PEG) solution (23).

(2) From soybean root nodules. Since we are interested in transferring nitrogen fixing genes from the legume to sugar cane, preliminary work of isolating protoplasts from soybean root nodules had to be done first. Field-growing soybean nodules (0.4 - 0.6 cm in diameter) were selected and thoroughly washed with water. The bacteroid-containing nodules could be cut into 0.3 - 3.0 mm pieces and incubated in an enzyme mixture (1% Driselase + 1% Onozuka R-10 + 0.5% pectinase + 0.5% hemicellulase). After 2 - 3 hours, a large number of protoplasts could be isolated. The isolation rate is 0.1 - 0.5 x 10^6 per gram fresh weight. The protoplasts retained roughly the same elongated shape as in the nodule's tissue, but had become larger by 36 - 54% as a result of absorbing water (Liu and Chen, unpublished).

(3) From sugar cane suspension culture. Protoplast fusion between different sources can be enhanced by using PEG solution (18). However, whether the fused product is able to divide in culture is the critical point in somatic hybridization studies. In order to establish suitable conditions which encourage the fused product to divide, working on medium composition, exploring the suitable growing stage, etc. for protoplast isolation must be thoroughly investigated.

We found that sugar cane protoplasts could be enzymatically released from sugar cane clone 65-28 suspension culture by using a combination of two enzymes : 5% (v/w) Onozuka R-10 + 1.50% (v/v) Glusulase, and shaking (160 rpm) for 2 - 3 hours at ca. 26°C. Approximately 10% of the original cells in the suspension culture were recovered as protoplasts, after three washings in a culture medium, by centrifuging at 90x g for 10 minutes. The stage of cell growth was critical for successful conversion of a high percentage of cells to protoplasts. It was found that the exponential phase was much better than the stationary phase in this respect. The higher the pH value of the enzymatic solution, the lower the percentage of the cells which were converted into protoplasts, pH 4.0 being most suitable for protoplast isolation (6).

6. Anther Culture and its Significance in Crop Improvement. Haploids are important to plant geneticists and breeders for at least two reasons: (1) homozygous plants can be obtained by using colchicine treatment; (2) mutant genes are not subject to the effect exerted by either dominant or recessive alleles, thus selection of a superior character is facilitated. However, for practical use in a breeding program they are required in large numbers. The conventional methods employed by plant breeders for haploid production are cumbersome and laborious. With the introduction of tissue culture method as suggested by Guha and Maheshwari (12) and Nitsch (40), excised anthers and isolated pollen can be induced to form haploid embryos and/or plants (see also Chapter 8, this volume). Tsay (50) has reviewed this topic and reported that more than 37 genera covering 70 species of higher plants can be cultured to generate callus, embryos and/or whole plants. However, the successful species have not included sugar cane. It is apparently a difficult task. One of the problems with sugar cane anther culture is that young anthers contain a high concentration of phenolics, the oxidation products of which quinones, are toxic, and thus cause death to the isolated anthers. One of the approaches used to curtail the formation of quinone is to culture the whole spikelets in liquid medium as tried in our laboratory.

Part of the immature sugar cane inflorescence with pollen at pollen-mother-cell or tetrad stages were cultured in the liquid MS basal medium, supplemented with 2.5 mg/l 2,4-D and 6% of sucrose. After 30 - 40 days shaking, some of the spikelets produced callus. A pre-treatment of placing the unemerged inflorescence in water for two days before spikelet inoculation seemed to enhance callus formation. There were differences in the response of various sugar cane genotypes in culture. Potato medium (1), in combination with a thermal induction from individual spikelets (Liu, Yang, and Chen, unpublished). In tracing the origin of the callus, it has been found that it arose from the vascular bundles which are underneath the ovary of the sugar cane spikelet (Liu, Yang and Chen unpublished).

C. *Conclusions*

The potential of sugar cane tissue cultures for crop improvement involves taking advantage of the capacity of callus cells to undergo genetic change in culture. Because the changes usually originate from the deletion or addition of a few chromosomes, or a portion of one, the mutant plant usually resembles the donor variety in all but a few characteristics. If the selection pressure is directed toward the deficient characteristics such as have been done in the case of smut-resistance and sucrose content, the indigenous commercial varieties could become better adapted and competent.

The problem with tissue culture breeding methods, as with other mutation programs, is that the induced genetic change is undirected and most often the derived clones are inferior to their progenitors. For screening an agriculturally useful genotype, the derived population must be large enough to improve the probability of finding an altered genotype. Therefore, the difficulties confronting the tissue culturist in screening a variety to be improved for one or two characters, while still holding its high-yielding ability, are the same as those faced by the conventional plant breeder. More efficient techniques in detection must be worked out before the tissue culture breeding method can be used widely.

Recent advances in the field of plant cell and tissue culture have increased the use of the many approaches afforded by these techniques for practical exploitation in agriculture. In our own work, we have been using these approaches since 1970. We have been able to achieve some positive results, and these have been outlined in this chapter. Similar progress is being made elsewhere, especially in Hawaii and Fiji. Thus, it appears that tissue culture will continue to play an important role in the improvement of sugar cane.

ACKNOWLEDGMENTS

The information presented in this report is a summary of work done by the author and his colleagues at this Institute and he wishes to thank them for the same. Thanks are also due to the National Science Council, Taiwan, China, for financial support for some of these studies.

REFERENCES

1. Anonymous, Research Group 301, *Acta Genet. Sin. 3*, 30 (1976).
2. Artschwager, E., *J. Agric. Res. 30*, 197 (1925).
3. Chen, W.H., M. Phil. Thesis, University of Leicester, U.K. (1978).
4. Chen, W.H., and Liu, M.C., *Rep. of Taiwan Sugar Res. Inst. 64*, 1 (1974).
5. Chen, W.H., and Liu, M.C., *Rep. of Taiwan Sugar Res. Inst. 71*, 1 (1976).
6. Chen, W.H., and Liu, M.C., *In* "Annu. Rep. Res. Dev. Council," Taiwan Sugar Corporation, p.29 (in Chinese) (1980).
7. Chen, W.H., Liu, M.C., and Chao, C.Y., *Can. J. Bot. 57*, 528 (1979).
8. Cocking, E.C., *Nature 187*, 962 (1960).
9. Galston, A.W., *In* "Propagation of Higher Plants Through Tissue Culture" (K.W. Hughes, R. Henke, and M. Constantin, eds.), p.200. DOE Conf. 7804111, Springfield, Virginia (1979).
10. Gamborg, O.L., Ohyama, K., Pelcher, L.E., Fowke, L.C., Kartha, K., Constabel, F., and Kao, K., *In* "Plant Cell and Tissue Culture," (W.R. Sharp *et al.*, eds.), p.371. Ohio University Press, Columbus (1979).
11. Giles, K.L., *In* "Frontiers of Plant Tissue Culture 1978" (T.A. Thorpe, ed.), p.67. University of Calgary Press, Calgary (1978).
12. Guha, S., and Maheshwari, S.C., *Nature 204*, 497 (1964).
13. Heinz, D.J., and Mee, G.W.P., *Crop Sci. 9*, 346 (1969).
14. Heinz, D.J., Krishnamurthi, M., Nickell, L.G., and Maretzki, A., *In* "Applied and Fundamental Aspects of Plant Cell, Tissue and Organ Culture," (J. Reinert and Y.P.S. Bajaj, eds.), p.3. Springer-Verlag, Berlin (1977).
15. Helgeson, J.P., Kemp, J.D., Haberlach, G.T., and Maxwell, D.P., *Phytopathology 62*, 1439 (1972).
16. Jagathesan, D., *In* "Improvement of Vegetatively Propagated Plants and Tree Crops through Induced Mutations," p.69, Wageningen: FAO/IAEA, Vienna (1976).
17. Jagathesan, D., and Ratnam, R., *Theor. and Appl. Genet. 51*, 311 (1978).
18. Kao, K.N., and Michayluk, M.R., *Planta 115*, 355 (1974).
19. Krishnamurthi, M., and Tlaskel, J., *Proc. Congr. Int. Soc. Sugar Cane Technol. (South Africa) 15*, 130 (1974).
20. Leu, L.S., and Teng, W.S., *Proc. Congr. Int. Soc. Sugar Cane Technol. (South Africa) 15*, 275 (1974).

21. Liu, M.C., *Taiwan Sugar 18*, 8 (1971).
22. Liu, M.C., and Chen, W.H., *Proc. Congr. Int. Soc. Sugar Cane Technol. (South Africa) 15*, 118 (1974).
23. Liu, M.C., and Chen, W.H., *Int. Soc. Sugar Cane Technol. Sugar Cane Breeders' Newslett. 37*, 39 (1976).
24. Liu, M.C., and Chen, W.H., *Euphytica 25*, 393 (1976).
25. Liu, M.C., and Chen, W.H., *Euphytica 27*, 273 (1978).
26. Liu, M.C., and Chen, W.H., *In* "Annu. Rep. Res. Dev. Council," Taiwan Sugar Corporation, p.23 (in Chinese) (1978).
27. Liu, M.C., and Chen, W.H., *In* "Annu. Rep. Res. Dev. Council," Taiwan Sugar Corporation, p.34 (in Chinese) (1979).
28. Liu, M.C., and Chen, W.H., *In* "Annu. Rep. Res. Dev. Council," Taiwan Sugar Corporation, p.26 (in Chinese) (1980).
29. Liu, M.C., Huang, Y.J., and Shih, S.C., *J. Agric. Asso. China 77*, 52 (1972).
30. Liu, M.C., Shang, K.C., Chen, W.H., and Shih, S.C., *Proc. Congr. Int. Soc. Sugar Cane Technol. (Brazil) 16*, 29 (1977).
31. Liu, M.C., Chen, W.H., and Shih, S.C., *Proc. Congr. Int. Soc. Sugar Cane Technol. (Philippines) 17*, (in press) (1980).
32. Melchers, G., Sacristan, M.D., and Holder, A.A., *Carlsberg Res. Commun. 43*, 203 (1978).
33. Murashige, T., and Skoog, F., *Physiol. Plant. 15*, 473 (1962).
34. Nadar, H.M., and Heinz, D.J., *Crop Sci. 17*, 814 (1977).
35. Nadar, H.M., Soepraptopo, S., Heinz, D.J., and Ladd, S.L., *Crop Sci. 18*, 210 (1978).
36. Newcomer, E.H., *Science 118*, 161 (1953).
37. Nickell, L.G., *Hawaiian Planters Rec. 57*, 223 (1964).
38. Nickell, L.G., *Proc. Congr. Int. Soc. Sugar Cane Technol. (Puerto Rico) 12*, 887 (1965).
39. Nickell, L.G., and Maretzki, A., *Physiol. Plant. 22*, 117 (1969).
40. Nitsch, C., *C. R. Acad. Sci. Paris, Series D 278*, 1031 (1974).
41. Ohyama, K., Pelcher, L.E., and Schaefer, A., *In* "Frontiers of Plant Tissue Culture 1978" (T.A. Thorpe, ed.), p.75. University of Calgary Press, Calgary (1978).
42. Price, S., *Proc. Congr. Int. Soc. Sugar Cane Technol. (Puerto Rico) 12*, 583 (1962).
43. Schenk, R.U., and Hildebrandt, A.C., *Can. J. Bot. 50*, 199 (1972).
44. Scowcroft, W.R., *In* "Advances in Agronomy" Vol. 29, (N. C. Brady, ed.), p.39. Academic Press, New York (1977).

45. Spiegel-Roy, P., and Kochba, J., *In* "Plant Tissue Culture and Its Bio-technological Application," (W. Barz, E. Reinhard and M.H. Zenk, eds.), p.404. Springer-Verlag Berlin (1977).
46. Steel, R.G.D., and Torrie, J.H., "Principles and Procedures of Statistics," p.78-79; 132-156; 370-371. McGraw-Hill Book Co., Inc., New York (1960)
47. Taiwan Sugar Corporation. "Cane Sugar Handbook" (in Chinese). Chapt. 3(1). (1956).
48. Taiwan Sugar Corporation. "Taiwan Sugar Statistics," 1976/1977, p.49 (1976).
49. Thom, M., and Maretzki, A., *Hawaiian Planters' Rec. 58*, 6 (1970).
50. Tsay, H.S., Ph.D. Thesis, College of Agriculture, National Taiwan University, Taipei, Taiwan (1978).
51. White, P.R., "A Handbook of Plant Tissue Culture," Jaques Cattell, Lancaster (1943).

IN VITRO METHODS APPLIED TO COFFEE

M.R. Sondahl
L.C. Monaco

Department of Genetics
Agronomic Institute
Campinas, São Paulo, Brazil

W.R. Sharp

Campbell Institute for Research and Technology
Cinnaminson, New Jersey

I. INTRODUCTION AND REVIEW

A. *Importance of Coffee*

The consumption of coffee as a stimulatory beverage began in Arabia (presently North and South Yemen) in the 14th century. The first public coffee house was inaugurated in Turkey in 1554 and in London in 1652. They appeared in Venice in 1615, in France in 1644, in Vienna in 1650, and in the USA in 1668. Soon after, the drink made of roasted coffee beans was adopted throughout the world despite some opposition due to its stimulatory effects. Coffee bean is a typical product of a tropical perennial plant.

The world production of coffee in 1977/78 and in 1978/79 was 69,615 and 74,544 million 60 kg bags respectively. In 1977 ca. 75% of the coffee production was sold in the international market generating ca. 12 billion US dollars. There are several countries that depend heavily upon the revenue of coffee exportation in order to buy modern goods and advanced technology; Burundi (94.2%), Ethiopia (75.4%), Rwanda (72.7%), Columbia (62.6%), El Salvador (62.2%), Ivory Coast (47.2%),

ISBN 0-12-690680-7

Haiti (45.1%) Kenya (41.7%), Tanzania (41.1%), and Costa Rica (40.2%). Coffee is also a very important source of revenue for small farmers, contributing to the development of rural areas around coffee plantations. The ten leading coffee producers in the world in the past 5 years (74/75 to 78/79) have been Brazil, Colombia, Ivory Coast, Mexico, Indonesia, Ethiopia, El Salvador, Uganda, Guatemala and India (12).

Despite the fact that it is one of the major agricultural products in the international trade market and has economic importance for many tropical countries, coffee plantations have not benefited from the same technological developments as other cash crops. It is expected that new research accomplishments pertaining to coffee will give rise to higher and stable world production and consequently more stable prices.

B. *Literature Review of in vitro Methods*

In vitro techniques can be applied in many aspects of the improvement of cultivated coffee. The pioneer work with coffee tissues was published by Staritsky (29). He was successful in inducing callus tissues from orthotropic shoots of *Coffea canephora*, *C. arabica* and *C. liberica*, but somatic embryos and plantlets only from *C. canephora*. Colonna established cultures of embryos of *C. canephora* and *C. dewevrei* with better growth and development from *C. dewevrei* (5). Callus from endosperm tissues of *C. arabica* was induced by Keller *et al.* with the objective of studying caffeine synthesis (16). Sharp *et al.* cultured somatic and haploid tissues of *C. arabica* obtaining callus growth (petioles, leaves, green fruits), proembryo formation (anthers) and shoot development (orthotropic shoots) (22). Monaco *et al.* were able to induce abundant friable callus from perisperm tissues of *C. arabica* and *C. stenophylla* (18). This tissue proliferated rapidly in the absence of auxin suggesting that it was autonomous for auxin. With the purpose of producing coffee aroma from suspension cultures, Townsley established liquid cultures of coffee cells from friable callus derived from orthotropic shoots of *C. arabica* cv. El Salvador (30). These same suspension cultures were used to analyze caffeine and chlorogenic acid contents (3) and to compare unsaponifiable lipids in green beans with those in cell suspensions (31). Callus formation (leaves) and shoot development (orthotropic shoots) were reported by Crocomo *et al.* (7). Anthers were placed in culture from several *Coffea* species (*C. liberica*, *C. arabica*, *C. racemosa*, *C. canephora* and *C. canephora* (4x) x *C. arabica*) however, rapid and friable callus development was observed only with *C. liberica* (19). Herman and Hass reported the presence of "organoids" from leaf explants of *C. arabica*; the normal plantlets were rooted and transferred to soil after 7 months (15). Solid cultures from stem explants of *C. arabica* cv. Bourbon Vermelho were established for analysis of caffeine production and possible commercial utilization of *in vitro* synthesis of such alkaloids (13). The distinction of high and low frequencies of somatic embryo induction from cultured mature leaf explants of *C. arabica* cv. Bourbon were described by Sondahl and Sharp (23, 24, 25). High frequency somatic embryos have also been obtained from mature leaf cultures of *C. canephora*, *C. congensis*, *C. dewevrei* cv. Excelsa, *C. arabica* cv. Mundo Novo, cv. Catuai, cv. Laurina, and cv. Purpurascens (24). Protoplasts have been isolated from leaf-derived callus tissues of *C. arabica* cv. Bourbon and callus regeneration was obtained in ca. 30% of the cultures (29). The histogenesis of callus induction and early embryo formation from *C. canephora* stem explants was described by Nassuth *et al.* (21). It was observed that the parenchymatous cells of the cortex contribute to formation of callus tissues. Embryo-like structures were present in the callus zone after 14 days of culture. Cultures of nodal explants of *C. arabica* led to the development of ca. 2.2 plantlets from arrested buds of leaf axils (9). Similarly, shoot development has been obtained from arrested buds of Arabusta (F_1 hybrid of *C. arabica* x *C. canephora* 4X) as a means of vegetative propagation (10).

Plant regeneration with coffee tissue has been accomplished from solid cultures of orthotropic shoots and mature leaves covering four species of the genus *Coffea* and five *C. arabica* cultivars. Shoot development from bud culture has been described for *C. arabica* and Arabusta. A summary of *in vitro* cultures using coffee tissues is presented in Table I.

II. LABORATORY PROTOCOL

Refer to Chapter 1 for required equipment and facilities and Chapter 2 for chemicals, supplies and media preparation.

For medium preparation, stock solutions, 50 X concentration of the mineral salts, vitamins, amino acids, and inositol are prepared previously and kept in 500-1000 ml dark glass bottles in the refrigerator or freezer. From each stock solution, 20 ml are taken to prepare one liter of final medium. Alternately, media can be prepared and frozen as 10 X concentrate. The mineral salts can be stored frozen in Whirl pak bags. The growth regulators (GR) are prepared just before using. Usually a concentration is prepared so that 1 ml of

TABLE I. Current Success of in vitro Culture of Tissues of the Genus Coffea

Success	Species	Explants	Reference
Undifferentiated callus tissues:			
Haploid	C. liberica	Anther	19,15
Somatic	C. arabica	Stem	29,13
	C. liberica	Stem	29
	C. stenophylla	Perisperm	18
	C. salvatrix	Leaf	24,25
	C. bengalensis	Leaf	24,25
	C. liberica	Leaf	24,25
	C. arabica	Petioles	22
		Leaf	8,22
		Green fruits	22
Triploid	C. arabica	Endosperm	16
Suspension cultures	C. arabica	Stem	3,30,31
Embryo culture	C. dewevrei cv. Excelsa	Embryo	5
	C. canephora	Embryo	5
Proembryo	C. arabica	Anther	22
Shoot development	C. arabica	Stem	7,9,10,22
Organoids	C. arabica	Leaf	15
High frequency somatic embryos	C. canephora	Stem	21,29
	C. arabica cv. Bourbon	Leaf	23,25
	C. arabica cv. Mundo Novo	Leaf	23,25
	C. arabica cv. Laurina	Leaf	23,25
	C. arabica cv. Purpurascens	Leaf	23,25
	C. dewevrei cv. Excelsa	Leaf	23,25
	C. canephora cv. Kouillou	Leaf	23,25
	C. congensis	Leaf	23,25
Protoplasts	C. arabica	Callus	28

the GR solution added to 1 liter of medium will give a final concentration of 5 µM. All growth regulators are dissolved in 0.1 or 1.0 N KOH (in 5, 10, or 25 ml volumetric flasks), or alternately some GRs (e.g., gibberellic acid) may be dissolved in ethanol. A small volume of DMSO can also be used for dissolving the GRs and it is especially recommended for use in the preparation of solutions of chemical mutagens (EMS, MMS, colchicine, etc.).

When a new cultivar or species of *Coffea* is subjected to *in vitro* study, a series of diallel experiments are programmed in order to find the best medium for growth and differentiation. Generally the salt solutions of Murashige and Skoog (20) provide a balanced composition for initial growth of coffee tissues. The specific requirements and concentration ratios of the growth regulators are compared by a diallel experiment. Such a 5 X 5 diallel experiment with kinetin and 2,4-dichlorophenoxyacetic acid is given in Table II.

To prepare the 5 X 5 diallel experiment one uses five 1 liter Erlenmeyer flasks with complete medium omitting 2,4-D and agar. Each liter flask (for example, the one containing 2.5 µM kinetin) is subdivided into 200 ml portions using 250 ml Erlenmeyer flasks. A total of 25 Erlenmeyers are used, each one coded with the respective treatment number (1 to 25). The 2,4-D is then added to each 200 ml treatment and the pH is adjusted to 5.5. Finally the agar (1.6 g/200 ml), is added to each treatment, then melted and poured into the culture bottles. Using square bottles, of ca. 50 ml capacity, 10 ml of melted medium is added to each bottle. This 5 X 5 diallel experiment as described provides 25 treatments with 20 replicates.

TABLE II. An example of a 5 X 5 diallel experiment with kinetin and 2,4-D

	2,4-D (µM)				
KIN (µM)	1.0	1.5	5.0	10.0	20.0
2.5	1[a]	2	3	4	5
5.0	6	7	8	9	10
10.0	11	12	13	14	15
20.0	16	17	18	19	20
40.0	21	22	23	24	25

[a] treatment number

A. Culture Procedure

1. Leaf Culture of C. arabica cv. Bourbon. Mature leaves from orthotropic or plagiotropic branches of greenhouse plants are surface sterilized in 1.3% sodium hypochlorite (25% commercial bleaching agent) for 30 minutes and rinsed three times in sterile double distilled water. If the coffee plants are growing in the field, the surface sterilization is much more difficult. The following procedure has given some degree of success for field material: 2.6% sodium hypochlorite for 30 minutes, rinse in sterile water, incubate in sealed Petri dishes overnight followed by another exposure to 2-6% sodium hypochlorite for 30 minutes and then rinse with sterile water. It has been found that 70% ethanol and solutions of $HgCl_2$ are toxic to coffee leaves; on the other hand, solutions of sodium or calcium hypochlorite are effective and non-toxic. Before immersing the leaves in the sterilizing solution, it is advisable to wash them by hand with 1% detergent solution, and rinse in distilled water. If ozone water is available, a 5 minute immersion is recommended.

Leaf explants of about 7 mm^2 are cut, excluding the midvein, margins, and apical and basal portions of the leaf blade. The elimination of the midvein (midrib) also excludes the domatia and consequently increases the number of sterile leaf explants. Domatia are deep pores (ca. 0.18 mm diameter) located in the acute angle formed by the midvein with secondary veins on the abaxial side of the coffee leaf. These leaf cavities, which the sterilizing solution will not penetrate, carry bacteria, fungi, and small insects. Leaf sections are usually cut in a sterile saline-sugar liquid medium. This was found to reduce oxidation at the cut edges and so favors subsequent growth. However, when dealing with highly contaminated material (such as from the field) it is better to cut the leaf explants on top of sterile filter paper (or paper towel) which is changed frequently to avoid cross contamination. If the explants are cut in liquid medium, each explant should be dried on sterile filter paper discs, before placing onto agar plates. All sides of a leaf explant must be cut since callus proliferation occurs only from cut edges. Histological study has demonstrated that the callus tissue originates from mesophyll cells of the leaf explant (27).

The leaf explants are placed on 20 X 100 or 20 X 150 mm Petri dishes containing the solidified saline-sugar medium (half-strength MS salts with 2% sucrose). The incubation of the leaf explants can be in the dark or in the light (no difference has been detected between these two conditions) for ca. 36 hours. This pre-experimental incubation period has been found to be very useful for the selection of viable explants and the elimination of contaminated explants. The abaxial surface of the leaf explants, which is clearly distinguished by a dull pale green coloration, in contrast to the shiny dark green coloration of the adaxial surface, is always placed upwards. French square bottles of ca. 50 ml capacity containing 10 ml autoclaved basal medium (BM) made up of Murashige and Skoog (MS) inorganic salts, 30 μM thiamine-HCl, 210 μM L-cysteine, 550 μM meso-inositol, 117 mM sucrose, and 8-10 g/l Difco Bacto agar are used. In primary culture, a conditioning medium containing a combination of kinetin (20 μM) and 2,4-D (5 μM) is used, and the bottles are incubated in the dark at 25 ± 1°C for 45-50 days. The composition of this "conditioning medium" was found to be ideal for high-frequency somatic embryogenesis (60% of the flasks) in leaf explants of *C. arabica* cv. Bourbon (23).

Callus proliferation of cultured leaf explants of *C. arabica* cv. Bourbon was initially studied in diallel experiments with a single cytokinin, kinetin, and several auxins [indole-3-acetic acid (IAA), indole-butyric acid (IBA), naphthalene acetic acid (NAA) and 2,4-D]. Evaluations were made with 50-70 day old primary cultures. Growth indices were evaluated based on the following grading scale: No. 1 (ca. 15 mg fresh weight), No. 2 (ca. 100 mg f.w.), No.3 (ca. 600 mg f.w.), No. 4 (ca. 1200 mg f.w.), and No. 5 (ca. 2400 mg f.w.).

Kinetin (0, 4.5, 9.0, 18, 36, and 72 μM) and IAA (0, 6.0, 12, 24, 48, and 96 μM) at all possible combinations were found to be ineffective for callus proliferation. All 36 combinations of kinetin (0, 4.5, 9.0, 18, 36, and 72 μM) and IBA (0, 5.0, 10, 20, 40, and 80 μM) were also ineffective in the promotion of substantial proliferation. The highest growth index, 1.4, was obtained with a kin/IBA concentration ratio (c.r.) of 4.5/20 μM.

Kinetin (0, 4.5, 9.0, 18, 36, and 72 μM) and NAA (0, 5.5, 11, 22, 44, and 88 μM) promoted excellent callus proliferation. Growth indexes of 2.9 and 2.8 were obtained with kin/NAA c.r. of 9/22 μM, and 9/44 μM, respectively. A lower range of kinetin concentrations (0, 2.22, 0.45, 0.9, 1.8, and 3.6 μM) and NAA (0, 0.22, 0.45, 0.9, 1.8, and 3.6 μM) was tested. A growth index of 2.4 was observed at the kin/NAA c.r. of 0.45/ 2.0 μM, which coincides with the 1/2 or 1/2 optimum c.r. found in the previous experiment.

Combinations of kinetin (0, 4.5, 9.0, 18, 36, and 72 μM) and 2,4-D (0, 4.5, 9.0, 18, 36, and 72 μM) gave the highest growth rate at the kin/2,4-D c.r. of 9/4.5 μM (3.3), 18/4.5 μM (3.0), and 18/9 μM (2.9). A lower range of concentrations was also tested with kinetin (0, 0.22, 0.45, 0.9, 1.8, and 3.6 μM) in order to tabulate information pertaining to the minimal and maximal concentration tolerance of coffee leaf tissues. The maximum growth index of 3.0 was found with a kin/2,4-D c.r. of 1.8/3.6 μM and 3.6/3.6 μM.

In order to test the effectiveness of other sources of cytokinins in the callus induction from leaf explants, diallelic experiments with a single auxin (2,4-D) versus 6-BA (6-benzylaminopurine), 2-iP [6-(y-dimethylallylamino)-purine] and zeatin were designed. 6-BA X 2,4-D and 2-iP X 2,4-D combinations were tested in the range of 1, 5, 10, 20, and 50 μM concentration, whereas zeatin X 2,4-D in the range of 0.1, 1.0, 5.0, 10.0, 20.0, and 50.0 μM for zeatin and 1, 5, 10, 20, and 50 μM for 2,4-D. None of these growth regulator combinations were more effective than kin X 2,4-D (or NAA). The maximum growth rates recorded were the following: 20/1 μM 6-BA X 2,4-D (2.1), 10/10 μM 2-iP X 2,4-D (1.26), and 1.0/5 μM zeatin X 2,4-D (1.84).

It is interesting to note that out of 296 treatments tested, a broad range of combinations of kinetin, 6-BA, or zeatin with 2,4-D or NAA were effective in promotion of callus proliferation. This suggests that in the coffee leaf explant system, no sharp sensitivity exists pertaining to callus growth within a wide range of growth regulator concentrations. Even though each experimental treatment was designed in such a way that subsequent concentrations doubled the previous concentration, the response of growth presented smooth transitions. In all growth regulator interactions tested, the necessity of having both a cytokinin and an auxin source was clearly demonstrated. To achieve a minimum growth index of 1.0 either 2.0 μM of NAA or 0.9 μM of 2,4-D, must be combined with 0.45 μM of kinetin. In the case of NAA, a ratio of about 1/2.5 or 1/5 was necessary for good callus proliferation. However, with 2,4-D, ratios of 5/1 and 2/1 were effective at high concentrations (more than 4.5 μM), whereas ratios of 1/2 and 1/1 were effective at lower range concentrations (less than 3.6 μM). The higher the auxin concentration the more friable the callus became; conversely, at higher kinetin concentrations callus tissues appeared more compact.

In another diallelic experiment, six concentrations of sucrose were tested with six concentrations of MS inorganic salts and it was found that half-strength MS salts with 60 mM sucrose provided best growth. Concentration ratios of KNO_3 and NH_4NO_3 were also studied at six diallel combinations and higher growth rates were observed with KNO_3 at the 2X MS concentration (18.8 mM) and NH_4NO_3 at the 1.0 X MS concentration (20.6 mM).

2. *Morphogenesis in C. arabica cv. Bourbon.* Secondary cultures are established under conditions of a 12-hour light period at 24-28°C by subculturing 45- to 50-day-old tissues onto an "induction medium" containing half-strength MS organic salts, except KNO_3, which is added at 2X concentration, 58.4 mM sucrose, kinetin (2.5 µM), and NAA (0.5 µM). Following transfer to the induction medium, the massive parenchymatous type of callus growth ceases and the tissues slowly turn brown. Two sequences of morphogenetic differentiation have been characterized in secondary cultures of leaf explants in *Coffea*; low frequency somatic embryogenesis (LFSE) and high frequency somatic embryogenesis (HFSE). Adopting the standard culture protocol described for *C. arabica* cv. Bourbon, LFSE is observed after 13-15 weeks and HFSE after 16-19 weeks of secondary culture. LFSE appeared 3-6 weeks before the visible cluster of HFSE in more than 5000 cultured flasks (23, 26, 27).

LFSE is observed by the appearance of isolated somatic embryos developing into normal green plantlets in numbers ranging from 1 to 20 per culture. The occurrence of HFSE follows a unique developmental sequence: a white friable tissue containing globular structures develops from the nonproliferating brown callus cell mass; the globular structures appear to develop synchronously for a period of 4-6 weeks. The proembryogenic globular mass gives rise to somatic embryos and finally to a plantlet, but this latter developmental process lacks the synchrony of the earlier stage, probably because of nutrient competition. The size of this pro-embryogenic tissue varies, but on the average about 100 somatic embryos develop per cluster of globular tissue. To speed up development and increase the percentage of fully developed plantlets, it is advisable to excise the proembryogenic tissues and grow them under light conditions at 26°C in 5-10 ml of liquid induction medium devoid of kinetin for 4-6 weeks. NAA can be replaced by IAA (10 µM). After this period, the torpedo-shaped somatic embryos and young plantlets can be plated onto saline-agar medium containing 0.5-1% sucrose in the presence of light. Individual plantlets are removed from the agar medium, gently washed, and immediately transferred to small peat pots inside a humid chamber. After a hardening period of 1-2 months, they can be exposed to normal atmospheric humidity and transferred to a greenhouse. Another way of hardening the plantlets is the use of ca. 100 ml of saline-agar medium without sucrose in 250 Erlenmeyer flasks closed with cotton plugs and paper. The Erlenmeyer flasks are exposed to sunlight in a shaded portion of the greenhouse (usually protected with plastic screens that filter ca. 60-70% of the sun's rays and placed underneath the bench). After ca. 2 months, the plantlets have good leaf and root development and can be transferred to small peat pots with sand, soil, and vermiculite or perlite balls in a proportion of 1/1/1.

3. *Leaf Culture of Other Coffea spp.* Callus proliferation was studied in other *Coffea* species with diallel experiments using kinetin and 2,4-D within a concentration range of 2-20 μM during primary culture. Callus was subcultured onto the "induction medium" used for cv. Bourbon. Callus proliferation was evaluated just before subculturing according to the grading scale mentioned earlier. Plants 1.5 to 2 years old of *C. arabica* cv. Catuai, Laurina, Mundo Novo, Purpurascens, *C. bengalensis*, *C. canephora* cv. Kouillou, *C. congensis*, *C. dewevrei* cv. Excelsa, *C. liberica*, *C. racemosa*, and *C. salvatrix* were used as sources of leaf explants.

A series of diallel experiments was designed to study the effectiveness of kinetin (2, 4, 6, 10, and 20 μM) and 2,4-D (2, 4, 6, 10, and 20 μM) in promoting callus proliferation of leaf explants of several *C. arabica* cultivars and diploid species. A great deal of variation in the extent of callus growth was observed among these cultures. In general, *C. arabica* required concentrations of 2,4-D lower than 6 μM, *C. canephora* required a 1/5 kin/2,4-D c.r., and *C. bengalensis* tissues underwent poor growth at all concentrations tested. The best callus growth indices (GI), recorded from cultured leaf explants are given in Table III. The optimum concentrations of kinetin and 2,4-D obtained in the above study (Table III) can serve as a guide for future experiments. After subculture onto the induction medium, the frequencies of somatic embryo formation should be compared with the concentration ratio of kin/2,4-D used during primary culture (conditioning medium). In this way, the best culture medium for each coffee species or cultivar can be selected.

4. *Stem Culture.* Soft internodes of orthotropic shoots are an excellent source of explants for coffee cell proliferation. Explants from plagiotropic shoots do not grow well and a higher percentage of cultures exhibit phenolic oxidation. Stem pieces can be surface sterilized with a saturated solution of sodium or calcium hypochlorite (ca.

TABLE III. Optimum Kinetin and 2,4-D Concentrations for Callus Induction and Growth of Leaf Explants of Various Species of Coffea, (excluding C. arabica cv. Bourbon)

Species	GI	Kin (μM)	2,4-D (μM)
C. arabica cv. Catuai	3.0	4.6	1.8-3.6
C. arabica cv. Laurina	1.8	10.0	6.0
C. arabica cv. Mundo Novo	2.4	2.0	2.0
C. arabica cv. Purpurascens	2.5	10.0	2.0
C. bengalensis	0.8	6.0	10.0
C. canephora cv. Kouillou	2.8	2.0	10.0
C. congensis	2.6	2.0	4.0
C. dewevrei cv. Excelsa	2.0	10.0	6.0
C. liberica	2.9	10.0	2.0
C. racemosa	3.0	2.0	2.0
C. salvatrix	1.8	20.0	4.0

5.7%) for 15-20 minutes (13, 30, 31). Also 1% sodium hypochlorite has been used for 10 minutes followed by two rinses with sterile water. To reduce phenolic oxidation, subsequent rinses with sterile cysteine solution (100 mg/l), 95% ethanol and sterile water can be used (21). Larger stem pieces are more suitable for sterilization and culture. Stem pieces should be washed with 1% detergent solution for 5-15 minutes, with agitation, before placing into the sterilizing agent. $HgCl_2$ solution (0.01-0.5%) is an effective disinfectant agent but residual mercury is very toxic. Use of 70% ethanol increases oxidation in soft stem internodes. Immersion in ozone-rich water for 5-10 minutes before surface sterilization is a good practice. The material to be sterilized can be rinsed with sterile double distilled water containing an autoclaved solution of cysteine (50-825 μM), PVP-40 (2.5 μM), or a filter sterilized solution of ascorbic acid (284-568 μM). These anti-oxidants can also be included in the liquid medium used for cutting the explants. The same plating technique described for leaf explants is also recommended for stem pieces.

Callus proliferation and somatic embryos were obtained from stem explants of *C. canephora* by Staritsky using the following medium: MS inorganic salts, thiamine (3.0 μM), L-cysteine (82.5 μM), meso-inositol (555.1 μM), kin (0.5 μM),

2,4-D (0.5 μM) or NAA (5.4 μM), sucrose (87.6 μM), and agar (10 g/l) (29). Sharp *et al.* used the same medium with the following modifications: thiamine (11.9 μM), NAA (10.7 μM), sucrose (58.4 μM), and coconut water (5%), (22). Additions of benzyladenine (0.4 μM), IBA (2.5 μM), and casein hydrolysate (100 mg/l) to the original medium have a beneficial effect on the development of embryos and plantlets (Staritsky, personal communication). The medium described by Staritsky (29) was also used by Frischknecht *et al.* (13) to induce callus on orthotropic internodes (5-10 mm long) of *C. arabica* cv. Bourbon Vermelho. A recent modification of the culture medium for stem explants of *C. canephora* orthotropic shoots has been given by Nassuth *et al.*: MS inorganic salts, thiamine (3.0 μM) cysteine (82.5 μM), glycine (26.6 μM), adenine (7.4 μM), meso-inositol (555.1 μM), IBA (24.6 μM), benzyladenine (4.7 μM), sucrose (87.6 mM), casein hydrolysate (100 mg/l), agar (8 g/l) (21).

The initiation of callus growth is enhanced in the absence of illumination at a temperature of 25-28°C. Growth is initiated from both cut edges on horizontally positioned stem segments. Longitudinal sections through primary explants revealed that callus originates mainly from the cambium and cortex of the stem (13). A transfer schedule of 35 days is suggested but not all the tissues survive regular subcultures. Staritsky described two types of callus tissues: one having a white spongy appearance with elongated cells and a second type, which is more compact with isodiametric cells (29).

Morphogenesis can occur in the same medium in the first to fourth subculture callus tissues growing in the dark. Somatic embryos develop from the tissue with isodiametric cells: friable conglomerates of multicellular yellowish globules readily dissociate on the uncovered surface of the nutrient medium. Two leaf primordia develop at one end of the globules and turn green after being transferred to light. If a high auxin medium (NAA 10 μM or 2,4-D 5 μM) is used to initiate callus growth, the following secondary medium is recommended: half-strength MS inorganic salts, sucrose (58.4 mM), thiamine (15 μM), meso-inositol (550 μM), NAA (0.5 μM), KIN (2.5 μM); light or dark.

5. *Liquid Culture.* Fast growing callus tissues (30-40 days old) derived from leaf or stem explants can be used to establish liquid cultures of coffee cells. The PRL-4 medium (14) has been used for several years for coffee suspension cultures on a 1-2 week subculture schedule (3, 31, 32). Rapidly dividing cells have a light cream color which becomes darker as the incubation period increases. 2,4-D can be replaced by IAA and kinetin (30). The cells achieve expo-

nential growth between day 4-8. The maximum growth rate will depend on the size of the inoculum and on the composition of the medium (30).

In our laboratory, coffee suspension cultures have been established in the following medium: MS inorganic salts, thiamine (30 μM), L-cysteine (210 μM), meso-inositol (550 μM), sucrose (58 mM), casein hydrolysate (1 g/l) and 2,4-D (5 μM), pH (5.5). The cells do not grow well in the presence of both 2,4-D and kinetin. Cultures are established with 10-15 leaf explant calluses (40 day old) in 50 ml of the above medium in a reciprocal shaker (150 rpm). After one week, the cells are separated from the remaining of the leaf calluses (centrifuge, 300x g for 5 min.) and inoculated into fresh medium at an approximate density of 10^5 cells/ml. After 4 weeks, ca. 25% fresh medium is added to Erlenmeyer flasks. After a dense cell suspension is established, each culture plate is divided in half and 50% of fresh medium is added.

6. *Embryo Culture.* Coffee seeds are surface sterilized before embryo excision. Colonna *et al.* used 95% ethanol for 3 minutes (1 minute under vacuum) followed by 5 minutes in 0.7% hypochlorite solution (6). Before the embryos were removed, the sterilized seeds were soaked in sterile distilled water for 36-48 hours. In our laboratory a 75% Clorox[R] solution has been used with success (ca. 3.9% NaClO) for 30 minutes with continuous agitation (rotatory shaker - 150 rpm), followed by three rinses of sterile water. The sterile seeds are incubated 3-4 days on saline-sucrose agar plates before excision. Soaking the seeds in sterile water has given negative results because of cross contamination in the liquid environment. Both $HgCl_2$ and ethanol are deleterious to the seeds. In both sterilization procedures mentioned above, the parchment is removed from the seeds. The silver skin is almost completely eliminated with the agitation in the shaker. An alternative method to obtain sterile coffee embryos consists of soaking seeds (without parchment and silver skin) in water for 12 hours, excising the embryos and sterilizing them (1% NaClO - 6 minutes).

Solid and liquid Heller's salt solution plus sucrose (58.4 mM), meso-inositol (1.1 mM), thiamine (3.0 μM), cysteine (6.5 μM), calcium pantothenate (1.0 μM), nicotinic acid (8.1 μM), adenine (7.4 μM), pyridoxine (3.9 μM) and glycine (4.0 μM), has been used by Colonna *et al.* (6) and Colonna (5) with *C. canephora* cv. Robusta and *C. dewevrei* cv. Excelsa and cv. Neo-anddiana embryos. These studies showed that *C. cane-*

phora embryos had a slower rate of development than *C. dewevrei* embryos which gave maximum growth at the 60-70th day of culture. Subculturing was necessary to provide for continued development. The liquid medium provided longer root growth, but development was more regular in agar medium. The first pair of leaves in both varieties developed on the 45th day of culture and on the 75th day a second pair of leaves appeared in *C. dewevrei* (cv. Excelsa). The addition of 2 μM IAA to the culture medium promoted a 17% increase in elongation, while 10% coconut milk in addition to the IAA produced a 38% increase in elongation at the 40th day of culture. The cultures were maintained under a 12 hour photoperiod at 28°C day and 21°C night temperatures.

C. arabica cv. Mundo Novo and cv. Catuai embryos are cultivated in half strength MS inorganic salts, sucrose (58.4 mM), nicotinic acid (30 μM), thiamine (30 μM), meso-inositol (550 μM), pyridoxine (15 μM), 6-BA (1 μM), IAA (0-20 μM), agar (8 g/l), at pH 5.5 in the presence of light, and at a temperature of 24-28°C. Evaluations made after 60 and 90 days indicated that IAA (5-10 μM) is excellent for induction of roots but the subsequent development of roots is better in media containing 0-0.25 μM IAA. Good hypocotyl and leaf development was obtained using IAA concentrations in the range of 0-25 μM. Considering these results, the medium described above has been adopted for *C. arabica* with IAA (2-5 μM) for initial culture of the excised embryos. Since IAA is naturally degraded by the IAA oxidase system, a reaction enhanced by light, its concentration will be lowered as the culture ages. After 50-60 days, the culture bottles are exposed to light with higher intensity or even transferred to the greenhouse in the shade (70-80% of shade).

7. Anther Culture. Flower buds are washed with 1% detergent solution with agitation for 5 minutes and rinsed with distilled water. They can be surface sterilized with 1% sodium hypochlorite for 15-20 minutes (greenhouse plants) or using 2.5% calcium hypochlorite for 20-30 minutes (field plants), followed by three rinses with sterile distilled water. The anthers can be removed in liquid medium (using 50-100 mm diameter Petri dishes) under a dissecting microscope with the aid of sterilized stainless steel needles and forceps. To minimize phenolic oxidation, the flower buds can be dissected in a liquid solution of autoclaved cysteine (825.4 μM, PVP-40 (2.5-5.0 μM), or a filter sterilized solution of ascorbic acid (567.8 μM), using two glass needles. Each flower bud should be dissected in a Petri dish containing fresh medium to avoid cross contamination. If oxidation has not been very severe, one can use 20 X 100 mm Petri dishes filled with ca. 20 sterile filter paper discs. Two to three paper discs are removed after dissecting one bud. After excision, the anthers are placed immediately in the culture medium. Flower buds at the first mitosis stage of microsporogenesis should be selected. The most appropriate stage for culturing is obtained when flower buds are 1.2 mm for *C. arabica* and ca. 1.5 mm for *C. liberica*.

At present there are not enough data with *C. arabica* anthers to allow recommendation of a successful culture medium. The best results reported to date were by Sharp *et al.* with anther cultures of *C. arabica* cv. Bourbon Amarelo, using the following medium: MS inorganic salts, sucrose (116.9 mM), meso-inositol (555.1 μM), thiamine (11.9 μM), cysteine (82.5 μM), kinetin (0.5 μM), 2,4-D (0.5 μM), agar (8 g/l), and adjusted to a pH 5.8 before autoclaving (22). The cultures were maintained in the dark at ca. 25°C. In our laboratory, using this medium in liquid culture, callus proliferation and reduced phenolic oxidation has been observed during the first 2-4 weeks of culture.

C. liberica anthers have been successfully cultured using flower buds 1.85 cm in length (anther length of 0.92 cm) at the binucleated stage of microsporogenesis (19). The medium adopted was the same as described by Sharp *et al.* with the following modifications: cysteine (330.1 μM), and 2,4-D (15.8 μM) (22).

8. Bud Culture. Characteristically, coffee plants have multiple arrested orthotrophic buds and two plagiotrophic buds at each stem node. The plagiotrophic buds differentiate only after the 10th-11th node of a developing seedling, whereas the orthotrophic buds are present in the first node (cotyledonary node). The removal of the apical meristem results in the development of two orthotrophic shoots, one at each leaf axil, at the most apical node. At each leaf axil of plagiotrophic branches, there are 3-4 serially arrested buds. These buds will differentiate into flower buds under proper environmental conditions. Sometimes, the uppermost buds of that series at one or both sides of a node will develop as a vegetative bud instead of a floral bud leading to a secondary plagiotrophic branch.

The presence of these arrested buds have been explored as a means of vegetative propagation of coffee plants. Nodal explants of aseptically grown *C. arabica* plants have been cultured on MS medium supplemented with 6-BA (44 μM) and IAA (0.6 μM), under 16 hour photoperiod (2,000 lux) at 25±0.5°C (9). Shoot development occurs on the average of 2.2 per node after 2-5 weeks. The use of NAA (1.1 μM) in complete darkness encourages good rooting of these young coffee shoots. It is recommended at least 3 month old *in vitro* plants be used for excision of nodal sections and that the leaves be retained during culturing. Similar techniques have been used by Dublin with *Arabusta* plants. Shoot development was observed in medium supplemented with malt extract (400 mg/l) and 6-BA (4.4 μM) (10).

B. *Selection of Optimum Culture Medium for Callus Induction and Differentiation*

The best culture conditions for growth and differentiation for any particular coffee species or cultivar can be selected using the basic coffee medium in diallel experiments during primary or secondary culture (see II *A. 1* and II *A. 2*). It has been observed that callus proliferation is not directly correlated with frequency of somatic embryo formation. However, a primary culture that induces poor callus growth (grade 0 to 1) will provide insufficient tissue for embryo formation during secondary culture. One approach would be the variation of the primary culture (diallel design) and subculture on a simple secondary medium. After selecting the optimum medium for primary culture (grade 2, 3, or 4), a diallel scheme can be used for secondary culture. A grading system is convenient for evaluation of callus growth since the tissues can be subcultured afterwards. Attention should be given to the type of auxin and cytokinin used, since no two growth regulators provide the same results. It has been observed with coffee leaf explants that kinetin and 2,4-D, at certain concentration ratios during primary culture, affect the frequencies of somatic embryogenesis during secondary culture (23). Finally, it should be noted that the selection of leaves of the same physiological age collected from plants of well characterized coffee cultivars grown under uniform conditions, will provide less variation among the replicated bottles.

C. *Protocol, Evaluations and General Guidelines*

Coffee tissues can be cited as one of a few examples

among perennial plants to follow an embryogenic process of plant regeneration. Repeated protocols for somatic embryos have been presented for *C. canephora* stem explants (21, 29), and for leaf cultures of *C. canephora*, *C. dewevrei*, *C. congenesis*, and *C. arabica* cv. Bourbon, cv. Mundo Novo, cv. Catuai, cv. Laurina, cv. Purpurascens (24, 25). Somatic embryos from *C. canephora* orthotropic stem explants develop in the same primary medium during the first to fourth culture. Somatic embryos from coffee leaf explants differentiate after 3-4 months in secondary culture. There are two developmental processes during culture of coffee leaf explants: low and high frequencies of somatic embryo formation. The high frequency process is associated with the embryogenic tissue. This tissue is extremely important as an experimental material for induced mutation and protoplast isolation.

Further work needs to be done to determine if *C. arabica* stem explants have any advantage over leaf explants in terms of culture duration and morphogenetic capacity. It is known that it is easier to root *C. canephora* explants (stems, petioles) than *C. arabica*. In the nursery of the Agronomic Institute, stem pieces of *C. canephora* cut below the cotyledonary node form a small nodule (callus) around the cut edges and from this region several vegetative buds develop. This same result has not been observed with *C. arabica* stem pieces treated in the same way. This is a further indication of the regenerative capacity of *C. canephora* tissues. It should be noted that *C. arabica* is an autogamous plant with only ca. 10% out crossing whereas *C. canephora* is an allogamous plant. Self-incompatible *C. canephora* germplasm contains greater variability than *C. arabica*. So far a genetic analysis of morphogenetic capacity among plant species or cultivars has not been described, but there are several indications in the literature that *in vitro* plant regeneration has a genetic basis (for example, monoembryonic versus polyembryonic citrus tissues, tomato cultivars, *Nicotiana spp.*, etc.) (11). On the other hand, the advantage of culturing tissues of regenerated plantlets to achieve higher frequencies of plant differentiation is well known (1). This is another indication that during *in vitro* culture it is possible to select for cellular clones with high morphogenetic capacity. Within the genus *Coffea* a comparison of the morphogenetic capacity among different species and cultivars has not been described as yet. To perform such experiments, stem or leaf explants should be used. However, soft orthotropic shoots are limited to 1 or 2 shoots per plant.

Plant regeneration has not been accomplished from continuous coffee cell suspensions, haploid tissues or protoplasts.

The development of such protocols will allow the application of all techniques for genetic manipulation and crop improvement within the genus *Coffea*.

The demonstration that coffee plants derived from somatic embryos are exact phenocopies of the mother plant could have practical applications. This will open the possibility of using *in vitro* techniques for vegetative propagation of *C. canephora*, selected mutants of *C. arabica* and some interspecific hybrids of commercial interest.

III. APPLICATIONS OF *IN VITRO* CULTURE TO COFFEE BREEDING

As for all crop species, tissue culture represents a great potential to be explored as a tool for genetic breeding. Although conventional breeding techniques can be applied to coffee breeding and selection, there will be situations where tissue culture may be uniquely advantageous.

A. *Vegetative Propagation*

The two most important species - *Coffea arabica* and *C. canephora*, are propagated via seeds. *C. arabica* is autogamous and so progenies are quite uniform whereas *C. canephora* is self-incompatible and consequently highly heterogenous. Favorable genotypes may be cloned through tissue culture. In the case of *C. canephora* it is possible to propagate exceptional mother trees. In the case of *C. arabica* it is possible to rapidly increase certain combinations of favorable genes, for example, for leaf rust resistance, in order to face the rapid evolution of physiological races of the fungi.

B. *Induced Genetic Variation*

As has already been indicated, the genetic material of *C. arabica* represents a very narrow gene pool. Attempts by conventional breeding in some populations resulted in little genetic advance. Besides hybridization, which has been used mainly, induced mutations could offer a new range of variability for the breeder. However, the data available have indicated that cuttings, germinating seeds, and pollen grains of *C. arabica* are quite resistant to ionizing radiations such as gamma and X-rays (2 to 100 kR) and to chemical mutagens like ethylmethane sulfonate. Although a large number of samples of such material have been irradiated, only a few mutants have been isolated (Monaco and Carvalho, unpublished data). Moh suggested that the apical meristem of the *C. arabica* embryo is formed by a single apical cell (17). This would explain the reduced number of chimeric mutants in *C. arabica* in the generation first arising from the mutagenized cells. Isolated cell cultures and subsequent plant regeneration would therefore offer an excellent possibility of using *in vitro* mutagenic techniques to induce genetic variability for selection of desirable variants. The use of the embryogenic tissue described earlier (Section II, *C. 1*) would be of great advantage for mutation induction experiments since almost all cells result in plants.

C. *Embryo Culture*

Study of the genetic relationships among wild species and cultivated ones has been hindered because of the difficulty of obtaining interspecific hybrids. Species like *C. bengalensis*, *C. ebracteolata* and *C. humbertii* have been described as caffeine free, and *C. eugenioides* as having low caffeine content (0.38%) (4). *C. stenophylla*, *C. racemosa*, *C. kapakata*, and *C. dewevrei* are immune to coffee leaf miner (*Perileucoptera coffeela*) whereas *C. canephora*, *C. bengalensis*, and *C. dewevrei* are the best source for root nematode resistance or tolerance. Attempts to cross these species with the cultivated *C. arabica* have failed or had limited success. It is possible through culture of immature embryos to obtain viable interspecific hybrids and at the same time apply *in vitro* chromosome doubling techniques. These interspecific hybrids as well as others with species of the Sections Paracoffea and Mascarocoffea would open a new source of genetic variability for breeding purposes. Similarly, several species do not cross with the diploid species *C. canephora* which needs to be improved in several characteristics particularly in reducing the caffeine content, which can be as high as 3.3%.

D. *Meristem Culture*

The development of protocols for meristem culture of several species of the genus *Coffea* are necessary for the utilization of techniques of *in vitro* cryopreservation. This would give further guarantee of germplasm preservation at a low cost and with limited risks.

E. *Microspore Culture*

As already indicated, *C. arabica* is a self-compatible tetraploid species, whereas *C. canephora* and all other diploid species of the genus are self-incompatible. The production of haploid plants through microspore culture has reduced value for *C. arabica*. However, in the breeding for disease resistance it is possible to select from highly heterozygous hybrid populations haploid plants which carry the desirable major or minor genes for resistance to different pests or diseases. By means of duplication of such material it is possible to select the most productive ones which carry several genes for resistance to adverse environmental conditions in a homozygous state.

In the case of self-incompatible species like *C. canephora*, haploid plants obtained by microspore culture might be used to produce commercial hybrids. Mother plants with general combining ability would furnish microspores for the production of haploid plants. A large number of haploids would permit the selection of different S alleles, as indicated in Figure 1. No artificial pollination will be required. The lines with best specific combining ability are interplanted in rows and the commercial seeds collected. These double hybrids will be more variable but will have better population plasticity (2). A single hybrid could be used despite the risk of genetic vulnerability.

A third possibility of using microspore culture is related to interspecific hybrids which show reasonable fertility.

	Alleles			
Haploid	S_x	S_y	S_z	S_w
Diploid	S_xS_x	S_yS_y	S_zS_z	S_wS_w
Diploid F_1 hybrid	S_xS_y		S_zS_w	
Double cross hybrid	$S_xS_z:S_xS_w:S_yS_z:S_yS_w$			

FIGURE 1. Scheme for production of double cross hybrids starting from haploids obtained through microspore culture.

In the case of *C. arabica* X *C. canephora* (4X), the descendants of advanced generations have high frequency of abnormal phenotypes due to chromosome imbalance. This requires that a large number of generations and large populations be evaluated in order to select high yielding genotypes. During these selection cycles many favorable genotypes of the donor species are lost. The microspore culture would allow the production of haploids on a large scale which can be tested for disease resistance or other plant characteristics. The selected material then can be duplicated for yield tests.

F. *Protoplast Fusions and Plant Regeneration*

The studies of interspecific or intergeneric hybridizations have demonstrated that many desirable characteristics occurring in the genus cannot be transferred to cultivated species due to genetic barriers. Crosses between species of the Sections Paracoffea and Mascarocoffea with species of the Section Eucoffea would be a practical example. Nuclear and cytoplasmic traits conferring resistance to adverse environmental factors, low caffeine content, drought, resistance, etc., could be transferred to improve the cultivated species.

ACKNOWLEDGMENT

The authors wish to thank Dr. D.A. Evans for helpful comments.

REFERENCES

1. Bingham, E.T., Hurley, L.V., Kaatz, D.M., and Saunders, J.W., *Crop Sci. 15*, 719 (1975).
2. Bradshaw, A.D., *Advan. Genet. 13*, 115 (1975).
3. Buckland, E., and Townsley, P.M., *J. Inst. Can. Sci. Technol. Aliment. 8*, 164 (1975).
4. Charrier, A., and Berthaud, J., *Café Cacao Thé. 19*, 251 (1975).
5. Colonna, J.P., *Café Cacao Thé. 16*, 193 (1972).
6. Colonna, J.P., Cas, G., and Rabechault, H., *Compt. Rend. 272*, 60 (1971).
7. Crocomo, O.J., Carvalho, F.J.P., Carvalho, P.C.T., Sharp, W.R., and Bandel, G., 26th Congr. Nac. Botanica, Rio de Janiero (1975).

8. Crocomo, O.J., Carvalho, E.F.P., Sharp, W.R., Bandel, G., and Carvalho, P.C.T., *Energ. Nucl. Agric.* Piracicaba, *1*, 41 (1979).
9. Custer, J.B.M., Van ee, G., and Buijs, L.C., "*IX Intern. Colloquium on Coffee*", London, (1980).
10. Dublin, P., "*IX Intern. Colloquium on Coffee*", London, (1980).
11. Evans, D.A., Sharp, W.R., and Flick, C.E., *In* "Horticultural Reviews", Vol. 3 (Janick, ed.) AVI Press (in press). (1980).
12. Frederacion Nac. Cafeteros de Colombia. "Boletin de Informacion estadistica sobre cafe". Div. Inc. Economia, Dept. Inf. Cafeteria. *48*, 184 (1978).
13. Frischknecht, P.M., Baumann, T.W., and Wanner, H., *Planta Medica. 31*, 344 (1977).
14. Gamborg, O.L., *Can. J. Biochem. 44*, 791 (1966).
15. Herman, F.R.P., and Haas, G.J., *HortSci. 10*, 588 (1975).
16. Keller, H., Wanner, H., and Baumann, T.W., *Planta 108*, 339 (1972).
17. Moh, C.C., *Radiat. Bot. 1*, 97 (1961).
18. Monaco, L.C., Medina, H., and Sondahl, M.R., SBPC Meeting, Recife *26*, 240 (1974).
19. Monaco, L.C., Sondahl, M.R., Carvalho, A., Crocomo, O.J., and Sharp, W.R., *In* "Plant Cell, Tissue and Organ Culture", (J. Reinert and Y.P.S. Bajaj, eds.), p. 109, Springer-Verlag, Berlin, (1977).
20. Murashige, T., and Skoog, F., *Physiol. Plant. 15*, 473 (1962).
21. Nassuth, A., Wormer, T.M., Bouman, F., and Staritsky, G., *Acta Bot. Neerl. 29*, 49 (1980).
22. Sharp, W.R., Caldas, L.S., Crocomo, O.J., Monaco, L.C., and Carvalho, A., *Phyton 31*, 67 (1973).
23. Sondahl, M.R., and Sharp, W.R., *Z. Pflanzenphysiol. 81*, 395 (1977).
24. Sondahl, M.R., and Sharp, W.R., *Abstr. Int. Conf. Regulation of Developmental Processes in Plants, Halle* p. 180, (1977).
25. Sondahl, M.R., and Sharp, W.R., *In* "Plant Cell and Tissue Culture - Principles and Applications", (W.R. Sharp *et al.*, eds.), p. 527, Ohio State University Press, Columbus, (1979).
26. Sondahl, M.R., Salisbury, J.L., and Sharp, W.R., *Z. Pflanzenphysiol. 94*, 185 (1979).
27. Sondahl, M.R., Spahlinger, D.A., and Sharp, W.R., *Z. Pflanzenphysiol. 94*, 101 (1979).
28. Sondahl, M.R., Chapman, M., and Sharp, W.R., *Turrialba* (in press), (1980).

29. Staritsky, G., *Acta Bot. Neerl. 19,* 509 (1970).
30. Townsley, P.M., *J. Inst. Can. Sci. Technol. Aliment. 7,* 79 (1974).
31. van der Voort, F., and Townsley, P.M., *J. Inst. Can. Sci. Technol. Aliment. 8,* 199 (1975).

IN VITRO METHODS APPLIED TO FOREST TREES

Harry E. Sommer

School of Forest Resources
University of Georgia
Athens, Georgia

Linda S. Caldas

Departmento de Botanica
Universidade de Brasilia
Brasilia, Brazil

I. INTRODUCTION

The tissue culture of trees has been reviewed several times recently (3, 4, 5), as well as being the subject of numerous recent workshops, small symposia and symposium sessions (2, 6, 7, 10, 17, 20, 26, 30, 31, 34). For an extensive treatment of the literature these references should be consulted. Since the purpose of this book is to emphasize techniques, particularly the use of tissue culture as a method of propagation, we will present an outline of the procedures that we have found useful in propagating trees. Despite the fact that the tissue culture of trees is often set apart from the culture of other species, the majority of the principles are the same as presented elsewhere in this book. The two major exceptions are (1) most tree populations are highly heterogeneous resulting in a high variability in culture responses and (2) cultures from mature tissues are generally less responsive than those of juvenile tissue.

ISBN 0-12-690680-7

II. LABORATORY PRINCIPLES

A. *Objectives*

Many people have talked about using tissue culture plantlets for reforestation. At present even under the best of conditions, this is somewhat unrealistic. Estimates of the cost of producing planting stock by tissue culture exceeds that of nursery grown seedlings by a factor of 3 to 30 times (Brown and Sommer, unpublished). The mass propagation of superior trees by tissue culture is nevertheless a desirable objective, although its practical application still lies in the future. However, there are still many useful objectives that can be achieved by utilizing or extending current technology. Among these are the multiplication of individual trees for use in seedling seed orchards, progeny testing, conservation of rare, endangered or slowly reproducing species, multiplication of desirable genotypes, and propagation of horticultural varieties of ornamental and orchard trees.

B. *Selection of Materials*

In theory it should be possible to start a culture from any living cell, however, initially, actively growing tissue or tissue with the potential for active growth should be chosen. For instance Brown obtained callus from ray cells of slash pine and red maple immediately adjacent to the cambium, but was unable to obtain callus from rays beyond the cambial zone (personal communication).

The first factor to consider in choosing an explant is the condition of the source plant. This is probably the least understood and least researched aspect of tissue culture. A standardized system should be used for the production of the mother stock. When using embryos or other seed parts such factors as maturity of the embryo, imbibition time and temperature, degree of stratification and extent of embryo dormancy should be investigated and standardized. For instance, fresh longleaf pine seeds have no dormancy and start germinating upon imbibition and can immediately be put into culture, while Mehra-Palta *et al.* (18) found embryonic cotyledon cultures of *Pinus taeda* produced more buds if the seeds were partially stratified. Even more care needs to be taken with seedling material. Sterile seeds can often be germinated and grown on an agar medium. We routinely use Risser and White's medium for this purpose (22). The major advantage is that contaminating organisms that are not killed during sterilization of the seed may be detected. In addition, the use of potentially damaging sterilants directly on the seedling is avoided. Seedlings can also be grown in the greenhouse, but careful standardization is required. For example, Romberger has described in detail the procedures for growing seedlings of *Picea abies* which he uses as a source of apical domes (23). It is important to note that (1) the seedlings used for the experiment are the most vigorous and (2) the temperature of a greenhouse in warm weather, even when equipped with cooling pads and white washed, will primarily be controlled by the ambient temperature and humidity rather than any thermostatic control system, thus resulting in less uniform seedlings.

Materials may be grown or gathered in the field. However, problems can arise due to varying conditions of growth or physiological conditions, and the usually higher number of contaminating organisms found on and in field material. In all cases, regardless of source of materials, usually greater success is obtained if the explant is in a stage of active growth or can readily be forced into such a state.

C. *Sterilization*

Control of contaminating microorganisms is an important part of culture establishment. Tissues such as cambium and apical meristems may be free of microorganisms, but adjacent tissue is often heavily contaminated particularly in field-grown plants, and to a lesser extent, in greenhouse-grown plants. Roots, stems, and seedlings should be freed of loose soil and dead tissue, trimmed and washed. Surface sterilization of explants, seeds and buds can usually be accomplished with 5% sodium hypochlorite (often Chlorox^R or other commercial bleach containing 5% sodium hypochlorite). For softer tissues, a dilution to lower strength may be needed but anything below 0.5% may prove ineffective in killing the majority of the microorganisms. If the surface of the tissue does not wet readily with the hypochlorite solution a surfactant such as Tween 80^R added to the solution at 0.1% may help. Alternately, the material can be dipped in 95%-80% ethanol and/or a 1:500 dilution of Roccal^R can be used followed by an alcohol dip and then hypochlorite. The duration of such treatments needs to be determined empirically. Following surface sterilization the tissues should be rinsed with sterile water, then with 0.1N HCl and again with sterile water. In the case of seeds, we surface sterilize both before and after imbibition, and at the end of stratification. If the seedlot is expected to be heavily contaminated an additional sterilization is added during the first week of stratification. The length of time the tissue spends in contact with the sterilant is determined by experiment and experience. Other chemicals such a H_2O_2 and $HgCl_2$ can be used for surface sterilization, however the above procedure has proved most effective for us (25, 27).

Not all contamination is necessarily of surface origin. Internal contamination may occur either from organisms trapped within areas not reached by sterilizing solutions or due to diseased or necrotic tissue. Currently there is no satisfactory solution to this problem, particularly if the infection is systemic. In removing the explant care should be taken not to recontaminate it. In obtaining cambium explants, bark and phloem must be cut away possibly exposing pockets of contamination not reached by surface sterilization. Therefore it is necessary to make longitudinal cuts to remove the bark overlap. Likewise the surface of a bud may be sterile, but contaminants may be trapped on the interior bud scales, so care must be taken to insure that as the bud scales are removed the interior of the bud is exposed only to sterile surfaces, i.e., areas where the bud scales or bud had not previously touched.

D. *Establishment of the Culture*

Some workers precondition the explant before introduction into culture (8, 28). Usually this consists of placing the explant on a basal medium for up to a few days, then discarding any contaminated or waterlogged explants. Alternately, the bud is first treated with a weak solution of hypochlorite and placed on a basal medium, then followed by sterilization with a stronger solution of hypochlorite (15).

Typically in attempting to establish a new species in culture we use a screening system. Both explant source and several standard media are screened simultaneously. Explants tested whenever possible are embryos, shoot tips, axillary buds, stem segments, cotyledons, leaves, hypocotyl segments, and roots from sterile seedlings, nondormant, but unbroken buds, and anthers. Each type of explant is tested on several of the following media. Medium 1 and Medium 2 (11, 25, 27), Murashige's A (14), Murashige's B (14), Halperin and Wetherell's system (13), steps 1, 3, and 7 of Kohlenbach's system (16), Murashige's A and B with Blaydes' (32) salts substituted for Murashige's, and Anderson's rhododendron medium (1). From these screening experiments it should be possible to establish the salt solution compatible with the species. For instance sweetgum tissue dies on Murashige's salts (14, 19), but grows well on Blaydes' (32). It should be possible to establish those explants most responsive to culture. For example, sweetgum hypocotyls readily form a callus, while the roots do not. This procedure will also demonstrate if any of these systems will produce organogenesis or embryogenesis in the cultures. For example, on Murashige's A *Paulownia* readily forms numerous shoots on leaves while sweetgum leaves turn black. Other workers have found Schenk and Hildebrandt's medium (24) a very useful basal medium, such as in radiata pine (21).

Most of the species we have worked with have given only partially satisfactory results. The next step is to take the information gained and start to redesign the medium for better results, usually in terms of obtaining an optimum number of buds or embryoids per explant. Initially the salt composition is held constant and the hormones and their concentration varied in a latin square type design. Auxins usually are tested at 0-10 ppm, while cytokinins usually are tested at 0.1-64 ppm. To further refine the system the salts are tested by varying the concentration of the combined salt solution from 2X to 1/8X or of individual components, or by testing one or more standard salt solution and modifications against each other. Vitamins and carbohydrate sources are additional variables of interest. Once a satisfactory medium is established, in the case of embryogenesis the next step is usually suspension cultures in liquid medium, while if buds are obtained the next step is rooting. In some cases an intermediate step allowing for bud development and multiplication is necessary (see 30). Some workers find that charcoal (0.1 to 2%) in the medium is beneficial at this stage e.g., in *Pinus pinaster* (9).

E. Rooting

Rooting of angiosperm trees has been relatively easy in culture; while gymnosperms have presented more problems. A modified Morel's medium (29) with 0.1-10 ppm IBA has given us satisfactory results on the angiosperms we have tested. For conifers the situation is more complicated. Often two auxins (IBA and NAA) work better than one, sucrose at 1% is generally better than 2 or 3%, often a low concentration of cytokinin is helpful, and a temperature of about 20°C has proved beneficial at times (7, 18, Sommer and Brown unpublished). Half-strength MS salts are generally better than full strength. Part of the effect of these treatments is to reduce the callusing at the base of the shoots, otherwise the callus may grow faster than the root primordia and suppress outgrowth of the roots.

F. *Hardening Off of Plantlets*

The final step is the transfer of the plantlet to soil and its hardening off. A few general principles apply whether considering an embryoid grown up to size on agar or a plantlet. (1) There must be a reasonable balance between root and shoot. Each should be capable of supporting the other; neither should be predominant. (2) The plantlet will be going from a constantly high humidity situation to one of varying and lower humidity. Adjustment needs to be gradual. This can be accomplished by misting, a humidity tent that is gradually removed, or by removing the cover over the plantlet for intervals of increasing length over several days. Care must be taken in removing traces of agar from the plantlet as they may provide a substrate for pathogen growth. (3) When the plantlet is moved from the low light intensity of the laboratory to full sunlight of the greenhouse, some shading that is gradually removed may help prevent leaf burn.

III. PROTOCOL FOR PLANTLET PRODUCTION IN LONGLEAF PINE

The protocol used to obtain plantlets of longleaf pine (*Pinus palustris*) is given below. It is presented only as an example of the approach outlined above. The protocol is of historical importance as it represents the method used in the first report of successful plantlet formation of a conifer. The protocol is adapted from Sommer *et al.* (27).

(1) Surface sterilize viable seeds in half strength Chlorox^R for ca. 15 minutes.

(2) Rinse in sterile distilled water and imbibe seeds for 36-40 hours in sterile water.

(3) Immerse briefly in 1:3 (v:v) dilution of Chlorox^R, rinse in sterile water and place in aseptic petri dishes.

(4) Dissect out embryos from whole unbroken seeds.

(5) Place embryos on 20 ml of Medium A (Table I) in 25 x 150 mm test tubes for 4-5 weeks.

(6) Maintain cultures at 25± 3°C, under 8,600 lux (cool white flourescent light) in a 16 hour photoperiod.

(7) Excise and transfer adventitious buds (from cotyledons) to Medium B (Table I). In some transfers the primary needles will elongate and roots will form. (In

TABLE I. Composition of Nutrient Media used in Plantlet Formation in Longleaf Pine. Table Adapted from Sommer et al. (24)

Chemical	mg/liter	Chemical	mg/liter
Medium A[a]			
$(NH_4)_2SO_4$	200	$MnSO_4 \cdot H_2O$	10
$CaCl_2 \cdot 2H_2O$	150	$ZnSO_4 \cdot 7H_2O$	3
$MgSO_4 \cdot 7H_2O$	250	H_3BO_3	3
KNO_3	1,000	$CuSO_4 \cdot 5H_2O$	0.25
KCl	300	$NaMoO_4 \cdot 2H_2O$	0.25
KI	0.75	$CoCl_2 \cdot 6H_2O$	0.25
$NaH_2PO_4 \cdot H_2O$	90	Inositol	10.0
NaH_2PO_4	30	Thiamine·HCl	1.0
$FeSO_4 \cdot 7H_2O$	27.8	Nicotinic acid	0.1
Na_2EDTA	37.3	Pyridoxine HCl	0.1
Sucrose	20,000	NAA	2.0
Agar	7,000	6-benzyladenine	5.0
Medium B[b]			
$Ca(NO_3)_2 \cdot 4H_2O$	300	$MnSO_4 \cdot H_2O$	5
$MgSO_4 \cdot 7H_2O$	740	$ZnSO_4 \cdot 7H_2O$	2.5
KNO_3	80	H_3BO_3	1.5
KCl	65	KI	0.75
$NaH_2PO_4 \cdot H_2O$	165	$CuSO_4 \cdot 5H_2O$	0.01
Na_2SO_4	200	MoO_3	0.001
$FeSO_4 \cdot 7H_2O$	27.8	Inositol	10.0
Na_2EDTA	37.3	Thiamine HCl	1.0
Sucrose	20,000	Nicotinic acid	0.1
Agar	7,000	Pyridoxine HCl	0.1

[a] Medium A based on Gresshoff and Doy (11)

[b] Medium B based on Risser and White (22)

addition, degeneration of some buds will occur, in some cases only roots will develop, with the buds remaining inhibited).

(8) Transfer healthy buds, which have failed to form roots after 5-6 weeks, to Medium B supplemented with indole-3-butyric acid (10 ppm) for four weeks; and then transfer back to medium B (this treatment will promote rooting in some cultures).

IV. CONCLUSION

The procedures outlined in Section II have been kept very general since there are many differences not only between species, but even within species in their cultural requirements. The principal limitation in tissue culture is the researcher. He must have good sterile technique, the ability to search, read and critically evaluate the literature and must not get discouraged easily. In addition, he must be able to apply his knowledge of his chosen plant, his experience and his intuition for success. Creative research on the tissue culture of trees is needed more than the exact repetition of some standardized procedure. For instance, few of the current methods allow comparison of the growth of a seedling and its tissue culture progeny. Reports of the propagation by tissue culture using explants from mature trees are rare (12). Most of our important economic species have yet to be propagated. No extensive data is available on field survival of tissue culture propagated trees. Many other areas for research could be listed. But it is obvious that much more research effort is needed to develop tissue culture into a practical means for the propagation of trees.

REFERENCES

1. Anderson, W.C., "Progress in Tissue Culture Propagation of Rhododendron," OSU Ornamentals Short Course, Portland, Oregon, (1978).
2. Bonga, J.M., *N.Z. J. For. Sci. 4*, 253 (1974).
3. Bonga, J.M., *In* "Applied and Fundamental Aspects of Plant Cell, Tissue, and Organ Culture," (J. Reinert and Y.P.S. Bajaj, eds.), p. 93. Springer-Verlag, Berlin, (1977).
4. Brown, C.L., and Sommer, H.E., "An Atlas of Gymnosperms Cultured *in vitro*: 1924-1974," Georgia Forest Research Council, Macon (1975).
5. Button, J., and Kochba, J., *In* "Applied and Fundamental Aspects of Plant Cell, Tissue, and Organ Culture," (J. Reinert and Y.P.S. Bajaj, eds.), p. 70. Springer-Verlag, Berlin, (1977).
6. Cheng, T.Y., *Int. Plant Propagator's Soc. Combined Proceed. 28*, 139 (1978).
7. Cheng, T.Y., *In* "Plant Cell and Tissue Culture, Principles and Applications", (W.R. Sharp *et al.*, eds.), p. 493. Ohio State Univ. Press, Columbus, (1979).

8. Cheng, T.Y., and Voqui, T.H., *Science 198*, 306 (1977).
9. David, A., David, H., Faye, M., and Isemukali, K., *In* "Micropropagation d'Arbres Forestiers," Annales AFOCEL #12, p. 33, (1979).
10. Eriksson, T., Fridborg, G., and von Arnold, S., *In* "Vegetative Propagation of Forest Trees - Physiology and Practice," p. 17. Institute for Forest Improvement and Department of Forest Genetics, College of Forestry, Swedish University of Agricultural Sciences, (1977).
11. Gresshoff, P.M., and Doy, C.H., *Planta 107*, 161 (1972).
12. Gupta, P.K., Nadgir, A.L., Mascarenhas, A.F., and Jagannathan, V., *Plant Sci. Lett. 17*, 259 (1980).
13. Halperin, W., *Amer. J. Bot. 53*, 443 (1966).
14. Huang, L.C., and Murashige, T., *TCA Manual 3*, 539 (1976).
15. Jones, O.P., Hopgood, M.E., O'Farrell, D., *J. Hort. Sci. 52*, 235 (1977).
16. Kohlenbach, H.W., *In* "Plant Tissue Culture and its Bio-Technological Application", (W. Barz, E. Reinhard, and M.H. Zenk, eds.), p. 355, Springer-Verlag, Berlin, (1977).
17. McComb, J.A., *Int. Plant Propagator's Soc. Combined Proceed. 28*, 413 (1978).
18. Mehra-Palta, A., Smeltzer, R.H., and Mott, R.L., *Tappi 61*, 37 (1978).
19. Murashige, T., and Skoog, F., *Physiol. Plant. 15*, 473 (1962).
20. Pierik, R.L.M., *Acta Horticult. 54*, 71 (1975).
21. Reilly, K.J., and Washer, J., *N.Z. J. For. Sci. 7*, 199 (1977).
22. Risser, P.G., and White, P.R., *Physiol. Plant. 17*, 620 (1964).
23. Romberger, J.A., Varnell, R.J., and Tabor, C.A., "Culture of Apical Meristems and Embryonic Shoots of *Picea abies* - Approaches and Techniques," U.S.D.A. Forest Service Tech. Bull. 1409, (1970).
24. Schenk, R.U., and Hildebrandt, A.C., *Can. J. Bot. 50*, 199 (1972).
25. Sommer, H.E., *Int. Plant Propagator's Soc. Combined Proceed. 25*, 125 (1975).
26. Sommer, H.E., and Brown, C.L., *In* "Plant Cell and Tissue Culture, Principles and Applications," (W.R. Sharp *et al.*, eds.), p. 461. Ohio State University Press, Columbus, (1979).
27. Sommer, H.E., Brown, C.L., and Kormanik, P.P., *Bot. Gaz. 136*, 196 (1975).
28. Söndahl, M.R., and Sharp, W.R., *Z. Pflanzenphysiol. 81*, 395 (1977).

29. Start, N.D., and Cumming, B.G., *HortSci. 11*, 204 (1976).
30. Thorpe, T.A., *In* "Vegetative Propagation of Forest Trees - Physiology and Practice," p. 27, The Institute for Forest Improvement and Department of Forest Genetics, College of Forestry, Swedish University of Agricultural Sciences, (1977).
31. Winton, L.L., *In* "Frontiers of Plant Tissue Culture 1978", (T.A. Thorpe, ed.), p. 419, University of Calgary Press, Calgary, (1978).
32. Witham, F.H., Blaydes, D.F., and Devlin, R.M., "Experiments in Plant Physiology," p. 195. Van Nostrand-Reinhold Co., (1971).
33. Zimmermann, R.H., *Int. Plant Propagator's Soc. Combined Proceed. 28*, 539 (1978).
34. Zimmermann, R.H., Hammerschlag, F., and McGrew, J.R., (eds.), "Nursery Production of Fruit Plants Through Tissue Culture, Applications and Feasibility," U.S.D.A., SEA. In press, (1980).

BIOSYNTHESIS OF SECONDARY PRODUCTS *IN VITRO*

Otto J. Crocomo

Escola Superior de Agricultura "Luiz de Queiroz"
Universidade de São Paulo
São Paulo, Brazil

Eugênio Aquarone

Faculdade de Ciências Farmacêuticas
Universidade de São Paulo
São Paulo, Brazil

Otto R. Gottlieb

Instituto de Química
Universidade de São Paulo
São Paulo, Brazil

I. INTRODUCTION

Living organisms are extremely complex from the chemical point of view. They are composed of nucleic acids, proteins, carbohydrates and fats, which, along with other intermediates in the main stream of cell organization, receive the most attention from biochemists concerned with comprehending the function and survival of organisms. Organisms are composed also of alkaloids, non-protein amino acids, terpenoids and phenolics, which, along with other classes of relatively small molecules receive equal attention from researchers concerned with comprehending the functioning and survival of a community of organisms. It is not always possible to assign a natural

ISBN 0-12-690680-7

compound unequivocally to either of these categories, a difficulty enhanced by the fact that both these so-called primary and secondary metabolites are derived biosynthetically from the same primary precursors and are essential for survival. Primary metabolites essentially provide the basis for growth and reproduction, while secondary metabolites provide, to a large degree, basis for adaptation and interaction with the environment.

Man being part of this environment should also be influenced by at least some secondary metabolites. This is indeed the case, and therapeutic, flavoring, odorous, poisonous, hallucinogenic and colored principles from plants have been utilized since the dawn of civilization and continue to play an important role in human affairs. Why from plants? While animals may rely also on behavior for communication, plants are of course totally dependent on chemicals, a probable reason for the vast distribution and fantastic diversity of their secondary compounds. Over four-fifths of the about 30,000 known natural products are of plant origin. It is, consequently, most unfortunate that the gradual substitution of forests and grassland by housing and agriculture should make the exploitation, as well as the discovery of new substances, of such economically important commodities increasingly more difficult.

The problem is being felt more acutely in developing countries such as Brazil. In order to demonstrate the richness of its dwindling native flora a selection of those compounds for which at least some use is known is presented in Table I. Clearly many more secondary compounds have been isolated from these and other native species as well as from cultivated ones. Would it be possible to produce such chemicals without further sacrifice of the environment?

II. PLANT TISSUE CULTURE AS AN ALTERNATIVE TO WHOLE PLANTS

About 30 years ago it was suggested that plant tissue cultures could be used as an alternative to whole plants as a biological source of potentially useful compounds. This feeling came from demonstrations that plant cells can be grown in suspension in liquid medium in the same way as microorganisms, e.g., fungal production of medicinal compounds such as penicillin, coupled with the hope that the compounds would be extractable from the culture medium.

TABLE I. Selected Secondary Products and Their Properties or Uses from Wild or Semi-cultivated Brazilian Plants

Compound	Property/Use	Species	Plant Part
Aliphatics			
chaulmoogric acid	antilepra	*Carpotroche brasiliensis*	fruit
ichthyothereol	ichthyotoxic	*Ichthyothere terminalis*	leaf
rubrenolide	psychotropic	*Nectandra rubra*	wood
spilanthol	flavor	*Spilanthes oleracea*	leaf
Alkaloids			
anibine	analeptic	*Aniba duckei*	wood
berberine	pigment	*Berberis laurina*	leaf, bark
caffeine	cardiac stimulant	*Ilex paraguariensis*	leaf
caffeine	cardiac stimulant	*Paullinia cupana*	seed
canthinones	antibiotic	*Simaba cuspidata*	wood
cocaine	local anesthetic	*Erythroxylon coca*	leaf
emetine	emetic	*Cephaelis ipecacuanha*	root
20-epiheyneanine	antispasmodic	*Peschiera affinis*	leaf
glaziovine	anxiolytic	*Ocotea glaziovii*	bark
guatambuinine	antitumoral	*Aspidosperma longipetiolatum*	bark
harman	tranquilizer	*Passiflora alata*	bark, fruit
harmine	hallucinogenic	*Banisteriopsis caapi*	leaf, bark
harmine	hallucinogenic	*Cabi paraensis*	leaf
hyoscyamine	antispasmodic	*Datura insignis*	leaf
leurocristine	antileukemic	*Catharanthus roseus*	leaf, root
5-methoxytryptamine	hallucinogenic	*Piptadenia peregrina*	seed

TABLE I. (continued)

Compound	Property/Use	Species	Plant Part
6-methoxycarboline	hallucinogenic	*Virola theiodora*	bark
nicotine	insecticide	*Nicotiana tabacum*	leaf
nicodine	antitumoral	*Fagara* spp.	wood
olivacine	DNA intercallating	*Aspidosperma olivaceum*	bark
pereirine	sympaticolytic	*Geissospermum laeve*	bark
pilocarpine	miotic	*Pilocarpus jaborandi*	leaf
quinine	antimalaric	*Cinchona ledgeriana*	bark
quinidine	antifibrill	*Cinchona ledgeriana*	bark
reserpine	tranquilizer	*Rauwolfia pentaphylla*	bark
scopolamine	antispasmodic	*Datura stramonium*	leaf
theobromine	cardiac stimulant	*Theobroma cacao*	seed
toxiferine	muscle paralysant	*Strychnos toxifera*	bark
tubocurarine	muscle paralysant	*Chondodendron tomentosum*	bark
vincaleukoblastine	antileukemic	*Catharanthus roseus*	leaf, root
yohimbine	aphrodisiac	*Aspidosperma quebracho-blanco*	bark
Flavonoids			
carajurine	pigment	*Arrabidaea chica*	leaf
cinnamyl phenols	insect sterilants	*Machaerium* spp.	wood
dalbergione	spasmolytic	*Dalbergia nigra*	wood
flavonols	dermatitis causing	*Apuleia leiocarpa*	wood
methoxyflavones	antitumoral	*Zeyhera montana*	leaf
morin	pigment	*Chlorophora tinctoria*	wood
pterocarpins	fungistatic	*Osteophleum platyspermum*	wood
rotenone	ichthyotoxic, insecticide	*Derris urucu*	root
rutin	capillary strengthening	*Dimorphandra mollis*	fruit
vestidol	fungistatic	*Machaerium vestitum*	wood

Compound	Property/Use	Species	Plant Part
Phenolics			
anthraquinones	laxative	*Cassia speciosa*	wood
arylpyrones	CNS-depressant	*Aniba guianensis*	wood
bergapten	skin pigmentation stimulant	*Brosimum gaudichaudii*	root
biflorin	antibiotic	*Capraria biflora*	root
burchellin	antitumoral	*Aniba burchellii*	wood
cardol	vermifuge	*Anacardium occidentale*	fruit
centrolobine	antibiotic	*Centrolobium robustum*	wood
chrysarobin	laxative	*Vataireopsis araroba*	wood
cinnamamides	local anesthetic	*Ottonia jaborandi*	leaf
cinnamein	odorous	*Myroxylon balsamum*	wood
cotoin	astringent	*Aniba coto*	bark
coumarin	flavor	*Dipteryx odorata*	seed
coumarin	flavor	*Torresea cearensis*	seed
curcumin	pigment	*Curcuma longa*	rhizome
eugenol	flavor	*Dicypellium caryophyllatum*	bark, leaf
filicin	tapeworm expellant	*Dryopteris filix-mas*	rhizome
lapachol	antitumoral	*Tabebuia impetiginosa*	wood
maclurin	pigment	*Chlorophora tinctoria*	wood
1-nitro-2-phenylethane	odorous	*Aniba canelilla*	bark, wood
plumbagin	antibiotic	*Plumbago scandens*	leaf
safrole	flavor	*Ocotea pretiosa*	wood
surinamensin	cercaricide	*Virola surinamensis*	leaf
usnic acid	antibiotic	*Usnea aspera*	entire plant
xanthones	varied	*Kielmeyera coriacea*	wood

TABLE I. (continued)

Compound	Property/Use	Species	Plant Part
<u>Steroids</u>			
antiarin	cardiotoxic	*Antiaris toxicaria*	seed
cevadine	insecticide	*Schoenocaulon officinale*	seed
digitoxigenin	cardiotonic	*Thevetia ahoai*	bark, seed
diosgenin	raw material for hormone synthesis	*Dioscorea laxiflora*	tuber
glycosides	cardiotonic	*Asclepias curassavica*	leaf
solasodine	raw material for hormone synthesis	*Solanum lycocarpum*	fruit
<u>Terpenoids</u>			
andirobine	bitter	*Carapa guianensis*	wood, seed
ascaridole	vermifuge	*Chenopodium ambrosioides*	leaf, seed
bisabolol	antiinflammatory	*Vanillosmopsis erythropappa*	wood
bixin	pigment	*Bixa orellana*	fruit
cucurbitacins	laxative	*Luffa operculata*	fruit
14,15-epoxygeranylgeraniol	cercaricide	*Pterodon pubescens*	fruit
eremanthin	cercaricide	*Eremanthus elaeagnus*	wood
eucalyptol	odorous	*Licaria puchury-major*	leaf, wood, seed
genipic acid	antibiotic	*Genipa americana*	wood
genipin	pigment on skin	*Genipa americana*	wood
geniposide	laxative	*Genipa americana*	wood
germacranolide	amoebicide	*Calea pinnatifida*	leaf
glycyrrhizin	sweet	*Periandra dulcis*	root
jatrophone	cytotoxic	*Jatropha elliptica*	root
linalool	odorous	*Aniba duckei*	wood
linalool	odorous	*Croton cajuçara*	leaf
maytenine	antitumoral	*Maytenus illicifolia*	root

Compound	Property/Use	Species	Plant Part
micrandrols	antibiotic	*Micrandropsis scleroxylon*	wood
nerolidol	odorous	*Myroxylon balsamum*	wood
nor-diterpenes	cercaricide	*Velloziaceae* spp.	
phorbol	cocarcinogenic	*Euphorbia* spp.	leaf
plumieride	fungicide	*Plumeria lancifolia*	bark
quassin	amoebicide	*Quassia amara*	wood
ryanodine	insecticide	*Ryania acuminata*	wood,root
serjanosides	ichthyotoxic	*Serjanea caracasana*	bark
stevioside	sweet	*Stevia rebaudiana*	leaf
tetrahydrocannabinol	psychotropic	*Cannabis sativa*	leaf

[a] Sources: Benigini et al. (1), Correa and Penza (3), Gibbs (5), Gottlieb and Mors (6,7), Rizzini and Mors (10), The Merck Index (17), Uphof (18).

The major advantages expected for cell culture systems over the conventional cultivation of whole plants are that (1) useful compounds could be produced under controlled conditions, (2) cultured cells would be free of contamination from microbes and insects, (3) cells of any plant could be multiplied to yield specific metabolites, and (4) cell growth could be controlled automatically and metabolic processes could be regulated rationally; all contributing to the improvement of productivity and the reduction of labor and costs (14).

In order to be useful as an alternative industrial source of secondary compounds, a cell culture must satisfy several requirements. A good yield of final product is essential. Indeed, its accumulation in the cell or release into the medium should be rapid in comparison to its degradation. The cells must be genetically stable in order to give a constant yield of product. The production must be profitable including the inherent cost of the culture medium, and in the case of biotransformation precursors of products, and the extraction and purification procedures.

Clearly, in order to make cell cultures attractive as an alternative to whole plants, the yield in product expressed per unit of tissue must be at least comparable. Dougall gives data for tissue cultures from 16 plant species producing a variety of secondary products at levels approximating or exceeding those found in whole plants (4). High yields of specific medicinally useful compounds resulted, e.g., in the cases of chlorogenic acid (14-30 mg/g dry weight of tissue from *Haplopappus gracilis*), phenolics (0.33-13.3 mg/g fresh weight of tissue from *Acer pseudoplatanus*), anthraquinones (0-900 μmoles/g d.wt. of tissue from *Morinda citrifolia*), serpentine (0-0.8% d.wt. of tissue from *Catharanthus roseus*), anthocyanins (0.1-2.4 mg/100 ml culture of *Populus* sp.), nicotine (0-0.25% d.wt. of tissue from *Nicotiana tabacum*), carboline alkaloids (0-850 μmoles/g f.wt. of tissue from *Peganum harmala*).

In several instances, such as in cases of attempted production of alkaloids by cell cultures, yields in secondary metabolites gradually declined. Interestingly, however, the plants regenerated from these cell cultures recovered the patterns and levels of alkaloids characteristic of the species (4). During culture the cells thus retain the capacity for synthesis of specific compounds. On the other hand, callus cultures sporadically give rise to variant subcultures showing different concentrations of particular secondary metabolites. Thus cellular variation can regulate secondary metabolism and may lead to improved biosynthetic capabilities of culture strains. Free cells prepared from *Nicotiana rustica* callus tissue give rise to individual clones showing large differences in growth and nicotine production. One of these clones developed into a relatively stable, unorganized strain capable of producing a high yield of nicotine (ca. 0.3% d.wt.) (15).

Using chemical mutagens, such as N-methyl-N'-nitro-nitrosoguanidine, it is possible to obtain many variant clones of carrot cells showing a wide variation in their capacity to produce β-carotene and lycopene. Of interest is that the carotenoid content of the improved clones is increased 3-fold in comparison with the original strain and 4-fold in comparison with the intact root (8).

To develop cell cultures giving high yields of specific compounds, cultures should be derived from individual plants which give high yields of the desired compounds. It has been shown that the indole alkaloids, ajmalicine and serpentine, can be produced in tissue cultures established from high yielding and low yielding plants of *Catharanthus roseus*. The culture from the former give higher yields of alkaloids than the cultures of the latter (21). Furthermore, by the use of radioimmunoassay techniques, Zenk *et al.* were able to select stable high producing cell lines from populations of cells which produced insignificant amounts of the alkaloids (21). These findings led Zenk to conclude that the selection of stable variant cell lines, producing high yields of the desired natural compounds is presently the most promising step towards an industrial application of cell culture techniques (20).

III. CONTROL OF COMPOUND PRODUCTION IN CELL CULTURES

Yields of specific compounds in cell cultures may be influenced by environmental factors, such as light, precursors, and nutrients, including growth regulators, and by biological factors, such as growth-production patterns, morphological and chemical differentiation, and biosynthetic capacity (4,14, 20).

It has been shown for a number of plant tissue cultures that light stimulates the formation of such compounds as carotenoids, flavonoids, polyphenols, and plastoquinones. On the other hand, biosynthesis of other metabolites is not significantly affected by light and may even be inhibited. Thus, irradiation with white or blue light resulted in an almost complete inhibition of alkaloid biosynthesis (16).

An exogenous supplement of the biosynthetic precursor added to the culture medium may increase the yield of the end product. However, administration of a direct precursor does not always produce the desired effect and finding the most efficient compound may sometimes prove to be a problem. The cost of the precursor to be used is clearly relevant from the practical point of view.

The chemical composition of the media has to be adjusted for the optimization of cell growth. Some plant tissue cultures will grow on nitrate as the sole nitrogen source, others require a mixture of nitrate and ammonium ions and some require casein hydrolysate or a few amino acids. The addition of ammonium ions result in a nearly 4-fold increase in the caffeine content in tea suspension cultures (19). Sucrose is frequently used as a carbon and energy source and leads to the optimal growth rate of plant cell cultures. Mannose, galactose, glucose and raffinose have also been used for this purpose. Both the nature and concentration of the sugar generally affects the yield of the secondary product. The growth substances required by plant cells are represented by auxins and cytokinins, and again their nature and concentration alter both cell growth and yield in specific compounds. Alterations in the yield of secondary products may be determined by still other factors, such as the pH of the medium, the concentration of some gases, oxygen, carbon dioxide and ethylene, and the presence of yeast extract or coconut milk.

The biosynthetic activity of cell cultures in the direction of cell growth or secondary product formation seems frequently subject to some sort of control mechanism. According to data obtained in kinetic studies, growth-product patterns may be classified in the following way (14): (1) cell growth proceeds almost in parallel with product accumulation, e.g., nicotine, tropane alkaloids and morindin-type anthraquinones, (2) decline or cessation of cell growth precedes product formation, e.g., polyphenol, (3) the experimental curve is diphasic and cell growth slightly precedes product formation, e.g., diosgenin.

Some secondary products accumulate entirely in specific structures, showing so-called *morphological differentiation*. Thus essential oils are found in glandular scales and secretary ducts, latex in lactifers and tobacco alkaloids are primarily synthesized in roots. Thus, accumulation of such products occurs only if specific structures are present in the cultures. On the other hand, cases are also known in which unorganized tissue will produce compounds found exclusively in specific tissues of intact plants. An example of *chemical differentiation* without morphological differentiation has been examined by Sugisawa and Ohnishi who showed that cells of *Perilla frutescens* leaves in culture contained as much essential oil as the intact leaves (13).

Callus cultures can give rise to subcultures showing

different levels of a particular secondary compound. Variation of a culture strain thus influences biosynthetic activity and, hence, secondary metabolism (see 20).

IV. CULTURE SYSTEMS IN SECONDARY PLANT PRODUCT BIOSYNTHESIS

Secondary plant metabolites may be obtained from cell cultures produced by batch or, less commonly, by continuous processes. Current achievements have been reviewed by Street (12), Butcher (2), Staba (11), and Wilson (19).

The batch process is initiated by transferring pieces of undifferentiated callus to a liquid medium of constant volume. The suspension is incubated in a shaker maintained in an air-conditioned room at an optimized temperature. The biomass increases by cell growth and division. When a factor in the medium becomes limiting, the cells enter a stationary phase during which their mass declines. When such cells are subcultured, they pass successively through a lag phase, a short phase of exponential growth, a phase of declining growth rate and finally return to the stationary phase (12). Large scale batch cultures use borosilicate flasks and forced aeration. Dispersion of the cells is achieved either by spinning the culture vessels (80-120 rpm) or by magnetic stirring.

Continuous cultures may be operated in closed and open systems. In *closed system*, the cells are retained and cell density increases progressively while growth continues. Nutrients in excess of requirements are supplied by the continuous inflow of fresh medium, balanced by the continuous harvesting of spent medium (12). The *open system* involves the input of new medium and a balancing harvest of an equal volume of culture. This system allows the establishment of steady states of growth and metabolism, the study of the transition between steady states and the identification of the controlling factors. Regulation involves the use of either chemostats or turbidostats. In the former case, continuous input of fresh medium is set at a fixed rate and determines the nature of the resulting equilibrium. In the latter case, cell density is set at levels monitored by reading the absorbancy of the culture.

For products which accumulate most actively when the growth rate is declining it is possible, by chemostat culture, to maintain the cells at the growth rate consistent with the highest cellular content of the compound. A considerably higher level of the product will be built up in such steady state cells than can be achieved in batch culture. Variation in the medium during this steady state may further enhance

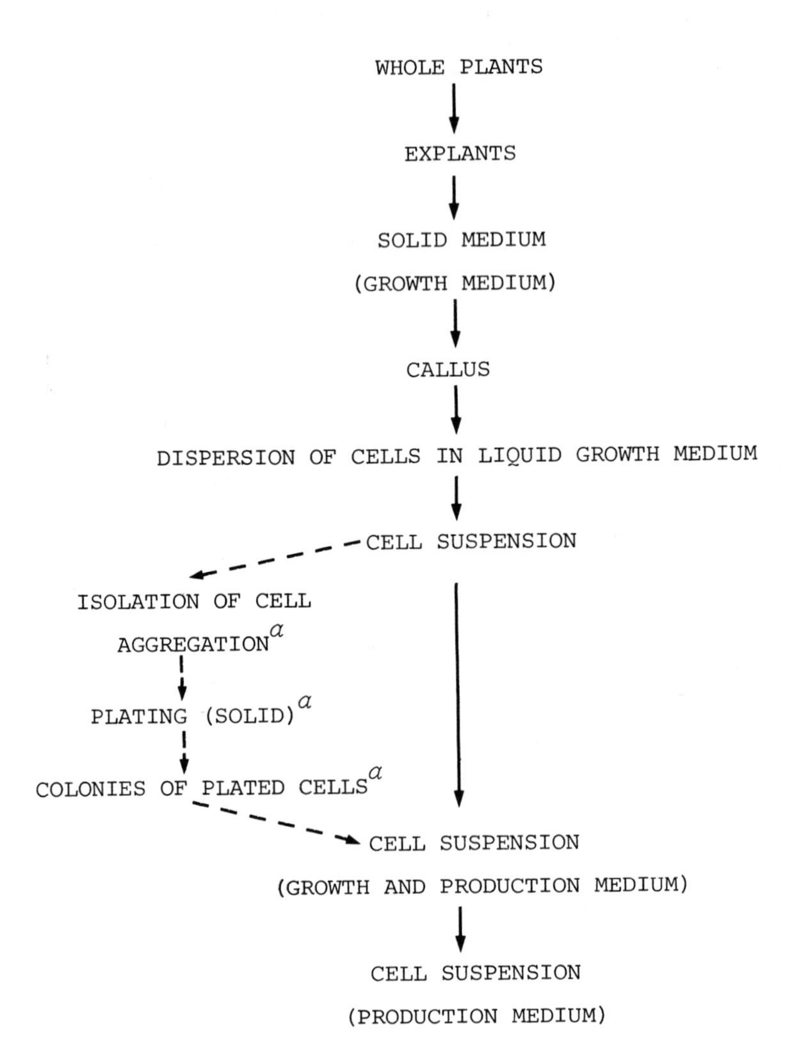

FIGURE 1. Flow chart showing stages and operation for the production of secondary products by cell culture.
[a]Steps for selection of high yielding strains (after 21).

biosynthesis of the secondary metabolite. Once the highest level of product is achieved consistent with the maintenance of cell viability the product can be harvested either through continuous run-off of surplus culture or through a single harvest of the whole culture. This latter alternative would probably be used where a relatively slow growth rate is necessary for optimal product accumulation (12).

The turbidostat system is used in cases where maximum levels of secondary product develop early during the growth cycle, either during or at the beginning of the decay of exponential growth. This enables studies to be made of the influence of growth regulators, physical factors, and precursors.

Closed and open continuous culture systems, illustrated by Street (12), make use of the procedure outlined in Figure 1. A similar flow chart of operations was used by Zenk *et al.* for the selection of high alkaloid yielding cell strains of *Catharanthus roseus* (21).

REFERENCES

1. Benignini, R., Capra, C., and Cattorini, P.E., "Plants Medicinali; Chimica, Farmacolagia a Terapia," 2 vols. Inverni E Della Beffa, Milão, Itãlia, (1964).
2. Butcher, D.N., *In* "Plant Cell, Tissue and Organ Culture" (J. Reinert and Y.P.S. Bajaj, eds.), p. 668. Springer-Verlag, Berlin, (1977).
3. Correa, P.R., and Penna, L.A., "Diccionario das Plantas Uteis do Brasil," 6 vols. Ministẽrio de Agricultura (IBDF), Rio de Janeiro, Brasil, (1975).
4. Dougall, D.K., *In* "Plant Cell and Tissue Culture - Principles and Applications," (W.R. Sharp, P.O. Larsen, E.F. Paddock and V. Raghavan, eds.), p. 727. Ohio State Univ. Press, Columbus, (1979).
5. Gibbs, R.D., "Chemotaxonomy of Flowering Plants," 4 vols. McGill-Queen's, Montreal, (1974).
6. Gottlieb, O.R., and Mors, W.B., *Interciência* (*Caracas*) *3*, 252 (1978).
7. Gottlieb, O.R., and Mors, W.B., *J. Agr. Food Chem.*, in press (1980).
8. Nishi, A., Yoshida, A., Mori, M., and Sugano, M., *Phytochemistry 13*, 1653 (1974).
9. Ogutuga, D.B.A., and Northcote, D.H., *J. Exp. Bot. 21*, 258 (1970).
10. Rizzini, C.T., and Mors, W.B., "Botânica Econômica Brasileira," EPU e EDUSP, São Paulo, Brasil, (1976).

11. Staba, E.J., *In* "Plant Cell, Tissue, and Organ Culture" (J. Reinert and Y.P.S. Bajaj, eds.), p. 694. Springer-Verlag, Berlin (1977).
12. Street, H.E., *In* "Plant Cell, Tissue, and Organ Culture" (J. Reinert and Y.P.S. Bajag, eds.), p.649, Springer-Verlag, Berlin (1977).
13. Sugisawa, H., and Ohnishi, Y., *Agr. Biol. Chem.*, *40*, 231 (1976).
14. Tabata, M., *In* "Plant Tissue Culture and its Bio-Technological Application" (W. Barz, E. Reinhard and M.H. Zenk, eds.), p. 3. Springer-Verlag, Berlin, (1977).
15. Tabata, M., and Hiraoka, N., *Physiol. Plant.* *38*, 19 (1976).
16. Tabata, M., Mizukami, H., Hiraoka, N., and Konoshima, M., *Phytochemistry* *13*, 927 (1974).
17. The Merck Index, 9th ed. (M. Windholz, ed.), Merck and Co., New Jersey, (1976).
18. Uphof, J.C.Th., "Dictionary of Economic Plants," Verlag Von G. Cramer, New York, (1968).
19. Wilson, G., *In* "Frontiers of Plant Tissue Culture 1978," (T.A.Thorpe, ed.), p.169. University of Calgary Press, Calgary, (1978).
20. Zenk, M.H., *In* "Frontiers of Plant Tissue Culture 1978," (T.A. Thorpe, ed.), p. 1, University of Calgary Press, Calgary, (1978).
21. Zenk, M.H., El-Shagi, H., Arens, H., Stockigt, J., Weiler, E.W., and Deus, B., *In* "Plant Tissue Culture and its Bio-Technological Application," (W. Barz, E. Reinhard, and M.H. Zenk, eds.), p. 27. Springer-Verlag, Berlin, (1977).

INDEX